Design and Problem Solving in Technology

Design and Problem Solving in Technology

John Hutchinson & John Karsnitz

Delmar Publishers Inc.

NOTICE TO THE READER

Cover photo by Larry Hamill
Cover design by design M design W

Delmar Staff
New Product Acquisitions Editor: Mark Huth
Developmental Editor: Sandy Clark Gnirrep
Managing Editor: Susan L. Simpfenderfer
Project Editor: Elena Mauceri
Production Coordinator: James Zayicek
Art Supervisor: Judi Orozco
Design Supervisor: Susan C. Mathews

For more information, write:
Delmar Publishers Inc.
3 Columbia Circle, Box 15-015
Albany, New York 12212-5015

Printed in the United States of America
Published simultaneously in Canada
by Nelson Canada,
a division of The Thomson Corporation

10 9 8 7 6 5 4 3 2 1 XX 99 98 97 96 95 94

Library of Congress Cataloging-in-Publication Data

Hutchinson, John, 1945–
Design and problem solving in technology / John Hutchinson & John Karsnitz.
p. cm.
Includes index.
ISBN 0-8273-5244-1 (textbook)
1. Technology. 2. Engineering design. 3. Problem solving.
I. Karsnitz, John R., 1945– . II. Title.
T47.H88 1994
600—dc20

92-21613
CIP

Contents

Chapter 1
The Technological Age

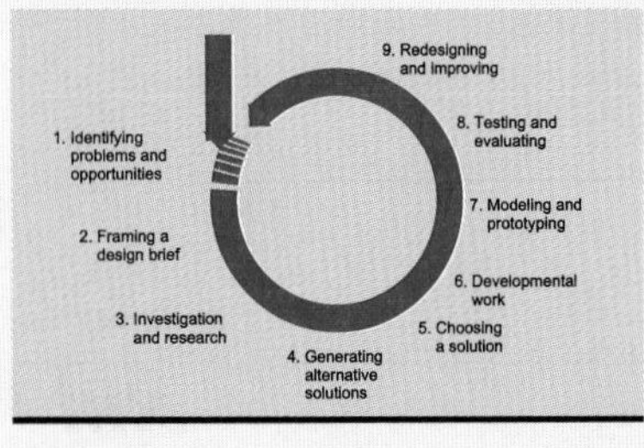

Chapter 2
The Design and Problem-Solving Process

Chapter 3
Problems and Opportunities

Chapter 4
Documenting Your Design Work

Chapter 5
Investigation and Research

Chapter 6
Designing Structural Systems

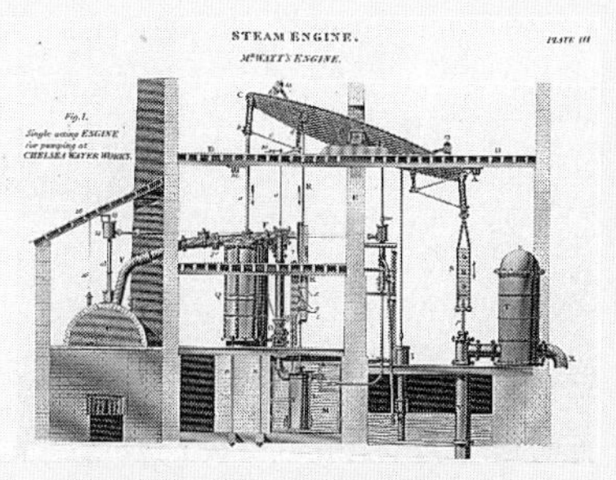

Chapter 7
Designing Mechanical Systems

Chapter 8
Designing Electronic Systems

Chapter 9
Designing Pneumatic Systems

Chapter 10
Generating and Developing Ideas

Chapter 11
Materials and Prototyping

Chapter 12
Testing and Evaluating

Dedication

This book is dedicated to our families.

To Harry K. Hutchinson, who was a wonderful father and the consummate craftsman.

Acknowledgments

The authors would like to thank the following individuals for their assistance in the development and preparation of this book: our wives—Suzanne, for writing the section on research and many hours of copy editing, and Patricia, for many creative illustrations and content advice; to Professor Bijan Sepahpour, ME, for suggestions in the mechanisms section and to Gregory Thomson for technical review; our friends and colleagues in the Department of Technological Studies at Trenton State College who helped provide the positive atmosphere and encouragement that makes creative work of this kind possible; the students in the Department of Technological Studies who make being an educator a worthwhile and fulfilling experience; the teachers and supervisors in New Jersey and other states, as well as in the provinces of Canada who are embracing the process-based technology education approach put forth in this text; and our Developmental Editor, Sandy Clark, who has been our advocate and has provided us with competent and enthusiastic support for this project.

The authors and Delmar Publishers Inc. would like to thank those individuals who reviewed the manuscript and offered invaluable suggestions and feedback. Their assistance is greatly appreciated.

Robert Berkemer
University of Wisconsin—Stout

Steve Brady
Rice Lake Middle School

Richard Buxton
Thomas Jefferson High School for Science and Technology

Ken Ford
Sauquoit Valley High School

Vivienne Frederick
Saratoga Springs High School

Charles Graham
Plymouth Meeting Schools

Thomas A. Hughes, Jr.
Foundation for Technology Education

Doug MacIntosh
Hudson Falls Schools

About This Book

This book is intended to provide students with knowledge, processes, and a framework for the development of skills necessary for the understanding of technology. It should be considered as a basic text for a major course in technology or as a supplemental text in an existing course in which the teacher wants to infuse a major new component in technological design and problem solving.

The book grows out of the belief that design and problem-solving skills can be learned by all students and that these skills can lead students to an in-depth understanding of technology, as well as other areas of study. This premise is grounded in the belief that the application of knowledge is preferable to a simple acquiring of facts. It is assumed that this text will be used in a course that is not the first experience in technology for most students; therefore, this text is aimed at a high school level. Advanced junior high school students, such as those in gifted programs or those with at least a full year of technology course work, may also find this book useful as either a course text or as a reference book.

The design process (problem solving) has been used as a framework for this book. Although the design process is not linear (and the chapters of this book are), it is useful to separate the major steps of the process into separate chapters in order to develop an understanding of the knowledge, skills, and thinking processes in each step. Examples have been used extensively to reinforce the principles discussed in each chapter.

Another important feature of this book is the structure of the technological content. Chapters are devoted to certain transferable "building blocks" of knowledge that cut across all of technology. An understanding of principles of structures will be transferable to all structures, including tables and chairs, as well as houses, bridges, and aircraft fuselages. Other areas of study include mechanisms, electronics, pneumatics, and human factors engineering. Students are encouraged to apply the knowledge that they have gained by solving problems in each of these chapters. These content elements are transferable to any division of technology, including the more traditional areas of communication, transportation, production, and construction.

It is intended that this book bring each student to the point at which he or she is capable of identifying and solving his or her own technological problem. Ideally, students should be involved with a major design project, as well as week-to-week narrowly defined problems to develop their knowledge and skill in specific areas.

Preface

The study of technology must be more than the accumulation of facts about hardware, materials, and processes. Emphasis should be placed on developing both knowledge and the ability to apply that knowledge. Through this book, you will be involved in developing a perspective of technology that will allow you to apply what you have learned to each new situation.

Because technology affects every aspect of our lives so deeply, it is important for basic survival that you understand the workings of this most powerful force. Technology has brought us the enjoyable material things around us, and it helps us meet our needs for food, clothing, shelter, and other basic requirements of life.

Technology has, however, also brought us reasons to despair. The weapons of war and mass destruction; the chemical pollutants that foul our rivers, oceans, and air; and at least some of the upheaval in our social system have also resulted from technology. If you are asked whether technology is good or bad, it would be wise to debate the question in your mind, for there is no easy answer.

That technology is necessary for survival cannot be questioned. We could no more eliminate technology from our society or our world than we could remove the veins from our bodies. Along with language, technology is considered a cultural universal, because all cultures develop both. While some cultures have developed what we in the West might consider less sophisticated levels of technology, they have developed technology just the same.

An educated person living in this last decade of the twentieth century must develop a technological capability to function adequately in the high-tech world of today: Nearly all jobs are linked to the changing technology; social, political, and environmental issues are linked to technology; even leisure time and sports are linked to the changes in technology. This book will help in the development of that capability.

About the Authors

John Hutchinson is a professor of Technology Education at Trenton State College in New Jersey. He holds a Ph.D. from The Pennsylvania State University. Dr. Hutchinson is an active member of the International Technology Education Association (ITEA), in which he has served as the chairperson of the International Relations Committee. He has authored numerous articles on technology education curriculum and philosophy and technical articles on microelectronics. He is an active member of the Technology Education Association of New Jersey and has served on the Commission on Technology Education for the state of New Jersey since 1984. Dr. Hutchinson has directed a number of funded technology curriculum development projects in New Jersey and has been a curriculum consultant to a number of states. He was a United States delegate to two NATO international conferences on Technology Education.

John Karsnitz is professor and chairperson of the Department of Technology Education at Trenton State College. He holds a Ph.D. from The Ohio State University and is the author of *Graphic Communication Technology, Second Edition,* another Delmar textbook. Dr. Karsnitz is an active member of the International Technology Education Association (ITEA) and the Council on Technology Teacher Education (CTTE), in which he has chaired two task force committees, as well as served on several standing committees. He is an active member of the Technology Education Association of New Jersey and has made many presentations to state regional and local groups concerning the transition to technology. He serves as a member of the college committee that is organizing a new general education core interdisciplinary course called "Society/Ethics/Technology." The Department of Technology Education is one of the premier institutions preparing technology teachers.

CHAPTER

1 The Technological Age

Introduction

Fig. 1-1

The artifacts of technology are all around us.
1-1a Courtesy of Sears Roebuck and Company
1-1b Courtesy of Sears Roebuck and Company
1-1c Courtesy of Eurosport
1-1d Courtesy of Sony Corporation
1-1e Courtesy of Trek Bicycle Corporation
1-1f Courtesy of Chevrolet

The study of technology is the study of the human quest for *solutions.* In recent years, technology has solved problems of survival in extreme conditions of cold, heat, pressure, vacuum, and radiation. The space suit developed for use by the shuttle astronauts is a marvel of technology, designed to withstand these extremes and yet be comfortable enough to be worn for hours while working in space. Yet, technology also gives us the tools and devices that make day-to-day life possible: the can opener, the microwave oven, the clothes on our backs. These are also the products of technology (Figure 1-1).

a. in the home

b. for personal use

c. play

d. to entertain

e. for exercise

f. to transport

Fig. 1-2

Biosphere is a project designed to see if a self-sustaining environment can support six people for two years. The results will have far-reaching effects on the future space program, such as a manned Mars mission.

Solutions are generated in response to problems. Buildings are usually constructed to enclose a space against the weather. In tropical environments, this may mean providing shade against the sun; in temperate climates, this may mean protection against cold, wind, and snow.

An interesting exercise is to look at something made by humans and find the problem it was intended to solve. For example, the Great Wall

Courtesy of Bettmann Archive

The Great Wall is over 1,500 miles (2,410 kilometers) long. Fifteen hundred miles is approximately the distance between New York City and Omaha, Nebraska. It is the only technological structure that can be seen from outer space, and it was built over 2,000 years ago. The walls are made of earth and stone, with a brick roadway on top for horsemen. The wall is 25 feet (8 meters) tall, with 35–40 foot (11–12 meter) towers every 200–300 yards (180–220 meters).

of China and Hadrian's Wall in northern England were both solutions designed to protect people who considered themselves civilized from people whom they considered barbarian.

Although technology is evident in all human history, advances in the last two centuries have provided us with almost all of what we take for granted today. The 1800s provided us with such inventions as the stethoscope (1816), refrigeration (1834), the reaper (1831), the sewing machine (1846), the elevator (1853), the telephone (1876), the incandescent lamp (1879), the skyscraper (1884), the automobile (1885), and the radio (1895). After the turn of the century, the first successful airplane was developed in 1903. Television, although it was invented in 1920, did not see commercial success until after World War II. In fact, you would probably find it difficult to find examples of products in use today that didn't come from these nineteenth and twentieth century inventions. Figure 1-3 contains photographs of a number of commercially available products. Can you describe the problem that each device was intended to solve?

What problems do the following items attempt to solve?

a. Hint: Used in the 18th century to process a food. The food was sometimes used to fatten pigs for market.

b. Hint: While this is a modern product, it could have been useful in the making of a Jules Verne movie.

c. Hint: Some people through natural or environmental causes lose full use of one or more human capabilities.

Fig. 1-3

1-3b Courtesy of Nikon
1-3c Courtesy of Golden Technologies

These new products and developments have changed the way people are born, the way we are brought up, the way we are educated, the way we work and live, and the way we spend our leisure time. Each new development, invention, and product has the potential to effect change in our society and our world—sometimes beneficially, sometimes not. Understanding the nature of technology and how we employ it to solve our problems is the subject of this text.

Defining Technology

Technology can be thought of as *process*, as *knowledge,* and as *artifact*. Because technology deals with solving problems, we can think of the process we use to arrive at these solutions as a technological process.

The knowledge we use to solve these problems, including information, theories, organization, and technical processes and skills, is also technology. Knowledge about working with materials has been accumulating through the ages, but more recently, information itself has become a resource to be manipulated and processed. The knowledge that deals with the processing of materials, energy, and information is also technology (Figure 1-4).

Fig. 1-4

Communications center

Actual hardware or systems are commonly referred to as "technology." For example, the mass media may call a new computer system "technology." Since things that we can see and touch are typically the solutions to the problems we want to solve, it is easy to see how they can be considered technology. In this book, however, we think that it is more important to focus on the process and knowledge needed to create the artifacts and products than to focus on the product itself. Viewing technology as process and knowledge keeps our focus on the future and on our ability to control that future. Viewing technology as artifact keeps our focus on the past and present—on what already exists.

Technology is **the human process of applying knowledge to satisfy our needs and wants and to extend our capabilities.**

Technology defined

Systems

Solutions to problems often take the form of systems. Systems can be divided into three parts: *input, process*, and *output* (Figure 1-5). Inputs to systems are called **resources**. These resources include materials, energy, information, people, tools and machines, and usually capital. These inputs are all necessary to drive the system.

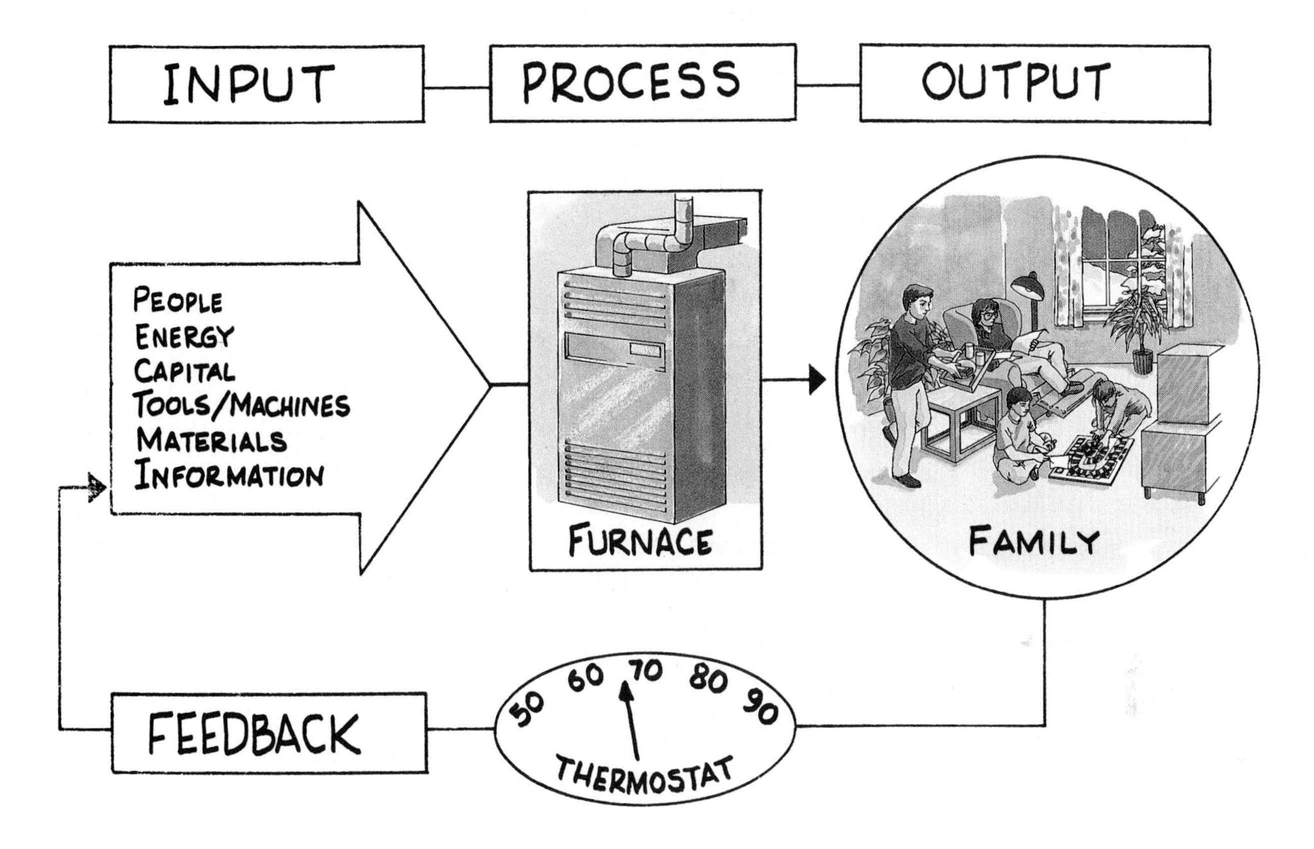

Fig. 1-5

Systems model

In a technological system, three things can be processed: *material, energy,* and *information*. In many systems, at least two of these are processed, one of which is energy. Can you think of a technological system in which energy is not processed?

Material Processing

Material is processed to change its size or shape, and to change its physical, chemical, or electrical characteristics. The material-based systems provide tangible products, ranging from pencils and paper clips to aircraft and computers. Material-processing technologies drove the Industrial Age, a departure from the agricultural-centered society it replaced.

Energy Processing

Energy is processed to change it from one form to another. Science tells us that energy can be found in six forms: mechanical, electrical, chemical, thermal, light, and nuclear. Technology provides us with devices that convert one of these forms into another form for our use.

Prime movers, such as steam engines and internal combustion engines, change fossil fuels into some form of mechanical energy. Smaller machines, such as the bicycle, convert human muscle power to mechanical energy used to drive the wheels. Energy processing is a part of every system.

Information Processing

Since the mid-1950s, the United States, Canada, England, and a number of other nations have entered the **Information Age.** This term is used because the primary economic activity in many Western societies has moved from the processing of materials to the processing of information. Information is processed to change its form in order to send, receive, store, or retrieve a message or a signal.

Computers have had an enormous impact on the Information Age because of their capacity to manipulate huge quantities of data. Management, medicine, and education are examples of fields that are information-driven and that rely on the processing of information. The mass media of television and newspapers is also an integral part of the Information Age.

System Outputs

Outputs from technological systems take a number of forms. The most obvious output of the system is the reason the system was created. It may be in the form of heat from a home central heating system or speed in a automobile cruise control system. This is called the *desired–expected output*. The matrix in Figure 1-6 helps clarify the four outputs of technological systems. The desired–expected output is considered a Type I output.

What is not so obvious are the other outputs of technological systems. While the home central heating system provides heat, it may also produce smoke and soot, carbon dioxide, and other less desirable by-products. These outputs are as much a part of the system as the desired output, heat, and they are planned for in the design of the system. A chimney is incorporated into the home heating system

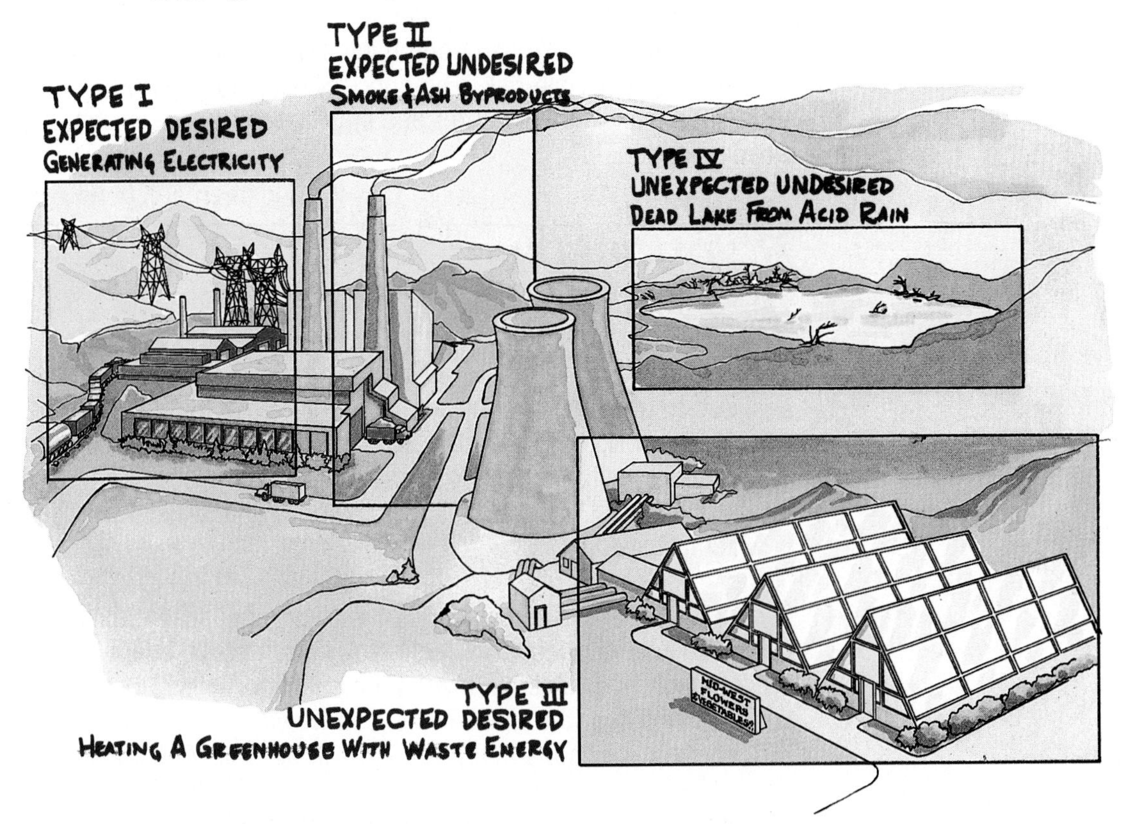

Fig. 1-6

Four outputs of technological activity

design to carry the *undesired–expected output* away from the home. This is an example of a Type II output.

The third or Type III output identified with technological systems is the *unexpected–desired output*. It is an advantage resulting from the use of certain processes or resources unique to that system. It is not harmful but rather a pleasant surprise. For example, it has been a problem to get rid of the waste heat from electrical generating plants. Some plants, however, are actually selling heat to local businesses, such as greenhouse nurseries, for winter heat.

The *undesired–unexpected output*, a Type IV output, is very important to the understanding and study of technology, because it may be the most influential factor affecting the adoption or elimination of a particular technological system. For example, burning fossil fuels for electrical generation and industrial process heat has been an accepted practice for over a century. But only recently has acid rain, which has been killing lakes and rivers and defoliating forests, been identified as a by-product of the combustion process. The undesired–unexpected output of this particular energy-processing system is having important environmental and political consequences.

Since all systems produce wanted and unwanted outputs, a **trade-off** exists between the things you desire and those you do not. All technological solutions then involve questions of trade-off and associated **risk**. As citizens we must become involved in questions of acceptable trade-off and possible risk. The concepts of trade-off and risk will be discussed later in this chapter and in Chapter 10, "Generating and Developing Ideas."

Relationship Between Science and Technology

Science and technology are not the same thing. *Science* seeks answers to questions about the natural universe. Science describes the natural world by asking the question "Why?" Why do things fall? Why do cells mutate? Why do some materials corrode? Science is important for our understanding of the natural and technological world in which we live.

Technology seeks different kinds of answers, because it asks very different kinds of questions. Technology attempts to adapt our environment by solving problems. It asks the question "How?" How can we keep warm during the winter? How can we we lift a heavy load? These are practical questions that deal with solving practical problems.

Fig. 1-7

Gutenberg and his printing press. Courtesy of Bettmann Archive

Scientific inquiry began during the Renaissance (about 1300 to 1600 A.D.). It was aided greatly by technological development, especially printing, which was developed in Western Europe in 1450 A.D. Up to this time, few people outside of the nobility and church possessed the ability to write and read, and books were rare and very valuable. With printing, discoveries could be recorded and shared with a great many people around the known world. The printing press made books widely available, and the number of literate people increased dramatically. The ability to share discoveries and information through books had a profound effect on science, as well as on literature, art, mathematics, and technology.

Technology has existed for as long as humans have inhabited the earth. Long before anyone knew about biology, people were cultivating plants and breeding animals. Before organic chemistry, there was wine making and other fermentation processes. The telescope and other optical instruments were developed long before the physics of light and optics were understood.

Technology uses knowledge from all disciplines, and it would be difficult to point to any one area as being more important than another. Knowledge and technique from art, design, engineering, history, language, mathematics, psychology, science, and others contribute to a technological solution to a problem.

Impacts of Technology

The four outputs collectively describe the impact of technological activity. If technology is finding solutions to problems, then the impacts are the new problems created by the solutions. It is often said that with technology come trade-offs and with each trade-off come risks. When you solve a problem, you also change the original conditions that led to the problem, so you have, in effect, changed the world. Trade-offs are the new problems created when you solved the problem in the first place.

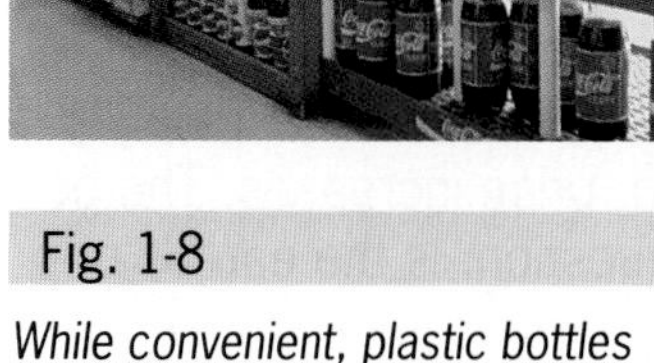

Fig. 1-8

While convenient, plastic bottles contribute to the waste stream.

An example of a trade-off can be seen by looking at the common plastic bottle. The plastic bottle was a solution to the problem of finding an inexpensive, nonbreakable, disposable, and easily manufactured container for liquids. At the time, petroleum (the primary raw material) was cheap and readily available. In addition, throwing away a bottle seemed so much more *convenient* than taking empty bottles back to the store. Unfortunately, we are coming to understand that the trade-off of convenience is burying us in garbage. It has been calculated that we are throwing away *91 billion* plastic bottles each year.

One billion seconds is equal to 31 years, 8 months. Ninety-one billion seconds is equal to about 2,882 years. You could pick up one plastic bottle each second for 2,882 years and only clean up the amount of plastic bottles we throw out for **1 year**.

There are individuals, communities, and companies that are now looking for solutions to the problem of all these disposed plastic bottles and other solid waste. Hopefully, as we study more about technology, we can look for solutions that will not cause yet greater problems.

Products

The convenience products of the 1980s are giving way to the performance products of the 1990s. More than 80 percent of the people in a recent survey stated that they would buy a product because it was environmentally safe. New products will: (1) be more compact, (2) use less resources, (3) have higher efficiency, and (4) be environmentally friendly. As a consumer-oriented society, we produce over 1,500 pounds of solid, nonindustrial waste per person per year. That is over 4 pounds per person every day! Our society must stop thinking only about production and consumption of goods and start thinking about impacts. We must think about the resources needed to produce the product; the impact of the product on the individual, society, and the environment; and the final disposal of the product after its usefulness is done.

Technology and the Earth's Resources

What do you believe about the earth's resources? Do you believe that there are sufficient world resources for the next century and beyond, or do you believe that we are rapidly using up these resources?

In 1980, as a way of settling this issue, Paul Ehrlich (author of *Population Bomb*) and Julian Simon (author of *The Ultimate Resource*) made a bet concerning the depletion of natural resources. The wager was based on the relationship between the supply and the market price of certain natural materials, sometimes referred to as supply and demand. Normally, as supply is reduced, price increases. The bet was thought to be a way of answering the question: Is the earth's population running out of natural resources? Five metals in quantities valued at $1,000 were chosen for the bet: chrome, copper, nickel, tin, and tungsten. In 1990, the price of the five metals was calculated and adjusted for inflation. Ehrlich had to pay Simon $576.07, the amount that the metals had lost in value over the ten-year period.

Paul Ehrlich subscribes to the **Malthusian** view of population and resources. Thomas Malthus (pronounced *MAL-thus*) (1766–1834) was an English economist. His essay on the "Principles of Population," published in 1798, predicted that world population would increase more rapidly than the food supply and that this would lead to war and disease. Although this prediction has proven true in many regions since that time, many problems of food and other resource shortages have been political rather than one of availability. While people in one part of the world starve, there are others who have more than enough.

The Malthusian view is that world population will eventually cause severe shortages of some resources. Although this particular bet proved Ehrlich wrong relative to material shortages in the short term, it remains inconclusive concerning long-range effects. As world population increases at an exponential rate, the competition for these resources will be fierce.

Julian Simon subscribes to the **Cornucopian** view that population growth is good and leads society to a better life. A cornucopia, also known as the "horn of plenty," is a symbol of nature's productivity. It is usually pictured as a funnel-shaped basket overflowing with fruits and vegetables. The symbol has its roots in Greek mythology. This view is based on the belief that people are the most important resource and human ingenuity is the key to solving problems, including those caused by population increases.

Simon was right in predicting that technological innovation would react to, and attempt to overcome, the problem of the high cost of a resource in short supply. Still, the question of long-term impacts must always be considered. Technological innovation is linked to human progress, as are the impacts of that progress.

Human progress is best characterized by change. Using resources and finding solutions to problems is a part of the change caused by technological activity. The Greeks, over 3,000 years ago, moved from the Bronze Age to the Iron Age because of a shortage of tin (needed to make bronze) caused by war. In the sixteenth century, a shortage of timber in Great Britain caused the country to convert to coal as a primary energy source. More recently in the United States, the rising cost of whale oil in 1850 led to the drilling of the first oil well in 1859.

The debate between the Malthusian and the Cornucopian views appears to illustrate that one-sided positions will most likely prove to be incorrect. In fact, important concepts can be learned from both world views. Today, new products are being designed that use less materials, thus conserving valuable natural resources. There is a Native American saying: "We do not inherit the earth from our parents—we borrow it from our children!" This is a view of our world that could help maintain a quality environment for all for us.

Technology: Curse or Answer?

People throughout history have wanted to stop change. The **Luddites** of eighteenth-century England, named after their leader Ned Lud, wanted to halt technological progress by smashing the new weaving machines being introduced into the textile industry of the time. Fabric had been produced by a decentralized **cottage industry**, but the new machines were eliminating jobs and changing the work of those that remained. The effect of the Luddites was minimal, and the weaving machines helped bring about the **Industrial Revolution** in Great Britain.

At the other extreme from the Luddites are people who today believe that any problem can be corrected through technological innovation. This **techno-fix** mentality seems to negate the need for careful consideration of issues and balanced human responsibility concerning the impact of technology on the individual, society, and the environment. This position also ignores the evidence that solving one problem inevitably leads to the creation of another problem.

As we all face the choices presented by technology, we must remember that these issues must be carefully weighed against the impact on the individual, society, and the natural environment. Products should be carefully designed, consumers must carefully choose, and citizens, through governmental regulations, must balance the trade-offs and risks of future technological progress for the betterment of all people.

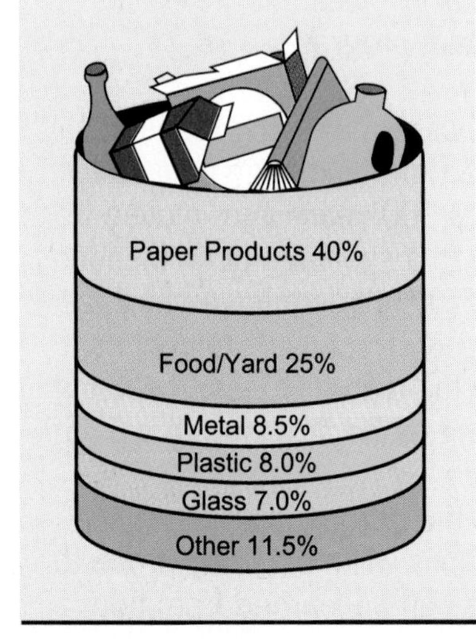

Fig. 1-9

What we throw away

The Waste Stream

The waste stream consists of those things that we throw away (Figure 1-9). In the United States, nearly 40 percent of our garbage is in the form of paper and paperboard products. These paper products are generated primarily by the publishing and the packaging industries. Other material in the waste stream, according to the Environmental Protection Agency (EPA), consists of 25 percent food and yard waste, 8.5 percent metals, 8 percent plastics, 7 percent glass, and 11.5 percent other. The waste stream can be controlled by carefully controlling the products we produce and by the way in which we choose to dispose of products when they are no longer useble. Germany has recently developed a plan for a recyclable car! Most societies deal with their "garbage" in one of four ways: landfills, incineration, recycling, or other alternatives.

Landfills

What does our society do with its waste? Currently, around 86 percent goes into landfills (Figure 1-10). A **landfill** is a specially prepared waste disposal site that is licensed and regulated by the government.

Fig. 1-10

Landfill. Courtesy of Waste Management Incorporated

Landfills are intended to be safe depositories for nonhazardous materials. Unfortunately, landfill space is becoming more expensive. Of the nearly 600 U.S. landfills, around one-half will be filled to capacity soon and three-quarters will be closed by the year 2000. In other developed countries, much less waste is put into landfills. For example, in Japan only 16 percent of their waste stream goes into landfills.

Incineration

An alternative that is gaining momentum as a way of dealing with our nation's waste is incineration. **Incineration** is the process of burning solid waste. The process reduces the volume of waste by around 90 percent and the weight by around 75 percent. Although this sounds great, the technological trade-offs are serious. Because it is nearly impossible to separate out all toxic materials from the waste to be burned, incineration poses a serious threat to the environment, both in terms of air quality and land pollution. If extremely high temperatures are not maintained during incineration, serious emissions, including cancer-linked dioxins, can be released. Even if the air quality is reasonably maintained, the ash has very high concentrations of **heavy metals.** Heavy metals, including lead, chromium, mercury, and others, can get into groundwater when the ash is disposed of in landfills. These heavy metals pose serious health risks to humans. In spite of these hazards, incineration has the potential of being an acceptable technological solution to the waste stream problem if all phases of the operation are carefully monitored, controlled, and maintained.

1-11a Solid arrows indicate that the product is made from recycled material.

1-11b Outlined arrows indicate that the product can be recycled.

Fig. 1-11

Recycle symbols

Recycling

A promising area of waste disposal is *recycling*. Recycling is the collection and reuse of valuable resources. Currently, around 13 percent of our waste stream is recycled in the United States. Many countries recycle considerably more material, including Japan, which recycles 50 percent of its waste stream.

Materials typically recovered are paper, metal, and glass. The *average value per ton* varies with the market, but some examples of 1992 prices are as follows: cardboard, $35; newsprint, $10; aluminum cans, $1,000; glass, $40; and clear plastic, $120. It should be remembered that the benefit of recycling is not only the value of the recovered material but also that natural resources are saved and the waste will not be entering a landfill.

A symbol that has become associated with environmental issues is the "recycle" circle (Figure 1-11). The three arrows represent the con-

Earth Day: Will the 1990s be the "Green" decade? The original Earth Day was held on April 22, 1970. Recycling has become more important, especially in the more populated states and regions of the country. Other environmental issues involving the air we breathe, the water we drink, and the land we inhabit appear to be of greater concern to society in the decade of the 1990s.

Courtesy of NASA

cepts of collecting, reprocessing, and remanufacturing of a natural resource. Originally, two variations of the symbol were envisioned: solid arrows for a product made from a recycled material and outline arrows to indicate that a material was recyclable. These variations are now widely used in the packaging industry.

Other Alternatives

Managing our waste stream is one of society's greatest challenges. One of the more recent trends is the design of new *environmentally safe* or *friendly* products. This trend includes the down sizing of products and the invention of new materials to replace those materials that damage the environment. In many Western European countries, a **Green Label** is used on a product to indicate that it has passed strict guidelines for environmental acceptability. This helps take the guesswork out of the many product choices and conflicting claims made by some manufacturers.

The reduction of product packaging materials is another important step in the reduction of waste. In 1991, a national retailing company announced its intention to reduce the materials used to package its products by 50 percent over a five-year period. Finding new ways to package consumer products that use less materials and other **source reduction** strategies can have a significant impact on the waste stream. Source reduction dictates that, if products and packaging are designed with less disposables, then less waste enters the waste stream.

Government Agencies and Regulations

Society shows many of its concerns through governmental regulations. The *Environmental Protection Agency (EPA)* is charged with enforcing environmental regulations and laws and with monitoring landfills, water resources, and atmosphere quality. The 1990 Clean Air Act and other tightened regulations have meant more monitoring and controls on medium and small businesses. Currently, the EPA is attempting to encourage less waste through source reduction and recycling programs.

Summary

Throughout history, humans have used technology to overcome problems. For thousands of years, these problems have been related mainly to food, shelter, clothing, and other basic needs. Until the end of the last century, most people worked in jobs related to agriculture and food production, and much of the technology centered on these issues, (Figure 1-12).

As the **Industrial Age** (1907–1957) began to take root, more and more people found themselves working in industries in which nonagricultural products were produced by the thousands or even millions. The fact that these workers were no longer working on the land to produce food was a significant departure from all of previous human history. For the first time, in many countries, more people were involved in producing goods than in producing food.

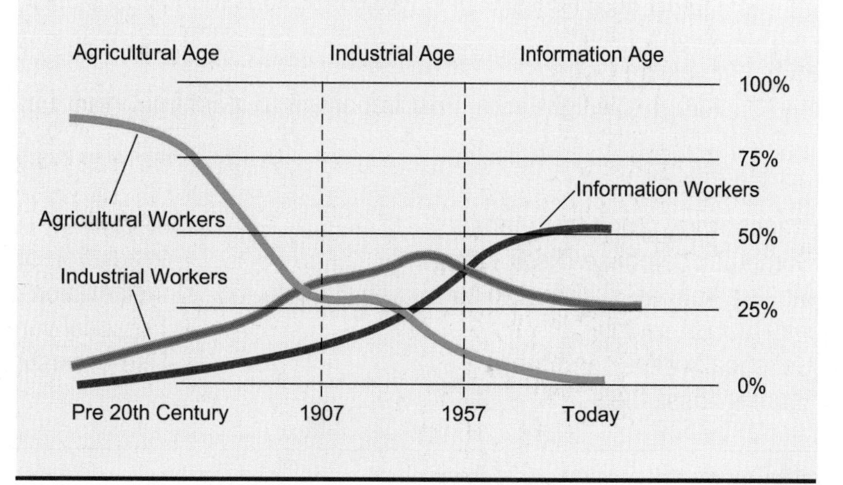

Fig. 1-12

Historical perspectives on technological development

For the next fifty years, America, Canada, and a number of other nations were in an Industrial Age, and in 1957 another significant milestone was reached. That was the year when the number of white-collar jobs outnumbered the blue-collar jobs in the United States. In other words, the people managing, coordinating, accounting, and otherwise dealing with information were in greater numbers in the work force than those who were directly involved in the production of a product. America had entered the Information Age.

The Agricultural, Industrial, and Information Ages have significantly different characteristics. For example, power in an Agricultural Age

Technological Dilemmas

The issues raised by the impacts of technology are not simple and often involve ethical dilemmas. An ethical dilemma occurs when moral reasoning or a position can be made to support multiple alternatives to solve a technological or other problem.

One way to deal with the ethical dilemma is to analyze the issue logically. Use the following five-step process to analyze the ethical dilemma created by technological activity.

Step 1 *Identify the Issue or Problem*
List ethical questions raised by this issue.

Step 2 *Identify the Impacts*
List all possible Type I, II, III, and IV impacts.

Step 3 *Identify the Options*
List the alternatives to deal with the problem.

Step 4 *Develop Arguments for the Options*
List all arguments for and against options identified in Step 3.

Step 5 *Evaluate Options*
Assign a weight from most important to least important for each option

Example Issues

- Transporting crude oil in tankers
- Computer database credit ratings
- Moving from an industrial to an information society
- Genetically developed foods
- Burning fossil fuels
- Recycling
- Irradiation of food
- Incineration of waste
- Personal versus mass transportation
- Use of pesticides and herbicides

A strategy for resolving technological dilemmas

comes from owning *land*; power in an Industrial Age comes from having *capital*; and power in an Information Age comes from having *knowledge*. Can you think of other differences?

Although the time in which we are now living has been called the Information Age, other terms have been used to describe this age of rapid technological progress: The Post-Industrial Age, the Super-Industrial Age, and the **Technological Age** are just a few. No one knows if there is another age to come in the years ahead. We can only recognize these changes in retrospect.

Observation/Analysis/Synthesis

1. Find an example of a technological system, and identify the inputs, what is being processed, and the expected–desired output.
2. Develop a questionnaire to assess the understanding of some group on how technology affects their lives. The questionnaire should have at least ten questions. Groups could be underclass students, parents, teachers, or others.
3. As a class or a small group, brainstorm a list of potential technological developments that could become reality in the next fifteen years. By consensus, give each development a rating of 1 (low) to 10 (high) in order to indicate the group's feeling about the likelihood of it becoming reality within this time period. Make a list of potential impacts of the development on the individual, society, and the environment. Some examples are: IQ-increasing drug; NY to LA Mag-Lev train.
4. Based on the world views of the Malthusians and the Cornucopians, prepare a debate on the issue of world resource consumption.
5. Develop a list of criteria for judging products as "environmentally friendly." Compile a list of products that meet these standards.
6. Develop a source reduction strategy by suggesting how a product could be packaged using less material but still be protected for shipping, or by providing an effective store display. Analyze existing packaging practices to gather ideas.
7. Develop a strategy for encouraging recycling in the school, home, or community.
8. Record the waste produced in your home by weight for seven days; graph the results.
9. Refer to the box on technological dilemmas.

CHAPTER

2 The Design and Problem-Solving Process

Introduction

Design is the *planned process of change*. Instead of something changing by accident, design demands that we plan change so that we end up with the results we want. It also means that we attempt to minimize trade-offs and control risk. Technology is all about design.

The Design Process

Creating solutions to problems involves the process of designing. Design is a broad term that is often associated with artistic expression, but it is best thought of as a thinking process involving planning with intention and purpose. We design everything from the look of a room to the vacation that we hope we will soon take. More important, design describes the process of developing solutions to problems.

Figure 2-1 shows a **design loop**, which was developed as a guide to help make problem solving in technology more effective. It provides a structure for thinking and doing, which is the essence of technological activity. There are many legitimate models for design and problem-solving work, and most have these themes in common. Other chapters in this book will provide depth in each step of the design loop. In addition, one important aspect of design is documenting your work. Chapter 4 details techniques of documentation to aid you in your work.

You should understand that designing is not a linear process; that is, when you design and make something, you do not think and act in separate, sequential steps. The creative process of designing is more like switching back and forth between a thinking–questioning–evaluating mode and an acting–doing mode. These modes have been called the "active" and "reflective" phases of design, and you are constantly moving between the two.

What this process of moving back and forth between the active and reflective phases means is that you will probably need to jump around

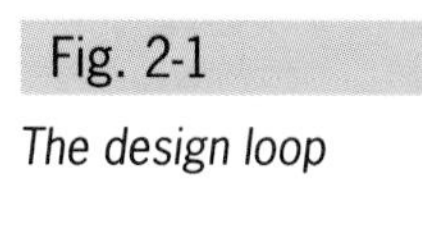
Fig. 2-1

The design loop

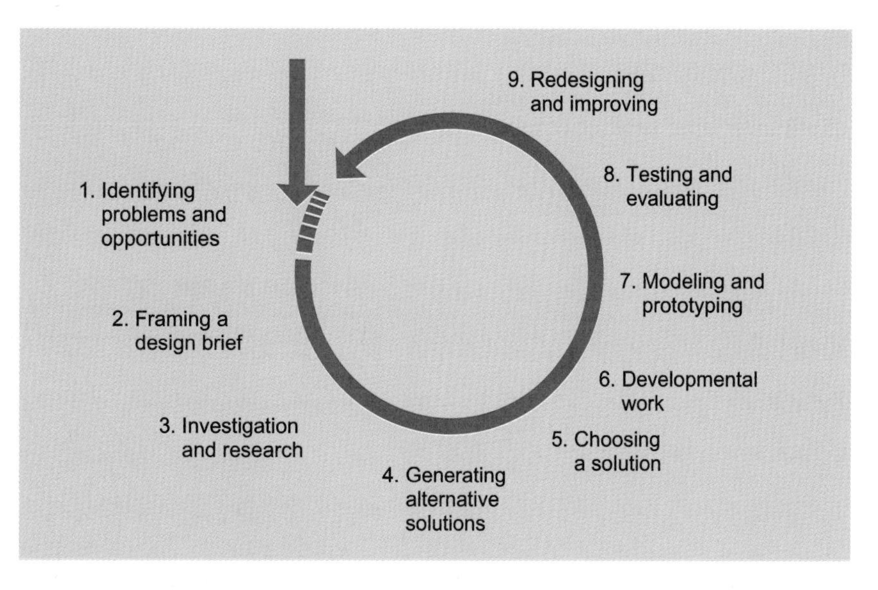

a bit during your designing. For example, if you get to the point of choosing a solution and it occurs to you that you need a bit more background, you will need to revisit the step "investigation and research." Or, if you are building the chosen solution and find a fatal flaw that will not allow you to complete that solution, you will have to go back to the step where you looked at "generation of alternative solutions" and pick another solution. There is nothing wrong with doing this. The design loop should provide you with a framework for your design and problem-solving work, and help you keep on track. The following section will provide a brief overview of the process. More detailed information on the process will be presented in subsequent chapters as you get involved in the actual design process yourself.

Steps in the Design Process

Step 1. Identifying Problems and Opportunities: *Analyzing Real-World Situations*

Central to the process of designing is the identification of a problem in need of a solution. On the surface, this appears to be a simple task, but it requires careful observation and a critical eye. Take, for example, the problems of the physically challenged person.

Some people, through injury, illness, or a birth defect, often cannot adequately function in a world built for more mobile people. This is sometimes referred to as the *technological situation.* A wheeled device may be a solution that comes into your mind, but there are

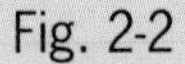

Fig. 2-2

Dysfunction resulting from spinal cord injury is determined by which vertebra is damaged.

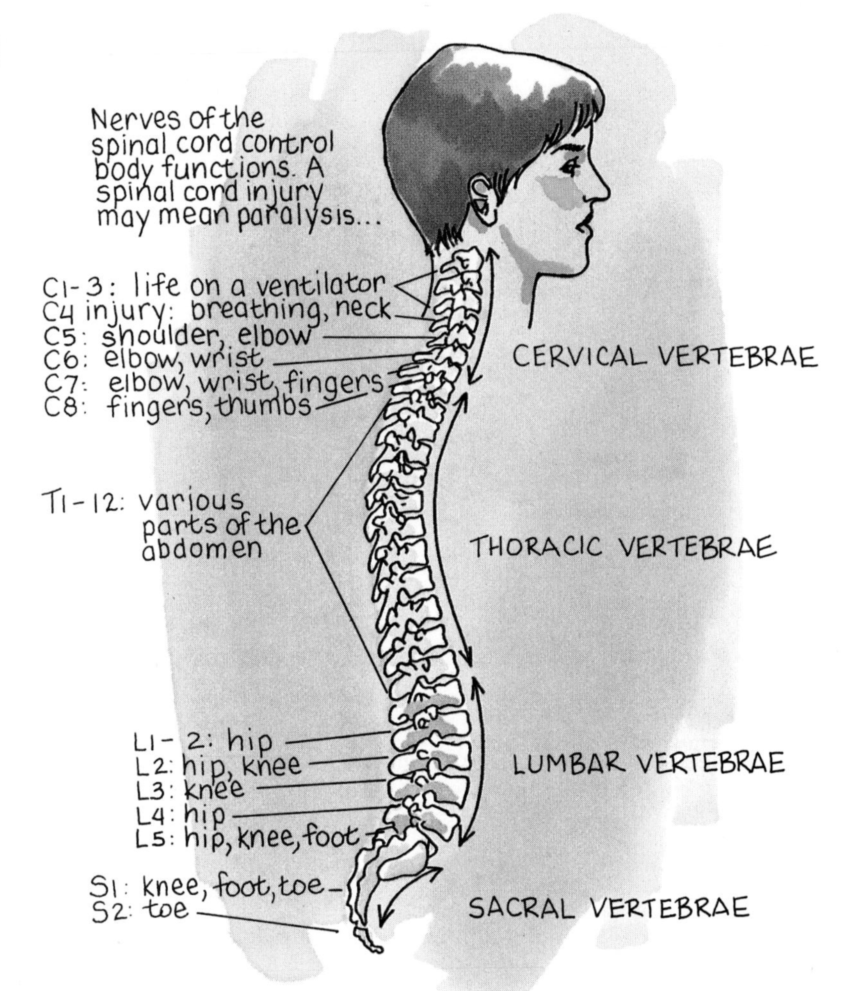

also other possibilities. Electrical stimulation of muscles and braced walking attachments for limbs are proving to be successful solutions for some mobility problems. There are a number of creative possibilities. Taking time to find the root problem is extremely important so that the ultimate solution works. Chapter 3 will help you develop skill in identifying problem situations.

Step 2. Framing a Design Brief: *Problem Clarification and Specification*

The second step in the process is one of clarification and specification. Here you will need to detail exactly what it is you intend to do. This can be difficult, and a good starting point is to ask yourself a number of questions. If the problem is centered around a mobility device for the physically challenged, some questions might be: Is it intended for an adult or a child? Will it be self-powered or motorized? Will it fold up? The specifications you generate will provide some expectations and limitations for the project, and can be referred to at a later time to see if the solutions you are considering address the important aspects of the problem.

Clarification begins with a **design brief**, which then guides further development.

Design Brief Ideas

- Design and develop a system to inform junior high school students about Halloween safely.
- Design and make a carrier for a fishing pole that will fasten to a bicycle or a motorcycle.
- Design a system that will make it easier for a person in a wheelchair to open a standard interior door.

Ideas for design briefs are limitless.

An example of a design brief for a device for the physically challenged follows:

> Design and make a device that will allow individuals with little or no use of their legs to move around the school between and within classrooms. The device should be easy to operate and convenient to use.

This design brief provides direction for the work of the designer, but it is valuable to think things through a bit further by specifiying in more detail the requirements of the solution. Here is an example of a set of specifications for the design brief just shown:

1. The device should meet all guidelines for such devices established by appropriate federal authorities and organizations representing the physically challenged.
2. The device should be capable of being easily stored in a 12"W × 24"D × 36"H space.
3. The device should be safe to use by an adult, and safety should be considered when choosing materials, finishes, and the shielding of possible moving components.

The specifications describe the limitations and requirements of the project's solution. Typical specifications set limits on time, capital, or materials and clarify questions of aesthetics and ultimate user popula-

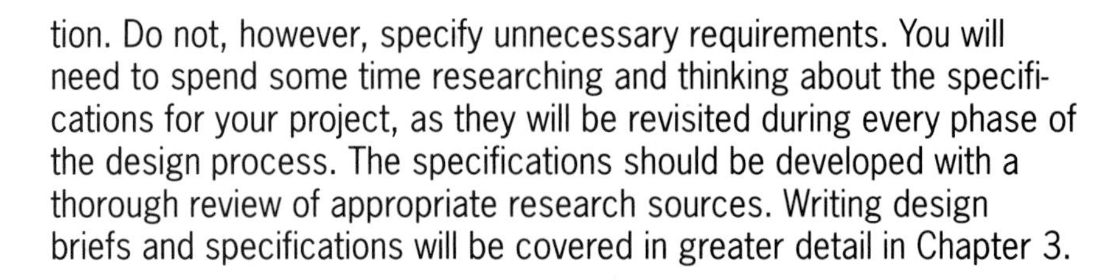

tion. Do not, however, specify unnecessary requirements. You will need to spend some time researching and thinking about the specifications for your project, as they will be revisited during every phase of the design process. The specifications should be developed with a thorough review of appropriate research sources. Writing design briefs and specifications will be covered in greater detail in Chapter 3.

Fig. 2-3

The library is a good place to begin investigating a problem.

Step 3. Research and Investigation: *Information Gathering*

In order to solve problems, all pertinent information must be gathered and documented for possible future reference. If the project involves humans, ergonomics, or human factors, engineering should play an important role. Ergonomics, discussed in Chapter 5, is the study of how human beings best interact with technology.

An example of this is the placement of controls on a bicycle. The distance must be correct between the seat and the handlebars, and where a person's hands fall on the handlebars to work the brake and gear controls correctly. These distances are designed so that most people can reach them comfortably. This also means that the placement of the controls is such that a person can exert the necessary force to stop the bike when his or her hands are in the position to reach the controls. Without ergonomic design, the controls would seem out of place, and using them would be more difficult, if not unsafe.

Beyond questions of the human user will be questions of various technological systems, such as those of structural or mechanical systems. Information about, and examples of, these systems appear in Chapter's 6 and Chapter 7. Sometimes information will be needed to answer questions involving control systems. Chapters 8 and 9 deal with electronic and pneumatic and hydraulic systems.

The importance of investigation and research cannot be overemphasized. Few solutions are new. Even most inventions involve many previously known principles and concepts. Effective designing means looking at existing devices and systems, as well as existing knowledge and solutions as a source of ideas for new solutions. Chapter 5 will provide you with information-gathering strategies.

Step 4. Generation of Alternative Solutions

Generating a number of alternative solutions is one of the most important steps and often the most difficult to do. Although it seems to be human nature to latch on to your first idea and try and make it work, designers know that the more ideas they have, the more likely it is that one of them will be a good one.

The solutions to almost any problem can take a wide variety of directions. Transporting people across a river, for example, can be accomplished by tunnel, ferry, bridge, helicopter, and so on. Likewise, within each of the many avenues of choice lie many alternatives. Choosing a bridge as a solution only begins the process of bridge design.

Generating Ideas. The generation of ideas for solutions is sometimes referred to as **brainstorming**. This term, however, is correctly applied to only a specific idea-generating technique that has firm rules. **Ideation** can also be stimulated and maximized by using techniques such as sketching and doodling, attribute listing, forced connection, and synectics. These techniques will be looked at more closely in Chapter 10.

Fig. 2-4

Brainstorming is more effective with a number of people.

Generating Solutions. Brainstorming and other techniques are useful methods of generating a lot of ideas. Of course, some ideas developed by using these techniques will be unworkable for various reasons, and this is the step in the design process where decisions are made to consider them or not. Ideas may be unworkable because of costs, technical know-how, or many other reasons, but these factors are not considered in techniques such as brainstorming. At some point, however, decisions must be made about the workability of the ideas generated. The workable ideas that emerge from ideation techniques must be further developed before they can be seriously considered, and it is during this phase that these possible solutions evolve.

With more advanced problems, each possible solution takes on its own complete loop in the design process. For example, the problem of transporting people across a river may have led to consideration of a tunnel under the river, a ferry system, a helicopter service, and a bridge. If a bridge is considered, what kind of bridge should be built? Suspension, cantilever, beam, and arch bridges all have unique characteristics, and selection of a particular type is based upon the situation. Most likely, however, the type of bridge would not be considered until that particular solution was selected from among the others.

Although you will probably not be called upon to design a river-crossing system, these principles apply to student design work as well. You will need to work out your alternative solutions to the point where you can see the further choices and problems that these solutions will generate. In some cases, this may mean the development, in detail, of a number of different directions for these solutions.

Developing a number of different solutions to a problem may mean that the size, shape, and materials will need to be addressed at this early stage. Here is where the application of sketching and drawing skills allow you to represent your ideas and begin the evolutionary process of design. You may even need to detail structural or mechanical systems to see if a solution can be made to work. Generating alternative solutions to problems is covered in Chapter 10.

Step 5. Choosing the Best Solution

Choosing the best among a number of ideas is less straightforward than it may appear. Two strategies for choice making are transferable to outside situations. The first involves listing the attributes (good and bad points) of the ideas and comparing them. The second strategy

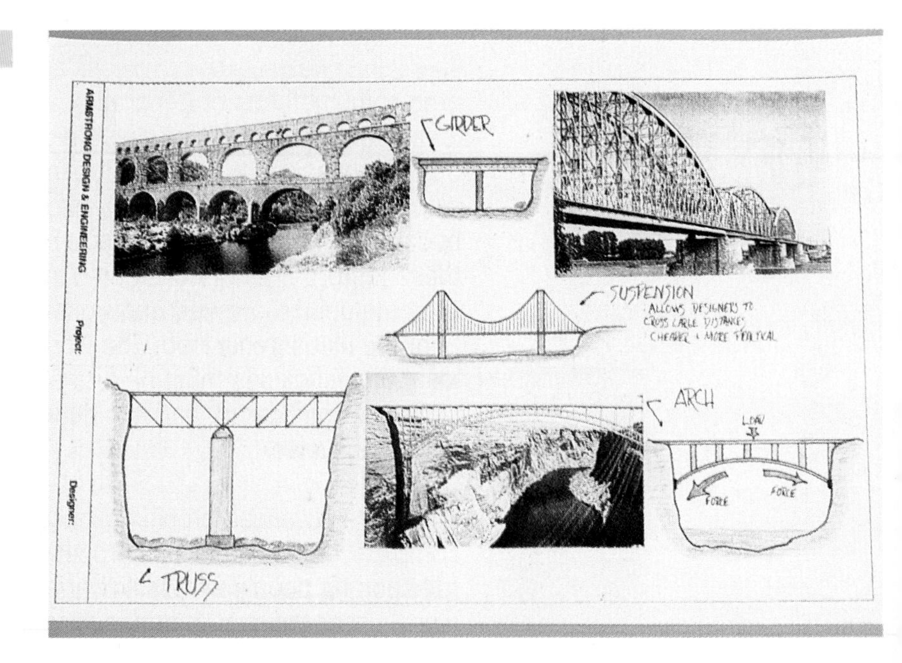

Fig. 2-5

Bridge types

involves the development of a matrix. On one axis of the matrix, you may list the specifications that were generated for the design brief (Figure 2-6). On the other axis, you list the alternative solutions. Checking off requirements and constraints met and scoring each solution is a way of organizing the data to evaluate each one. It is important that criteria be developed for making evaluations.

The evaluation process may indicate a way to combine features of several solutions into an optimum solution, that is, one that is best suited to the particular constraints imposed by safety, aesthetics, economics, and your school technology lab facility, among others. Documentation should summarize reasons for choosing a particular solution and relate them back to the original design brief.

Fig. 2-6

Choosing the best solution

5 - meets perfectly 4 - meets well 3 - meets somewhat 2 - meets minimally 1 - does not meet	Solution 1	Solution 2	Solution 3
meets organization guidelines?	4	5	3
meets organization guidelines?	5	3	5
safe?	4	4	2
convenient?	3	1	3
'Totals	16	13	13

Step 6. Developmental Work

After choosing the solution most appropriate to the specifications, you are proceeding toward the construction stage. However, there may be a number of subproblems that need solutions (this is the case with most technological problems), so modeling, experimentation with different materials, and fastening techniques, shapes, and sizes will need to be done before actual construction of the final design is undertaken. A model may be constructed that will serve to overcome problems in the finished product or to try out specific ideas. This may seem to drag out the project, but more often it saves time. The saying "There is never enough time to do the job right, but there is always time to do it over" applies well here.

This is also the stage at which technical planning—engineering drawings, exploded-view drawings, cutaways, and blueprints—may be

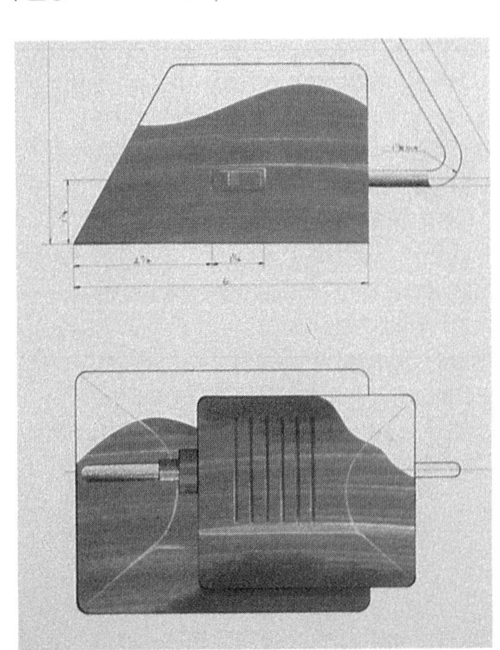

Fig. 2-7

Working drawings are necessary at the final design stages. Courtesy of Dan Davis

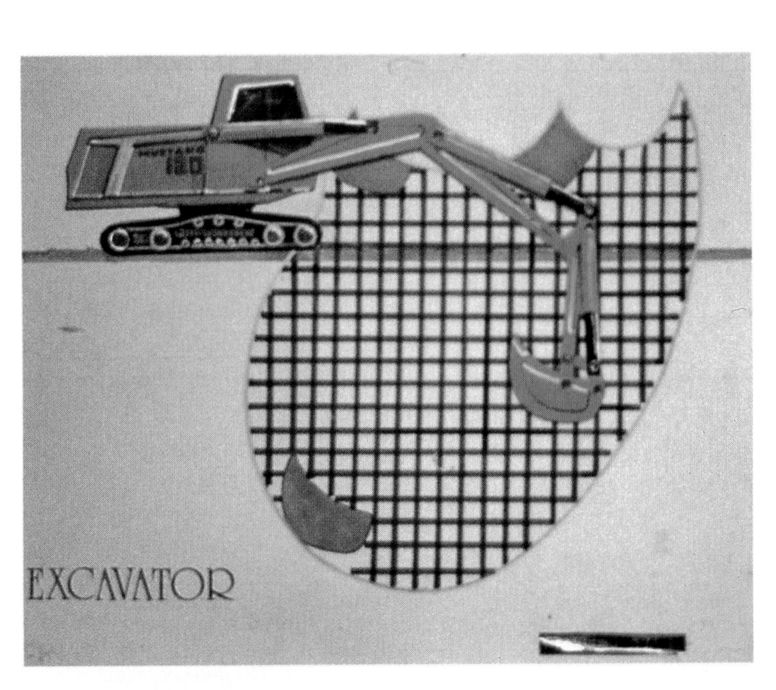

Fig. 2-8

Low-relief modeling is used to work out mechanical operations.

employed (Figure 2-7). Clarifying and communicating your ideas may include color drawings and sketches, two-dimensional models constructed on cardboard or "homosote" that demonstrate or confirm mechanical action or movement of parts, and/or three-dimensional models that help you to work out size, shape, appearance, or other important factors in the design. Figure 2-8 shows a simple model made from card stock that was used to work out appearance and material layout problems. Chapter 11 will deal with the development of ideas into workable solutions.

This is also the stage in industrial planning and design at which ideas need to be "sold" to clients or management for further development by a firm. The need to sell the idea in the world of business and industry means that concepts, mechanics, appearance, and other factors must be represented in a way that is clear to the nontechnician decision maker. Chapter 4 will help you with both developmental and presentation techniques.

Step 7. Modeling and Prototyping: *Construction*

Modeling a solution may take a number of forms: Two-dimensional and Three-dimensional models, computer models, and mathematical models are commonly used. In large aerospace corporations, for example, **computer models** have become the standard by which aeronautical engineers "fly" new aircraft designs on the computer long before 3-D models or prototypes are developed. There are a number of other common **model** types:

Appearance models do not function but are intended to show what a product will look like when it is produced. The clay automobile model used to work out bodywork production is an appearance model (Figure 2-9). Often, photos of new products in catalogs are of appearance models.

Functional models may not look like the end product, but they are operational. These are used to work out the mechanical or electronic systems. For example, Figure 2-10 is a functional model of a pneumatic device.

Prototypes are models that function and look like the finished product but are usually hand built. They are often developed as a "proof of concept," such as a car of the future that appears in an automobile show (Figure 2-11). The term "prototype" is an appropriate one for the type of work often done in technology courses in which only one design solution is constructed. An overview of modeling and prototyping is found in Chapter 11.

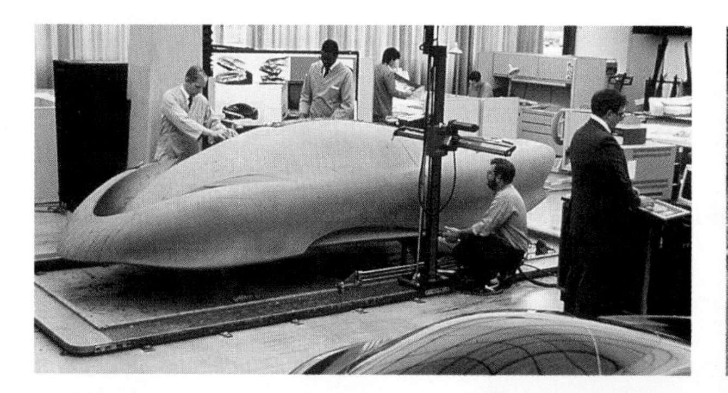

Fig. 2-9

An appearance model of an automobile design. Courtesy of General Motors

Fig. 2-11

Car of the future. Courtesy of Honda of America, Incorporated

Fig. 2-10

A pneumatic device is worked through until proper operation is achieved.

Fig. 2-12

Testing the final solution of a toy in the real world

Step 8. Testing and Evaluating

When the modeling or prototyping is completed, the testing and evaluation of the design solution begins. This may be as simple as applying the specifications to the end product to see if it does all the things that it is supposed to do. But more often it is performance testing, as in the case of a practical device. In the case of an aid for a disabled person, testing may involve giving the device to an appropriate individual to use for a time. An interview at the conclusion of the test would provide invaluable feedback on the performance of the device in the real world and could lead to possible improvements or modifications. This phase is important to the design process, because it tells you how well you have accomplished your goal and whether more development work is needed. Some techniques and ideas about testing and evaluation are found in Chapter 12.

Observation/Analysis/Synthesis

1. List the steps in the design loop, and briefly describe what occurs in each step.
2. Devise an appropriate test to measure the success of:
   a. a toy for a 5-year-old child
   b. a roadside emergency light
   c. a road map
   d. a backpack
   e. a word-processing system

Analyzing Existing Products

You will work in small groups and will have a number of products to choose from for the analysis activity. While each product was carefully designed for the marketplace, the actual details of the process are unknown.

Step 1 Write down the original problem the product was meant to solve.

Step 2 Write a design brief for the problem, including specific specifications.

Step 3 What questions had to be answered before the product could be manufactured? Make a list of as many questions as you can.

Step 4 What other alternative solutions to the problem that you have identified in Step 1 might be possible (besides the product that you are analyzing)? Make a list.

Step 5 Devise a test to assess the effectiveness of the product

CHAPTER

3 Problems and Opportunities

Introduction

What is a problem? If my car does not start, is that a problem? What if the fuel tank is empty? Is that the problem? Although the fact that the car doesn't start is a symptom of a problem, the lack of fuel is one possible cause of the problem. But is this example a technology problem? To answer this question, we will need to look at some definitions.

Faultfinding is the process of looking for the cause of a malfunction. If a car will not start, we might look at the electrical system and the fuel system. If we find, through our faultfinding process, that the car is out of gasoline, we would correct the fault and attempt to start the car. If the car starts, we have diagnosed and corrected the fault. Faultfinding is different from the design and problem-solving process described in this text. Faults are somewhat like math problems: There is only one right answer! Problems solved with technology have many answers; some solutions are better than others, but there are many alternatives just the same.

The search for **scientific truth** is another process that is important for us to understand. In Chapter 1, the differences between science and technology were outlined, and it was pointed out that technology seeks to meet our needs and wants and to extend our capabilities, whereas science seeks to explain and describe the physical world. Science uses the **scientific method** to gather evidence in order to confirm or refute a carefully developed hypothesis (guess) about the physical world. Meeting a need with technology can take many forms, but in science there can be only one true answer to a question such as "What is the acceleration of a body in Earth's gravity?"

Calculating acceleration in Earth's gravity

$$A = \frac{F\text{ (Force) in Newtons}}{M\text{ (Mass) in Kilograms}}$$

The problems we will try to solve with technology, therefore, are not malfunctions, and they are not answers to scientific questions. The

Sean's Assignment

Sean was given the assignment in his technology course to come up with an idea for his semester project. Because this was going to be a rather long and in-depth project, he wanted to do something he was interested in, not just pick the first idea that popped into his head. Using the techniques he had learned in class, Sean took a piece of paper and, in the center, wrote "INTERESTS" and drew a circle around the word.

He thought about the things he liked to do and began to write them down around the circle. First came "cycling" and then "camping." After that came "basketball" and "volleyball," followed by "canoeing." He was not one to sit around the house; he had a lot of interests. The list might get pretty long at this rate.

He made some spending and future college money by cutting lawns and he was looking forward to his 17th birthday when he could get a small, used pick-up truck to haul his lawn maintenance equipment. He'd have a sizeable contribution toward his college tuition by graduation. He put down "lawn maintenance" as one of his interests.

His plans were to pursue a career in an environmental field, but he wasn't sure whether it would be engineering, law, or something else. His camping, canoeing, and hiking experiences through the scouts and a community Sierra Club group had sparked first his interest and then his ambitions to work in preserving the natural environment. He wrote "ecology" as another interest on his paper.

Sean had a cousin with cerebral palsy, a condition that makes muscle control difficult. If he could do something in that area for a project, he thought, that would be pretty interesting, and he wrote that down. He would have to find out more about both his cousin and cerebral palsy before he could tackle such a project. He spent some time reading and made a few telephone calls.

Next, Sean sat back and thought about the problems and difficulties he encountered with his first interest, cycling. He knew that he wasn't going to redesign the bicycle, so he thought about the minor things that irritated him or caused him inconvenience.

He thought about people stealing his bike; it was a pretty good one. He had bought a hardened cable, one of the strongest you could buy, so he doubted that he could come up with a better one. Still, for every chain, there's a chain cutter, so he put down "prevent theft," drew a circle around it, and connected it to the word "cycling."

Then there are the bumps. It would be nice if there were a way to make the ride a little smoother on the back roads he usually traveled. He wrote that down and connected it to the word "cycling."

Sean worked for most of the evening, although he ran out to the garage to look at some feature on his tent and the grass trimmer, read a bit, and made more phone calls. He came up with ideas for problems to solve for five of his seven "interests." He spent some time looking over what he had done and finally decided that his semester project would be a problem that he could tackle in the time he had and would provide something worthwhile at the end.

problems that we are interested in are those that we can define in such a way that we will have alternative courses of action. The results of solving problems with technology will be *products* and *objects*, *systems*, and *environments*. For example, the problem of getting a message to someone on the other side of the world has a number of technological solutions: telephone, FAX, mail, radio, or television. Of course, you could always deliver the message in person, but that would involve a new set of problems.

Problems Versus Solutions

It is sometimes difficult to distinguish between problems and solutions. Whereas problems tell us the difficulties we must overcome, solutions tell us how to overcome them: *Design and make a bridge to get to the other side of the gorge* is clearly a solution; *design and make something to get to the other side of the gorge* describes what we must do but leaves us with alternatives. Although this may seem to be a simple point, it is a very important distinction. Starting with *design and make a bridge...* precludes any other options from being explored.

Inventions Versus Innovations

Invention is the development of something completely new. However, design work does not stop after a new product has been produced. The products of technology evolve over time. The invention of something is rare compared to the many improvements made in things already invented. These changes and improvements are called **innovations**. Most of the work done in technology classes may be categorized as innovation, such as the design and construction of a BMX bicycle light. The light bulb, battery, switch, and materials used as a container have already been invented, but, by putting these components together in a clever way, a student may come up with an innovation in bicycle lights.

It is conceivable, however, that a student working on a project to design and create a solution to a problem may invent something new. For example, if you are given an assignment to develop a design project based on a device to aid a physically challenged person, your work may lead to an original invention. There are quite a number of students who have received patents for design work they have done for a class assignment.

It is not important, however, that your design solutions be original inventions. The development of both inventions and innovations is a creative endeavor, and, commercially, innovations are probably just as profitable as inventions. A product that already exists is an excellent starting point for design work.

Profile of an Inventor

Dr. Percy L. Julian in the late 1940s invented substance S, a synthetic cortisone. Cortisone was known as a powerful pain-killing drug for arthritis sufferers. As Dr. Julian said, "every problem grows into a new problem" was certainly true for this new drug. Prior to his work, natural cortisone was extracted from oxen bile. Unfortunately, it took the bile of nearly 1,500 oxen to treat one patient for one year, making cost prohibitive for all but a very few of the millions of rheumatoid arthritis sufferers. This problem became Dr. Julian's opportunity. His invention of substance S made this beneficial drug available at a cost that everyone could afford.

Dr. Julian attended DePauw University in Indiana, where he graduated top in his class, earning membership in Phi Beta Kappa and Sigma Xi honorary societies. In 1931, Dr. Julian received his Ph.D. from Howard University. Most of his work was with products derived from agricultural products, especially soybeans. Some of his other products include Aero-Foam, a petroleum fire-fighting foam; synthetic male and female hormones, progesterone and testosterone. Progesterone is especially useful in saving babies by protecting mothers with problem pregnancies from miscarriage. In 1954, Dr. Julian founded the Julian Laboratory, Inc., where he directs the production of his many health-promoting products.

Percy Lavon Julian, soybean chemist. Courtesy of Bettmann Archive

Fig. 3-1

Mountain bike. Courtesy of Trek Bicycle Corporation

The bicycle is a good example of a product that has undergone many innovations. From the basic invention in 1839, the bicycle of today has only two wheels in common with the original. Contemporary mountain bikes use 21-speed transmissions, ergonomically designed handlebars and seats, "space age" alloy or composite materials used in the frame and gearing, and specially designed tires for rough terrain. Take one look at the huge number of accessories available to customize the bicycle, and it's easy to see that innovations are a valid design route.

Problems and Opportunities

Fig. 3-2a, b

Some of the products of technology used in the morning.

Opportunities for your own design work are all around you. For example, if you made a list of the things you use every morning before school, it might contain many of the items in Figure 3-2a, b. What are some of the appropriate questions you could ask about these items? You might ask if they are effective or if they give you value for the money you spent for each of them. Are they trouble-free? Are they attractive? Durable? And so on. Can you think of other questions you could ask?

These items can be found in most homes. Someone—maybe you—chose particular brands and models from among a wide variety of choices. A quick trip through a store will tell you that there are many choices for each: color, size, materials, style, quality, and so forth. But all these products can be improved, and, in fact, that is going on all the time in the companies that produce them. The improvement of an existing product is one possible route to a design problem.

Looking at Purpose

Another strategy to uncover possible design opportunities is to ask a completely different question: the purpose of the item. This can be addressed by asking "What is the problem that the item is trying to solve?"

For example, what is the problem that the toothbrush is attempting to solve? Is it attempting to solve the problem of "How do we brush our teeth?" Of course not—that would be too simple. The problem really is: Eating food causes small bits of that food to accumulate in places in the mouth which, over time, cause tooth decay and gum disease. Now, looking at the problem from this perspective, can you think of some other device that was developed, other than a toothbrush, to solve the same problem?

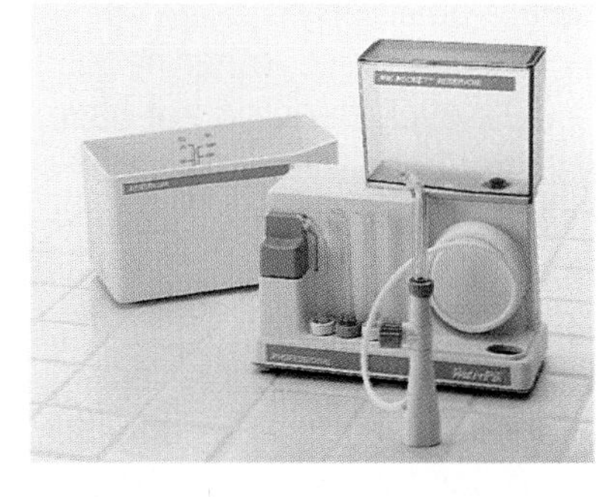

Fig. 3-3

The Water Pik® represents another solution to a common problem. Courtesy of Teledyne Water Pik®

If you said a water-jet device (such as the Water Pik®), then you would be correct. Just as wall-mounted hot-air dryers and paper towels will both dry wet hands, there are a number of different solutions to any given problem. The first step is to back up and ask "What is the problem this device is trying to solve?" This is a very useful strategy to use to uncover design opportunities.

Looking at Needs

Another approach to looking for possible problems to solve is to look at the problems common to all of us. A developmental psychologist, Abraham Maslow, has identified a number of needs that all humans have.

He has organized these needs from the most basic—the things that we all must have for simple survival—to higher-level needs—those that we have when our basic needs have been met. It is called **Maslow's Hierarchy.**

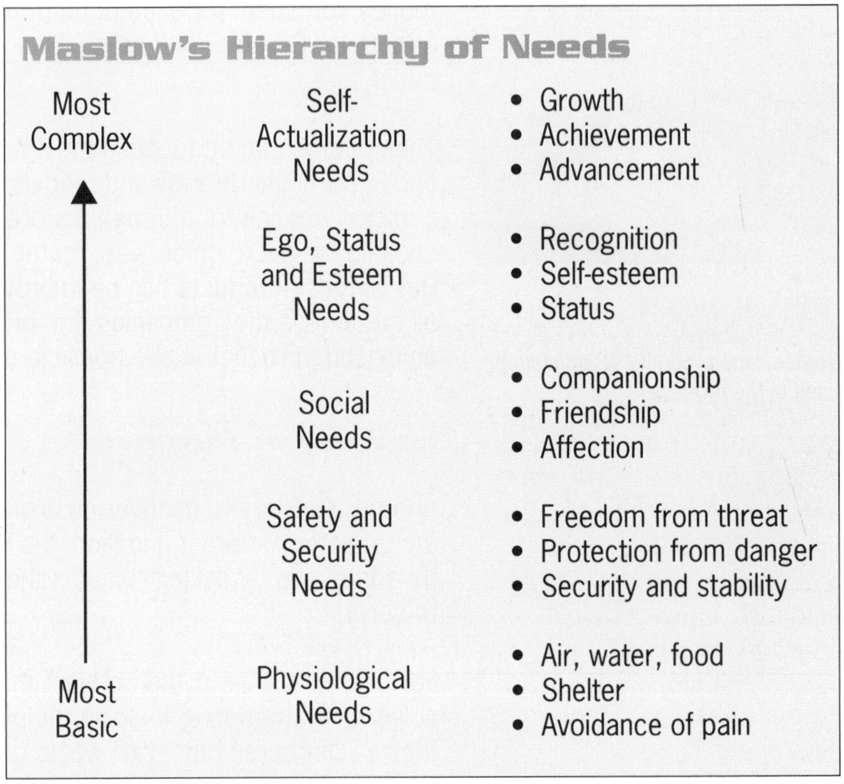

Maslow's Hierarchy of Needs

	Need	Examples
Most Complex ↑	Self-Actualization Needs	• Growth • Achievement • Advancement
	Ego, Status and Esteem Needs	• Recognition • Self-esteem • Status
	Social Needs	• Companionship • Friendship • Affection
	Safety and Security Needs	• Freedom from threat • Protection from danger • Security and stability
Most Basic	Physiological Needs	• Air, water, food • Shelter • Avoidance of pain

Maslow's Hierarchy of Needs

At the most basic level, we must all have *water, food, shelter*, and, according to the climate, *clothing*. These are basic **physiological needs** for our individual survival. If we don't have these, other needs simply do not matter. You do not need a CD player if you do not have food. After the basic physiological needs have been met, there are higher levels of needs. These are *safety and security needs, social needs*, and others.

Life in America at the end of the twentieth century is not representational of life for most of the rest of the world's population. As only a small percentage of this population, most Americans enjoy a standard of living in which their basic needs are met, (Figure 3-4). For almost 70 percent of the people in the world, the struggle to meet these basic needs is their entire life. If you look at the most basic needs, you may come up with ideas for development.

Solving the problems of people in the developing nations, satisfying their needs and wants, and extending their capabilities is most effective with appropriate technology. "Appropriate technology" is a term used to describe the sensible and sensitive selection and application of technology.

In developing countries, appropriate technology takes the form of using existing skills and resources to solve problems with technology. It has been found that, in most cases, industrialization has led to assaults on traditional occupations, local culture, and the environment, so ways are being found to use technology in order to create jobs with locally available resources, labor, and capital. The development of renewable energy resources, such as solar, hydro, and wind, are examples of such efforts. Other basic needs in the categories of shelter, food, and clothing offer many design opportunities in appropriate technology, both locally and with specific regions in mind. Investigating the needs and the resources in these regions will be an integral part of any design work undertaken.

5. Food Processing

Small improved tools which are still labour-intensive and can be easily manufactured by village craftsmen have found wide acceptance among small-scale farmers in developing countries. Shown here are four examples which have speeded up the processing of the major crops.

Groundnut Sheller

Source: Village-Level Technology. Bangkok (UNICEF)

This hand-operated groundnut shelling machine has an adjustable wooden paddle which cracks the nuts in the trough. The shelled nuts fall through the weld mesh into a receptacle placed underneath.

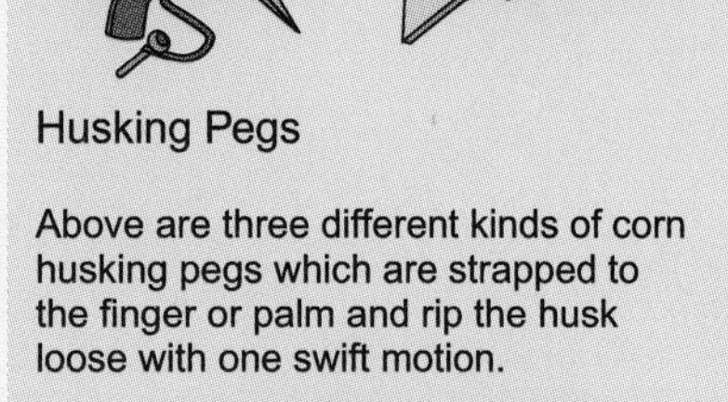

Source: VITA

Husking Pegs

Above are three different kinds of corn husking pegs which are strapped to the finger or palm and rip the husk loose with one swift motion.

Maize Sheller

This wooden, hand-held maize sheller uses the ridges to pull out the grain from the cob.

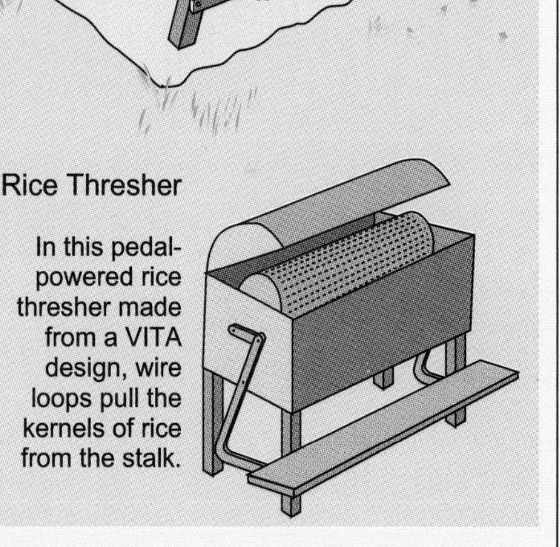

Rice Thresher

In this pedal-powered rice thresher made from a VITA design, wire loops pull the kernels of rice from the stalk.

Source: Village Technology in Indonesia (UNICEF)

There are other situations in which the basic human needs cannot be taken for granted, as in winter mountaineering. In this situation, shelter is crucial for survival in the harsh mountain environment, and designing for these conditions is no small task. Winds are heavy and will flatten some of the most sturdy tent designs. How might you go about designing a structure for survival in severe winter mountain conditions, (Figure 3-5)? Remember, it must be small and light enough to pack in with food and all the other necessary survival gear.

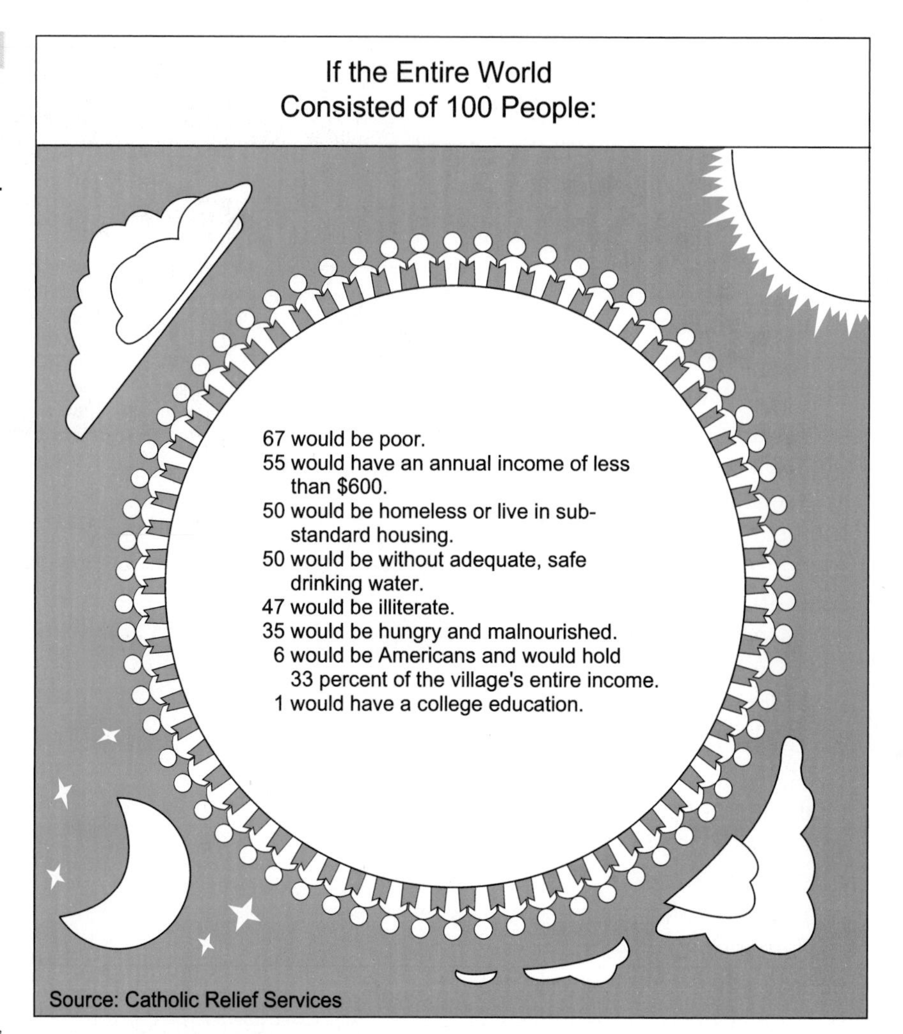

Fig. 3-4

Throughout much of the world, hunger and poverty are facts of life. Yet the quality of life for most people in America reflects far better conditions.

Fig. 3-5

Mountain conditions during the winter offer many design challenges. Courtesy of Johnson Camping, Incorporated

Looking at Real-World Contexts for Design Work

Opportunities for design work in technology are everywhere, but a good place to start are places with which you are most familiar. The *home*, already mentioned, is a good source of design problems. The *school* and *community* in which you live are examples of places about which you should also know a great deal. In addition, you probably have *interests*, *hobbies*, and perhaps *sports* you like to play.

School. Around a school there are many design opportunities. If people visiting a school often get lost trying to find a particular classroom or office, this is a problem waiting for a solution. Other problems might center around facilities in the technology or science labs or might address technical problems of theater production. Theater production, which ranges from producing stage plays to rock concerts, presents design opportunities in lighting, scenery structure, and moving stage sets, as well as a host of other areas (Figure 3-6).

Fig. 3-6

A theater production offers many problems and opportunities for design work. Courtesy of Nick Bryson

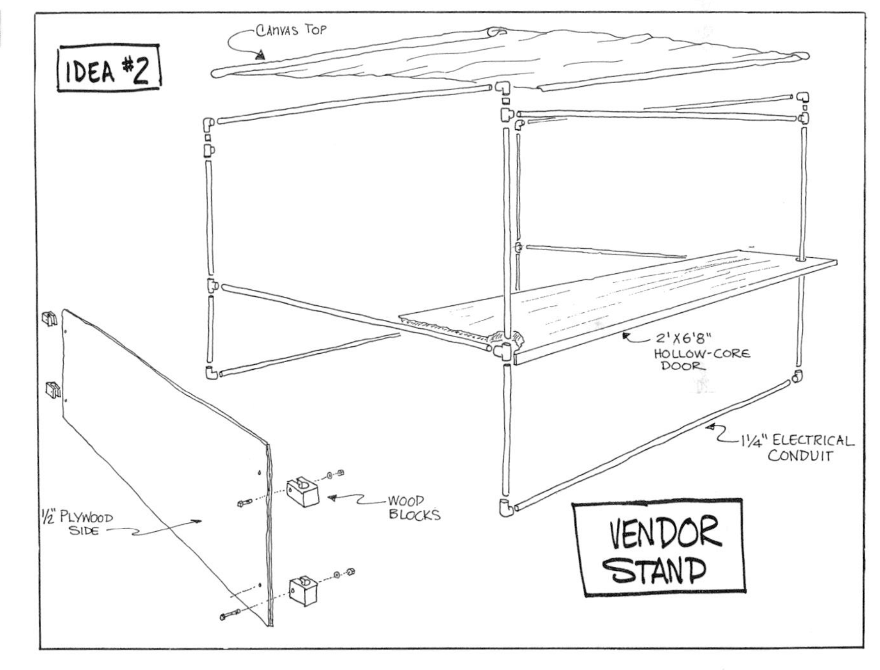

Fig. 3-7

Design ideas for a vendor stand for a civic organization.

Community. There are many design opportunities within your local community. Some groups who may need your expertise in design and technology are local volunteer and service organizations, clubs, and small businesses. If you sit down with a representative from almost any one of these groups, you will find that it has problems that make ideal design projects (Figure 3-7).

If you choose this approach, try to find a group that interests you. For example, let's say that you are interested in architecture and the outdoors. You could interview people at a community or county park.

Your interviews may point to the need for a new information center for the park. Your end product may be presentation renderings, architectural drawings, and a model for a proposed new information center.

If your interests lie in fishing, a hatchery may be a place to look for design problems; if it's in theater, a local playhouse group may be just right for you. The trick, again, is to ask the right questions so that you get at a problem that you can address in the time you have and the resources that are available.

Hobbies, Sports, and Other Interests. Your own interests and hobbies are some of the best areas for design opportunities: You are already familiar with the topic; learning about your own hobby doesn't seem like work; and, you can probably use the end product when the project is over.

People who engage in a particular sport or hobby for some time become very knowledgeable about the subject, and it is not uncommon for hobbyists to develop new inventions and improvements in existing products. Who is better qualified as a designer than someone who sees a need or an opportunity firsthand? Good examples of this can be seen in the developments in surfboards and skateboards. People who had a passion for these sports are the ones who have led the way in the improvements and innovations in equipment over the years (Figure 3-8).

Fig. 3-8

Sports interest and hobbies offer excellent design challenges. Courtesy of Bic

This is also a good example of the fact that you need in-depth knowledge about the area in which you are designing. If you try to do design work with only superficial knowledge, your solution will probably look naive to the educated eye.

Issues. As we grow older, our world expands to include the world outside of our own immediate experience. *Issues* are controversies facing society today, and they are good starting points for problems. For example, acid rain is an issue facing many industrialized and developing nations of the world because of the widespread burning of fossil fuels. It is poisoning entire lakes, damaging and killing trees and crops, and creating other environmental disasters. Although most issues will not be solved in technology classes, the development of monitoring systems, public awareness strategies, and other creative design work can be appropriate and meaningful. In addition, *future-oriented news items* and *speculation* about the future can lead to ideas for design projects.

Looking at Mind Maps

Sean, in the narrative found earlier in this chapter, decided to look at his own interests in choosing a design and technology project. He did this by developing a simple **mind map**, sometimes called a **concept tree**. This is a technique that may also help you to come up with design ideas. Mind mapping is putting thoughts on paper in a way that helps you to think through ideas.

Generating a simple mind map is not difficult, but, as you begin to put ideas down in detail, it takes time and thought. You may begin with either the word *interests* or with a specific interest, such as boating, if you have already limited your design work to that specific area. This word becomes the center of the mind map, and all other ideas will spin off from it. As your interests come to you, just write them down as "balloons" that radiate from the center word, as in Figure 3-9.

Fig. 3-9

Mind map of ideas for design work

As you begin to fill up the space around the center with other words, you will want to think about some of the problems you encounter in these areas. Sean's interest in cycling made him think of the possibility of losing his bike to thieves. Let's say that your interest is playing the guitar, which makes you think of how you would like to get rid of the wires going from the guitar to the amplifier because you are always tripping over them. Write down a key word or two describing the problem and connect it with a line to the appropriate interest. Soon, if you work at it, you will have a page covered with problem ideas. These are all potential design projects.

Mind mapping is not limited to individual effort. Figure 3-10 is a mind map developed by a class looking for a variety of technology problems around one theme. Because a class trip to Washington, D.C., was scheduled for later in the year, the students decided to use that theme as the focus for their mind mapping.

There are many different strategies for finding problems suitable for your design work, but they all attempt to get you to look more closely at common things around you. The inventor and the innovator are people who have a little different perspective of the world: They see the problems. This may be the most creative part of their work.

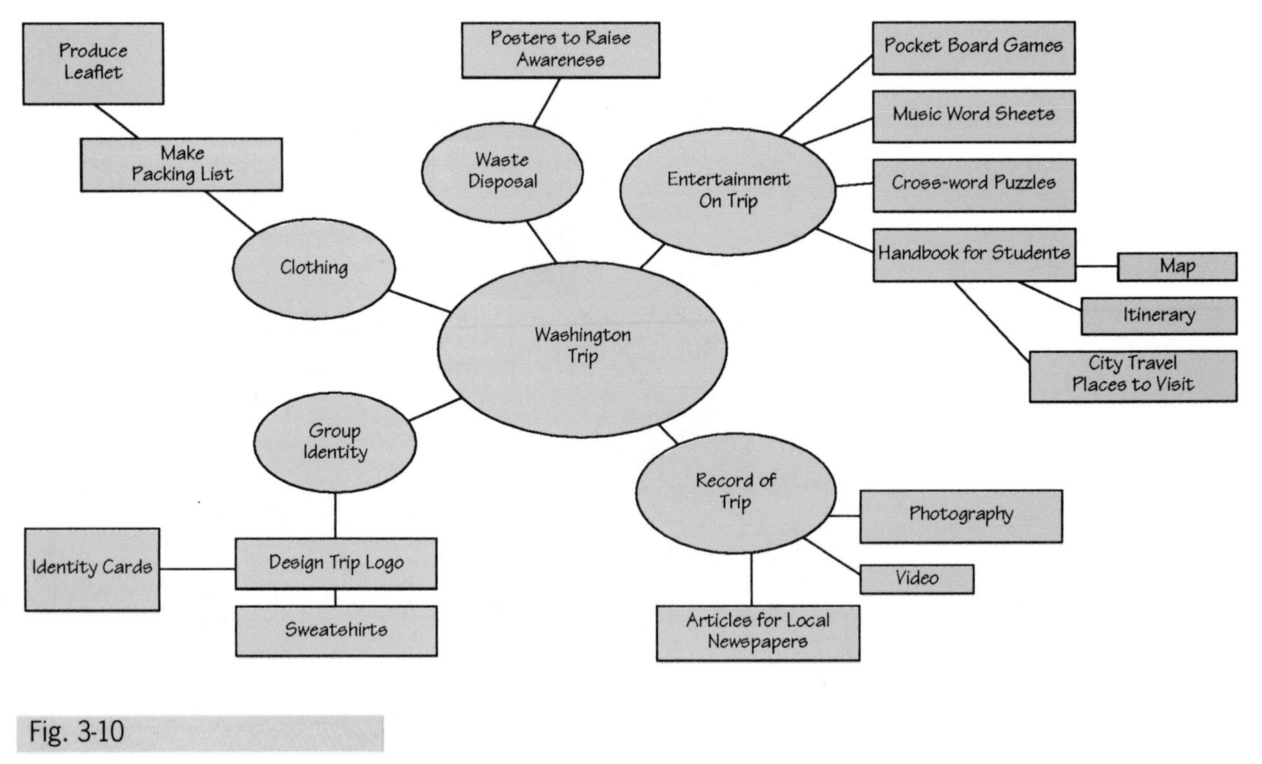

Fig. 3-10

Design opportunities centered around a class trip

You may often hear, when someone looks at a clever new product, "I could have done that!" and they may be right. But someone identified the problem, perhaps with a new perspective, which allowed that solution to be developed.

Developing a Design Brief

Once you have identified an area in which you want to work—let's say that you have a problem to solve—the next step is to clarify it so you have some direction. This is the development of the design brief.

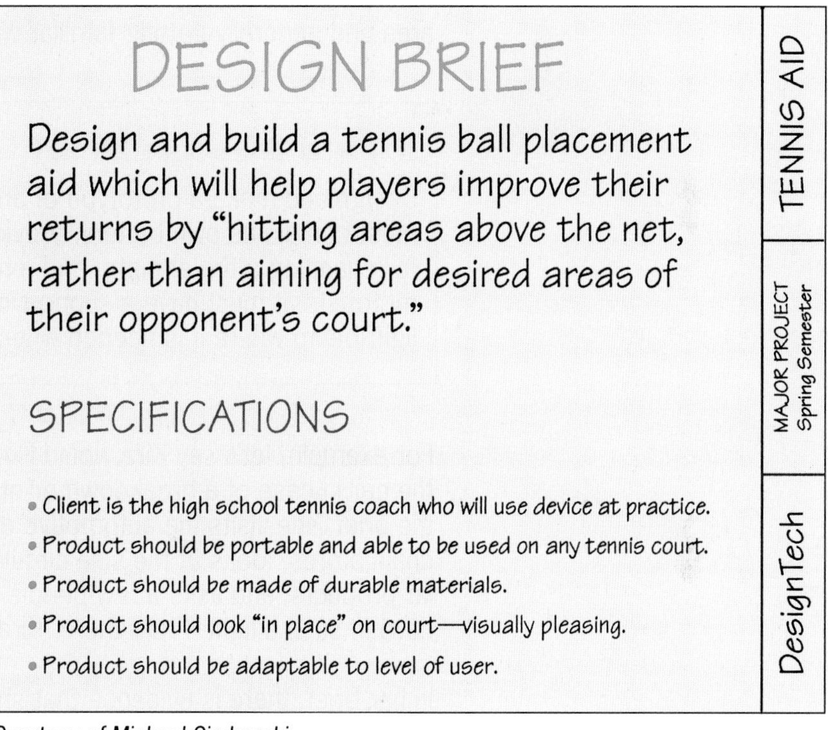

Courtesy of Michael Simborski

Writing a Design Brief

The writing of the design brief will make you think very precisely about the problem you are tackling. You will need to make it broad enough to open up a number of possible solutions so that you will have a chance to think creatively. The design brief is a statement with two parts: (1) what the designer (you) will do and (2) what the successful design will do.

What the Designer Will Do. The first part of the design brief, the statement about what the designer will do, is easy to write. Design briefs most often begin with such wording as:

- Design and make a device ...
- Design and construct a structure ...
- Design and build a prototype ...
- Design and develop a system ...

These statements are instructions for yourself.

What the Successful Design Will Do. The second part of the design brief, the statement about what the design solution will do, is the part that will take considerable thought. It is often necessary to gather information about your subject before this part of the design brief can be completely written. Gathering information may include such things as finding out what else has been done by others in the area and generally getting familiar with the topic.

Kira's Design Brief

Design and make a prototype of an automotive emergency light. It must contain its own battery, provide a bright white light suitable for changing a tire at night, and a red or amber light to warn other motorists. It must have a support or other means of aiming the light beam where it is needed when the unit is set on the ground.

For example, let's say Kira would like to make a light to be carried in the car in case of a breakdown or other emergency. Before writing the brief, she visits the automotive department of a few discount chain stores, looks at the sale circulars in the Sunday papers for similar products, and asks a few people what features they would like to have in such a light. From there, Kira develops this design brief.

In her brief, there is evidence that she thought about the situations in which the light might be used, such as warning other motorists, helping motorists change a tire, or light some other area. Kira may have seen some of these features on similar lights in stores or in catalogs, or these may have been suggestions by people with whom she discussed the project. Kira now has a beginning point and a direction for her design and development work. She also has some criteria for evaluating the success of the project.

Fig. 3-11

Student prototype

Notice that she specifically said that the device will be in the form of a prototype. This means that it must look and function as a finished product, but the actual construction might be different from that necessary for a mass-produced product. Chapter 11 will discuss prototyping techniques, which will include using a variety of materials to achieve a finished product effect.

Writing Specifications

Sometimes the simplest devices require the most developmental work. This can be seen in the products for aircraft where safety and reliability are the highest priority.

The requirements of most aviation products are that they be light, strong, simple for the pilot to use, and fit in as small a space as possible. These requirements make sense when you look at a small plane. Many single-engine airplanes weigh less than 1,200 pounds. Every ounce of additional weight is subtracted from the load that the plane can carry, because the amount of lift the wings can provide is fixed. That load includes the pilot and passengers, fuel, oil, baggage, and everything else.

The components of the light aircraft are subjected to terrific vibration and stress, so strength is important, even for seats, switches, and radios. A pilot flying a plane usually has many things to do at once, so instruments, avionics (radios and other aviation electronics equipment), and other control devices need to be simple and straightforward to operate. In addition to all these requirements, the small aircraft has very little room for accessories, so anything a manufacturer would like to sell the pilot for her or his plane must fit into a small space.

The requirements for aviation products are really specifications that a designer would want to consider in the development of almost anything for the aviation market. In the same way, the specifications for your design solution should help you clarify what it is you are going to do.

Aircraft components, such a gauge for monitoring the engine oil pressure, may cost three or four times as much as a similar automotive gauge. Much of the higher cost of the aircraft gauge is in the extensive testing and certification of the gauge to meet Federal Aviation Administration (FAA) specifications. Specifications help you end up with what you want.

With regard to Kira's emergency light, a bright, white light suitable for changing a tire was specified. This is a specification that is difficult to measure. If Kira were an experienced designer, she might have felt it was necessary to research the amount of light suitable for the tire-changing operation. This might have taken the form of looking up figures in a table for recommended light intensity to perform certain similar operations (light is measured in units of footcandle or lux), or she might have set up a few simple experiments to determine the light intensity suitable for tire changing at night.

Brief

Design and construct a portable work surface that will also act as a storage unit for pieces of drawing equipment while not in use. In use, the unit must be freestanding; however, it must fold to a smaller size to make movement more convenient.

"How much do I specify?" is always a question you will need to answer. The more experience you have in designing, the more precise your specifications should be. The following are some considerations you should think about when writing your specifications.

Functional Questions

1. *Are there guidelines for the development of products in this area?* National associations, independent evaluation groups such as Underwriters Laboratory (UL Listed), government codes, and so on provide guidelines.
2. *Are there size considerations?* Will it fit the human user, through a door, or in a pocket? Will it fit in an existing space?
3. *Is it a powered device?* Will it be battery powered, line-current powered, spring powered? Will it use a DC motor?
4. *What costs are involved?* Are you on a limited budget? What are the available resources?
5. *Can it be produced?* Do you have access to the materials, proper tools, and machines? Do you have, or can you develop, the skill to produce it?

Human Factors Questions

6. *Does it fit the human user?* Have you considered clearances, reach, size, and so forth?
7. *Are you specifying safety considerations?* Are there moving parts, high voltage, toxic by-products and so forth?

Environmental Questions

8. *Are you considering the environment?* Are the resource inputs and waste generated in manufacture environmentally friendly? What happens to the product when its useful life is finished?

Aesthetic Questions

9. *Is it aesthetic?* Does it have a pleasing appearance and style? Are materials used appropriately? What about finish?

Production Questions

10. *Is the product marketable?* Will there be consumer demand for it?
11. *Is the product producible?* Can it be efficiently mass produced?

The specifications describe the limitations and requirements of the project solution. They should be developed through a thorough review of appropriate research sources and are subject to change or updating as you become more knowledgeable about your subject. You will need to spend some time researching and thinking about the specifications for your project, since they will be revisited during the testing and evaluation phases. The success of your work will be judged in part against what you specify you want to do and what it is you come up with as the final solution.

Beginning design work, with simple problems, may not require specifications separate from the design brief. But, as you tackle more sophisticated problems, the design brief becomes cluttered with too much detail. The previous design brief may be written in the following format to call attention to the specifications:

Kira's Emergency Light

Design Brief

Design and make a prototype of an automotive emergency light. It should be suitable for changing a tire at night.

Specifications

- Must be powered by its own battery
- Must have a bright, white light capable of illuminating an area large enough to change a tire safely
- Must have a red or amber light to attract the attention of passing motorists for safety reasons
- Must be able to aim the white light when the unit is set on the ground

The format example just given, with accompanying specifications list, is most suitable for more complex design briefs. Simple briefs can be written with the specifications as part of the brief itself.

Summary

At first, finding a problem may be difficult, but, as you begin to look around, you will find an endless variety of problems and opportunities for design work. Begin in areas with which you are familiar: home, school, community, hobbies, and sports. You need knowledge about the subject to attempt any design work, so these areas are excellent starting points.

You can find design opportunities by defining the problem a particular product is trying to solve. Using this as your jumping-off point, your focus will be to look at other creative approaches for solving the same problem.

Look at the needs people have, especially people with special needs. Talk to people with physical challenges, and find out things they have difficulty doing that the rest of us take for granted. This is another good place to start when looking for problems.

When you have found an area on which you can focus your design efforts, you will need to develop a design brief and accompanying specifications. The design brief will help you to define what it is you intend to do and constraints within which you will do it. The design brief is a statement about what you as the designer will do and what the successful solution to your problem will do. It will be necessary to do some preliminary investigation into the problem to write a useful design brief.

Observation/Analysis/Synthesis

1. Write a sentence that describes the actual problems in the statements below:
   a. I need to build a bookshelf.
   b. I want a trailer for my bicycle to carry my surfboard.
   c. My friend needs crutches because of his accident.
2. Find a problem situation, and write it out in each of the following areas: home, school, community, church, leisure activities, sports.
3. Write a preliminary design brief for each of the problem statements below. You should only use your existing knowledge about the subject.

a. Because Uncle Walt has use of only one hand, it's difficult for him to open pop-top cans.
b. I can never find my keys.
c. The printer paper always ends up in a pile on the floor.
d. Books kept in the basement are damp and full of mildew.
e. I have a box full of photographs and slides, and I can never find what I am looking for.

4. Investigate the topics below, and then write a design brief with specifications for each problem.
   a. Light bulbs with sufficient candle power to illuminate the interior of the stomach give off too much heat to be used safely in noninvasive surgery.
   b. Many people do not have any idea of the quantity of waste we discard or where it ends up and therefore are not motivated to conserve or recycle.
   c. Individuals with arthritis in their fingers often have difficulty grasping small objects and grasping objects tightly.
   d. Barking dogs are annoying.

CHAPTER

4 Documenting Your Design Work

Introduction

Keeping a record of ideas and development work is an integral part of designing. The term "designer" is used for a wide range of occupations. The architect and the civil engineer are designers of structures; the graphic artist works with elements on a page, such as typeface, layout, and artwork to design print materials; the industrial designer works with product design; and the fashion designer works with clothes. These and many other occupations involve design, and a record of work is a necessary part of the job.

Fig. 4-1

Why Document Your Design Work?

In the world of business and industry, the *engineer's notebook* is a careful record of ideas, calculations, thoughts, and plans for a particular project. It is not unusual for the engineer to be pulled from the middle of a project, only to pick it up again six months or several years later. The documentation within the engineer's notebook saves countless hours of rediscovery when the engineer returns to a task.

In a similar manner, the records kept by the inventor may well be the legal basis for awarding a patent when several similar patent applications are submitted. Questions such as "Which applicant had a critical idea or design developed first?" can be answered with documentation evidence. Contrary to popular belief, ideas do not generally blossom forth fully developed; they evolve and grow. The documentation of the development of an idea over time is generally the collection of original sketches, notes, calculations, and other evidence.

The evidence of that evolution of thought must be preserved by the inventor, not just for legal purposes, but perhaps more importantly, so that the inventor can "backtrack" when the inevitable "false trail" is taken. Being sure about where you have been makes future progress possible. Because we all are forgetful, and designing is a very involved process, good documentation helps the designer keep track of important facts and information, ideas, and the details of design solutions.

The Portfolio

The designer generally keeps a **portfolio**. The portfolio is a collection of materials that document the thinking and physical work of an individual. An architect would show a portfolio to a prospective client so that that person can evaluate the experience of the architect and the style of the architect's work. If the prospective client sees evidence of quality work and style compatible with the project's requirements, then the deal may be "closed." The portfolio plays an important role in the designer–client relationship.

Like the architect, the engineer and the artist know the importance of the portfolio. The portfolio to the artist is usually a collection of original artwork, photographs of other artwork, and artwork from shows and exhibitions. The engineer and the inventor may develop a portfolio as well, describing projects and original work through photographs, sketches, illustrations, and publications. These are used for job promotion and job search, and may even be used to obtain funding for a project or to open one's own business.

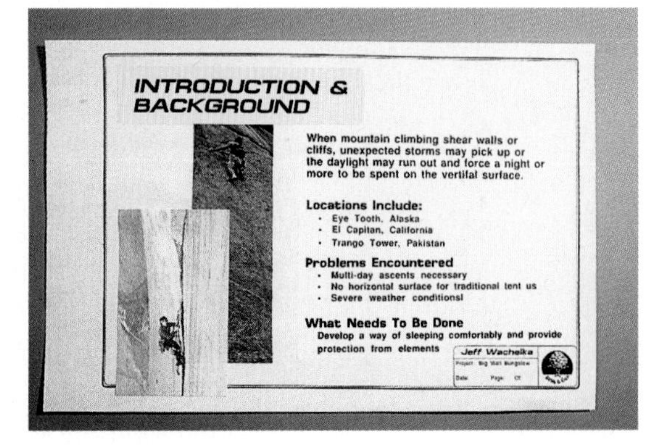

4-2a

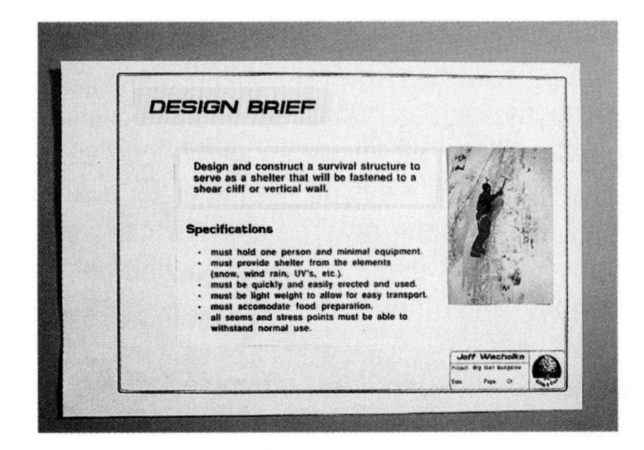

4-2b

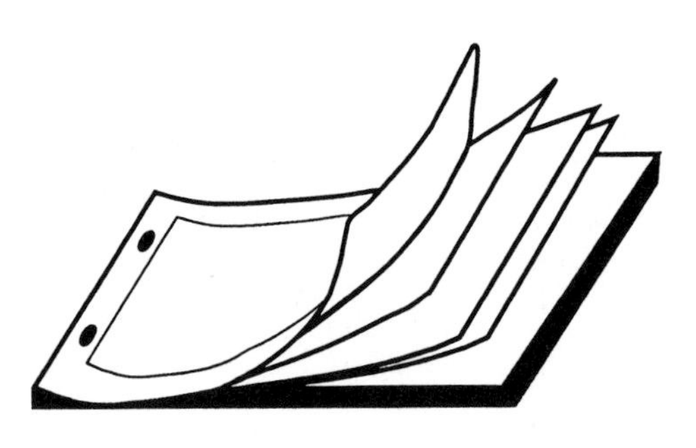

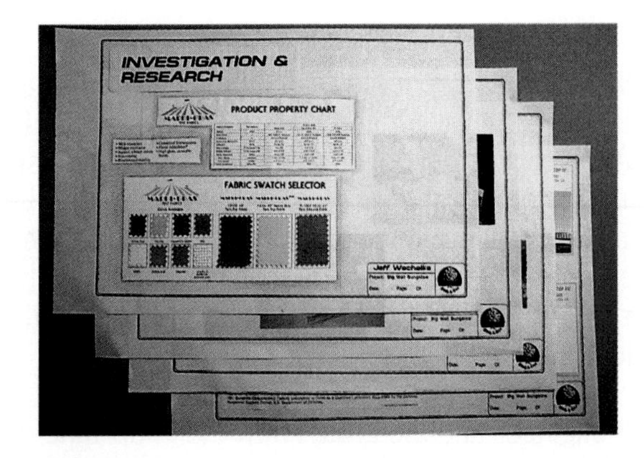

4-2c

4-2d

Figure 4-2a-h

Courtesy of Jeffery Wachelka

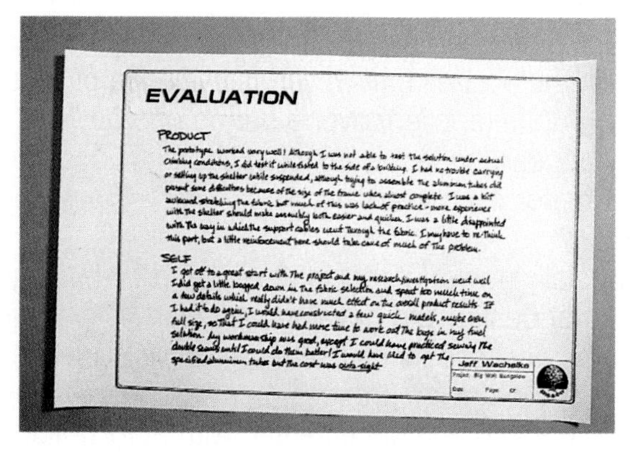

4-2h

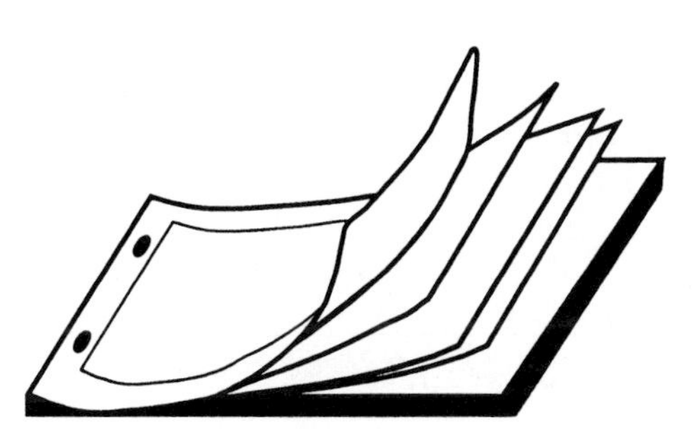

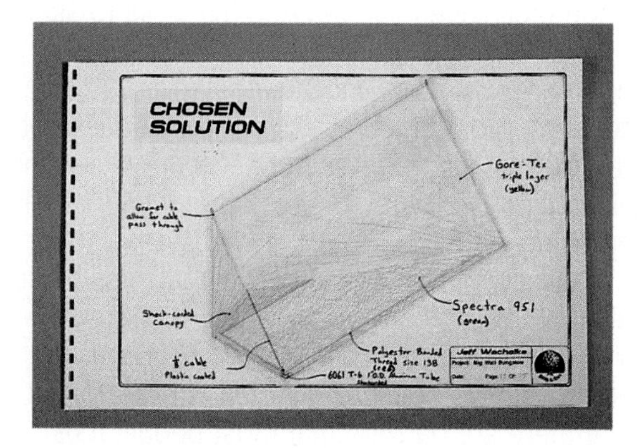

4-2e

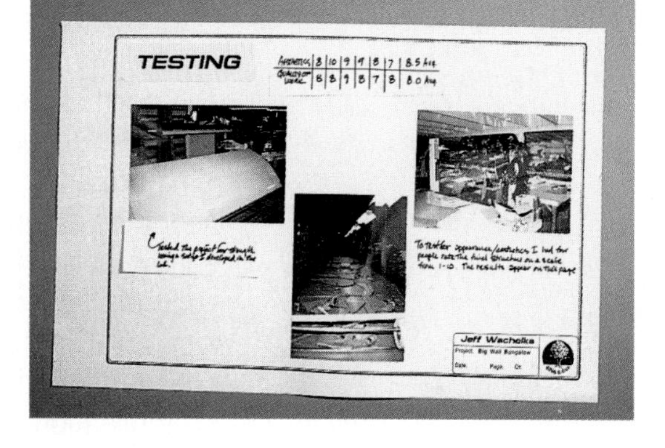

4-2g

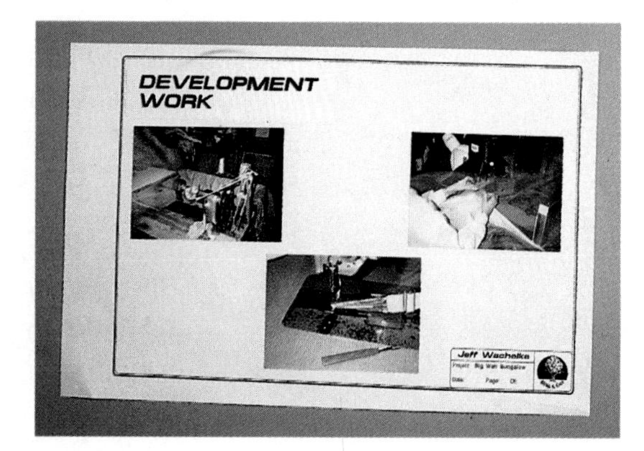

4-2f

In education, the portfolio is taking on new importance. Many critics of education are saying that paper and pencil tests are of only limited value when it comes to assessing what students have learned. Because tests only assess a student's knowledge at one point in time, they provide little evidence of what someone actually knows; we have all learned something for a test and then promptly forgotten it when the test was over. Within the portfolio, however, is evidence of what an individual has actually done (applied) over a period of time. This evidence does not simply appear the last week of school; it is accumulated over weeks, months, or even years.

The portfolio is an especially powerful tool in design and technology courses. Because much of design and development work is graphic in nature and because the portfolio lends itself to graphic evidence, it becomes a central part of the design process on two levels: First, each problem tackled is tracked and documented with a portfolio; second, the individual portfolio of each project is incorporated into a cumulative portfolio, which can act as evidence for application to a college and other further educational institutions, application for a job, or application for an award or honor.

The documentation of your design work is an integral part of both the study and understanding of technology. By means of the portfolio, you will gather evidence of your ideas and your creative and developmental work.

Documentation Basics

Sketching and drawing form a language that allows an individual to visualize, refine, resolve, and communicate ideas. It is a skill that can be learned. While some people have a natural talent for sketching, everyone can pick up the techniques and develop the skills to sketch.

You do not have to be an artist to draw and sketch. While a few people find it very easy to learn, most of us must practice the techniques and "tricks" to be effective visual communicators. Although sketching is useful for communicating ideas to others, it is often most useful when you are trying to clarify and develop your own ideas. Getting those ideas down on paper as quickly and effectively as possible is a major goal of sketching.

Before we discuss the techniques of sketching, it is important that you become aware that there are barriers that most people must overcome before they can learn to sketch and draw successfully. One barrier is the fear of not being able to sketch and draw, and looking foolish. Although it is rare that someone will admit they have this fear, just ask someone to sketch an idea or even a map, and the first words out of that person's mouth are usually excuses about why their drawings will look awful. Overcoming this barrier to sketching will take some work, but the results are well worth it. Sketching is an effective form of communication that has been said to be "worth a thousand words."

Sketching and Drawing

Preliminary Sketches. Sketching and drawing is about developing and presenting your ideas. The first ideas you put down on paper are usually in the form of **preliminary sketches** (Figure 4-3). These are done quickly so that you do not lose your train of thought or flow of ideas. This process should not be encumbered with straightedges, rulers, compasses, or other drawing instruments. These instruments only slow down the process that, by its very nature, is free and creative.

Annotated Sketches. The addition of notes about materials, fasteners, and other features to a drawing is called an **annotated sketch** (Figure 4-4). These remarks are reminders to yourself about what you were thinking at the time you developed the idea. Ideas are often lost if they are not recorded immediately.

Developmental Sketches. The next step in designing is the further development of the ideas in the preliminary sketches. These **developmental sketches** contain more detail as the ideas are refined (Figure 4-5). At the completion of the developmental sketches, the solutions should be workable, with features such as mechanical linkage or electronic circuitry worked out. Annotations are often a part of developmental sketches.

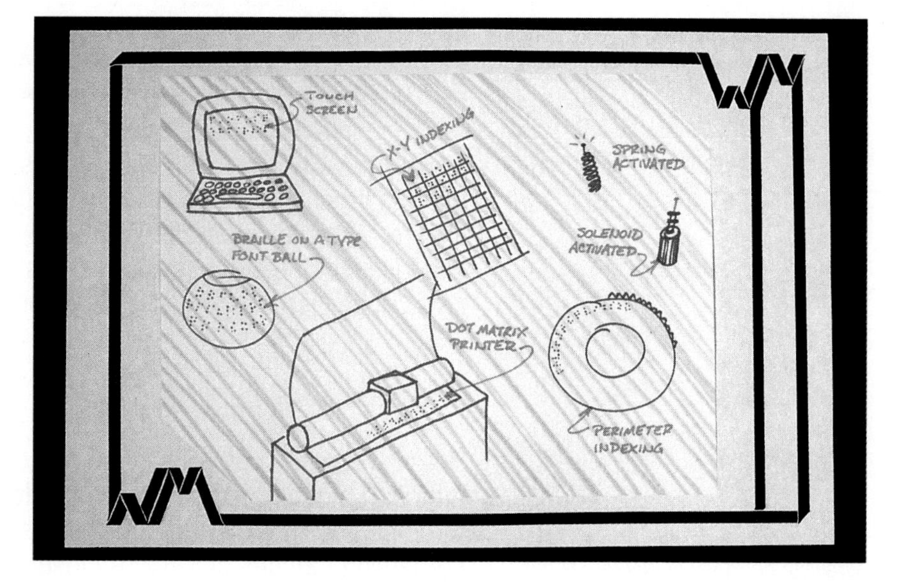

Fig. 4-3

A preliminary sketch of computer Braille reader ideas. Drawing courtesy of Wendy Snyder

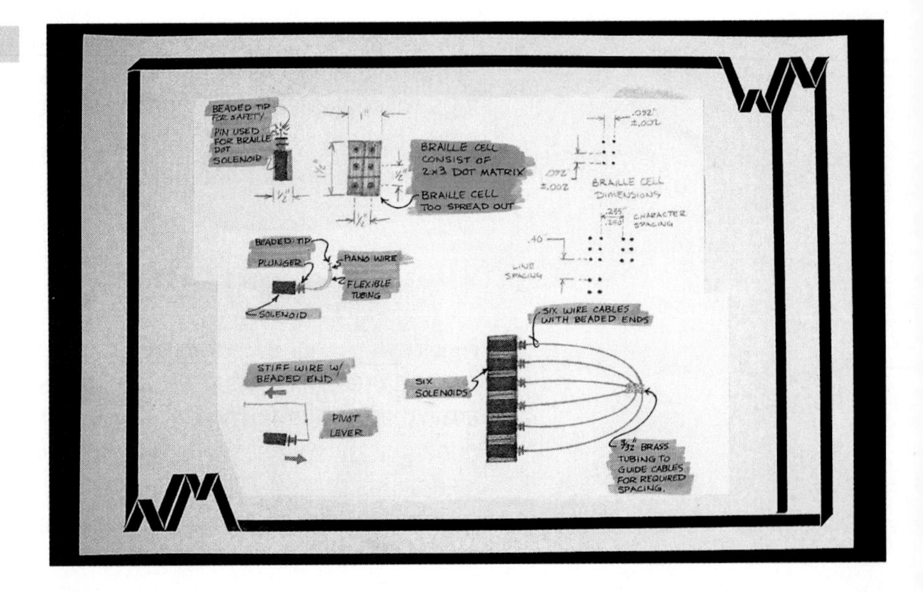

Fig. 4-4

An annotated sketch of computer Braille reader components. Drawing courtesy of Wendy Snyder

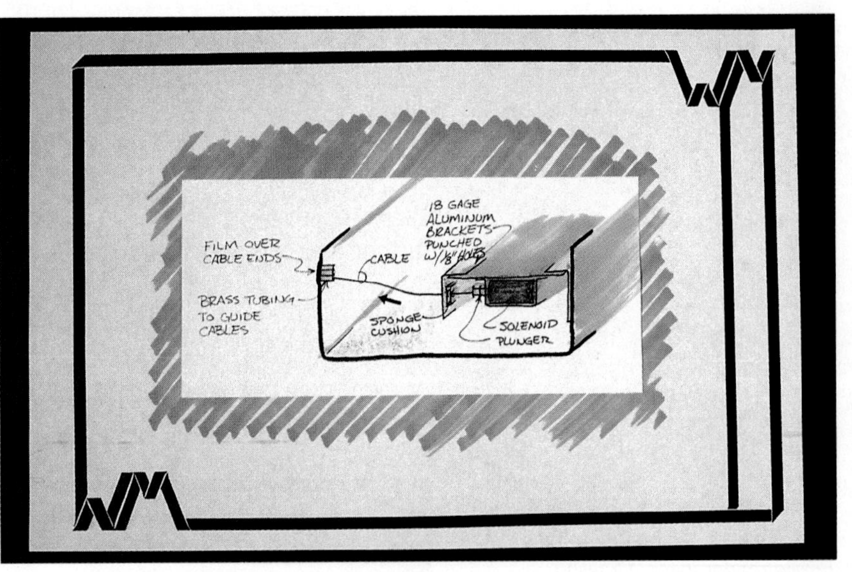

Fig. 4-5

A developmental sketch showing the solenoid control in the final Braille reader system. Drawing courtesy of Wendy Snyder

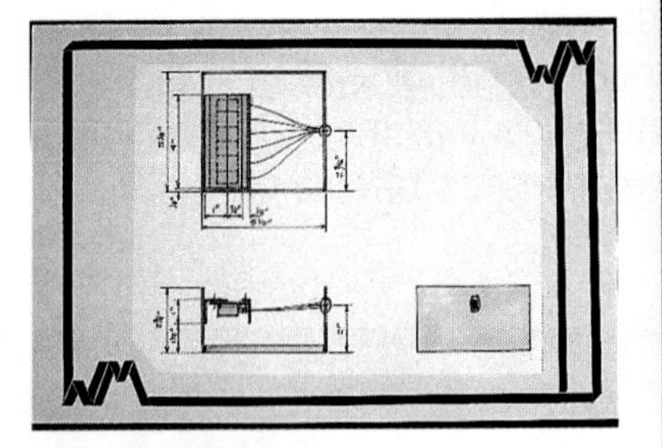

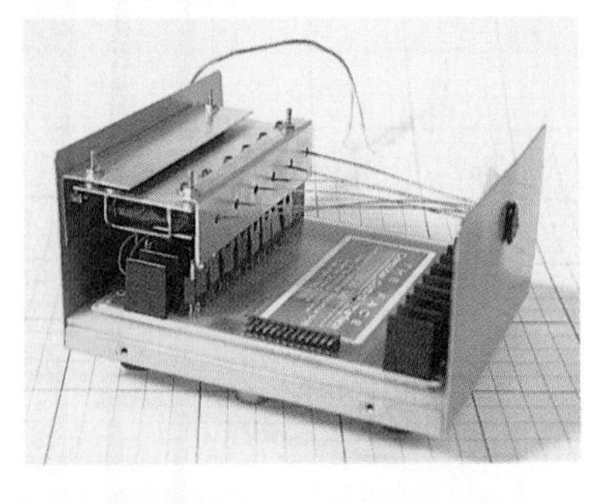

Figs. 4-6a, b

A working drawing for final placement of components for the Braille reader. Drawing courtesy of Wendy Snyder

Working Drawings. In the final stage, working drawings are developed that contain the information needed to actually make the solution (Figure 4-6). These are often drawn to scale, so that size, proportion, and location of features such as holes can be finalized. Working drawings will be discussed in a later section.

Seeing

Before you can sketch well, you must be able to see and visualize. Although most of us use our sight effectively to avoid tripping over furniture and running into doors, we seldom look at details unless something catches our eye. A closer look at everyday objects often

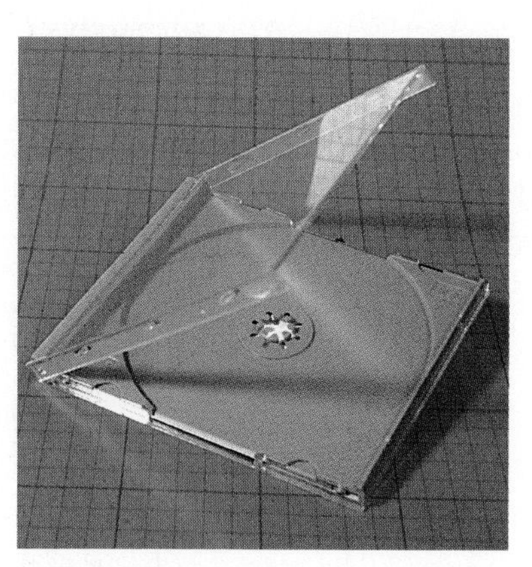

Fig. 4-7

A compact-disk case

reveals features we have missed. These features may be proportion, material, a design element such as line or texture, or the way in which one part fastens to another.

Let us say you want to sketch the plastic case for a compact disk (Figure 4-7). Have you noticed that it is made from a number of different parts; that it has raised lines molded into two edges; that it is hinged so that it can lie flat when opened; how it is held closed? Sketching a CD case demands a closer look at these and other details.

Sketching practice develops both the habit of looking more closely at objects and the ability to translate the features you have noticed into sketches of your own original ideas. In fact, sketching is an excellent method of analyzing an object, because you are forced to notice details.

Fig. 4-8

Line, Shape, and Form

Line. A line defined mathematically has only direction and length, but no width. In sketching, of course, the width of a line is not only real but also important to the appearance of the sketch. To achieve a good line, a soft, *HB pencil* is fine for most sketching, because it can be used to make faint lines or bold lines, according to how hard you press it against the paper (Figure 4-8). It will become blunt quickly, so it is important to maintain a sharp point.

Fig. 4-9

Shapes enclose a space.

Shape. Lines are used to enclose a *space*. The *shape* of an object is the two-dimensional space enclosed by lines. Shapes can be *natural* shapes, *geometric* shapes, or *free-form* shapes (Figure 4-9).

Form. When a shape is given a third dimension, it becomes a form. In sketching, this can be done by using lines, shading, and/or color. The geometric forms of the cube, cylinder, pyramid, and sphere are developed from the square, circle, and triangle. Many other geometric forms are also common, such as the tetrahedron and various prismatic shapes. These forms are the basis for many of the produced objects seen every day.

Fig. 4-10

You can view an object as being made from a combination of the basic shapes.

Fig. 4-11

The Xanadu house in Florida represents a possible direction for housing in the future. Courtesy of Xanadu House

Natural forms, sometimes called *organic forms*, are also important. In addition to geometric forms, these are also used as a basis for creating products and structures, such as the house pictured in Figure 4-11.

Sketching demands the ability to produce a likeness in three dimensions. There are a number of drawing styles that allow you to do this, including axonometric, isometric, oblique, and perspective drawing (Figure 4-12). Each of these styles has unique advantages and disadvantages, but perspective drawings make objects look more realistic.

Fig. 4-12

A television remote control sketched in different styles of visualization.

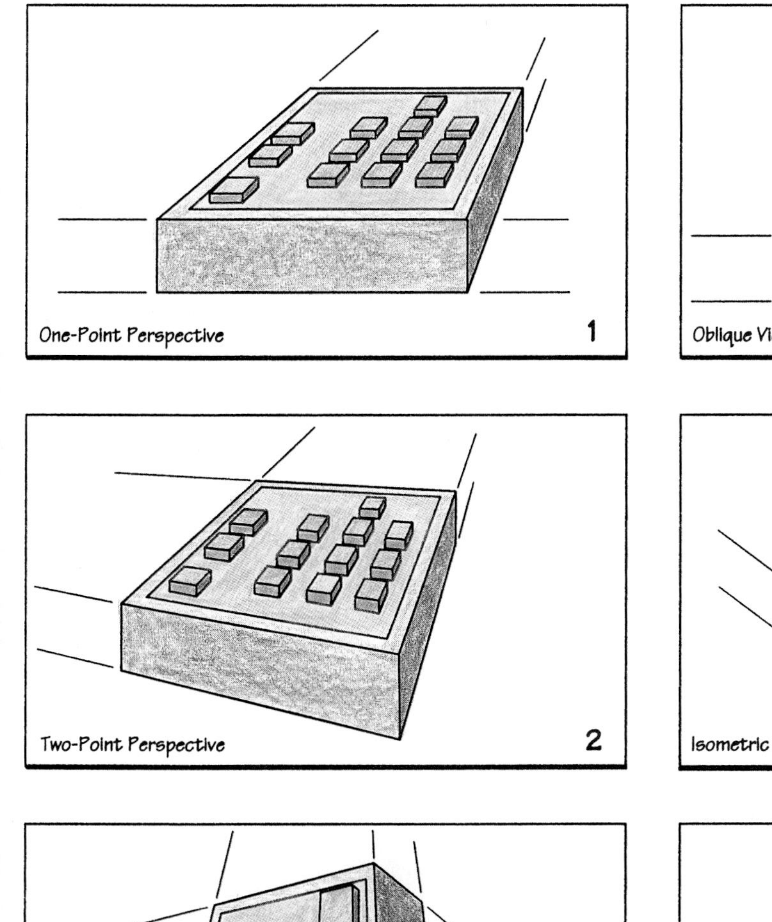

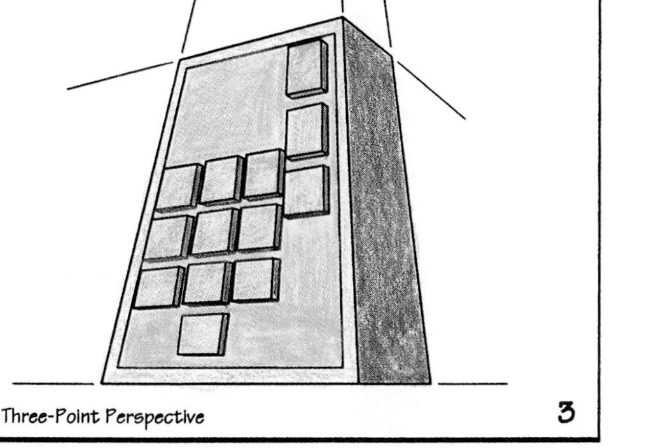

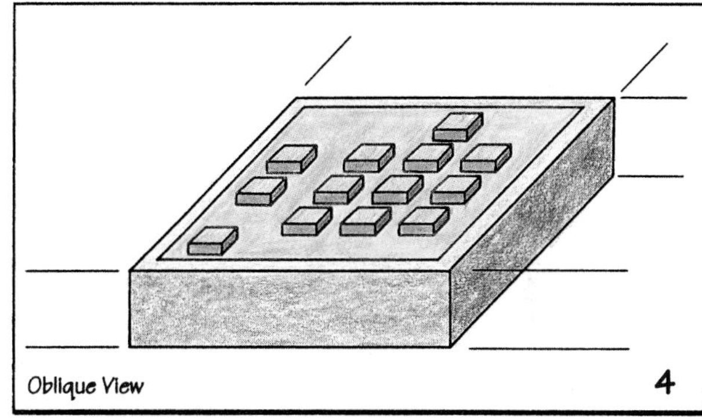

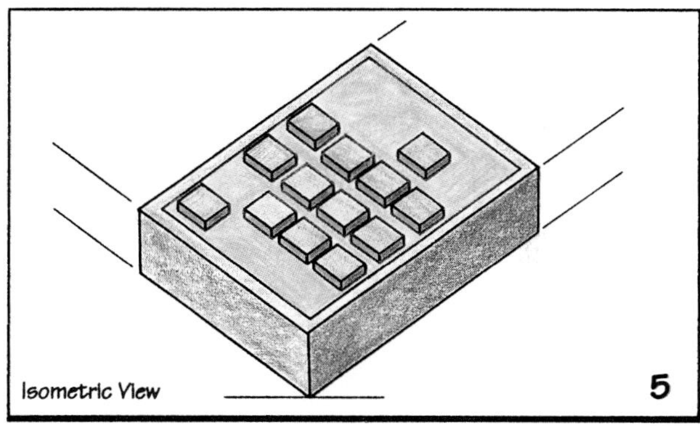

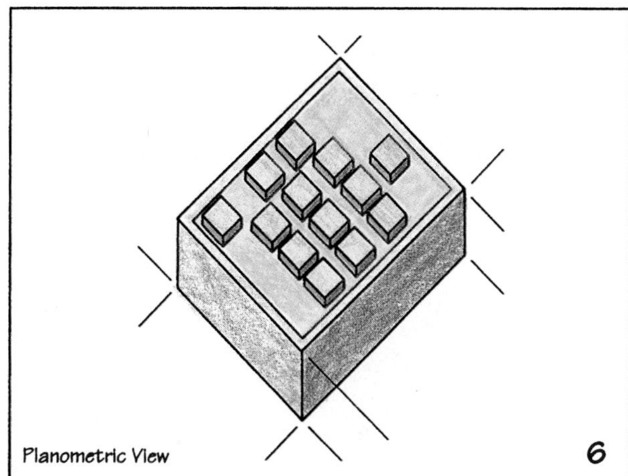

Perspective Drawing

Perspective drawing means taking into account that things seem smaller the further away they are from you. A car in the distance appears very small; as it moves closer, it appears much bigger, even though it is the same car. We often do not take notice of this phenomenon, because we take it for granted.

The same principle applies to a long building (Figure 4-13). Standing near one end, we see the full height of the near wall. The other end of the building, in the distance, seems smaller, and the roof connects the two by slanting down from the near wall to the far wall. At the same time, the ground at the bottom of the near wall almost seems to rise to meet the bottom of the far wall.

When we draw a line along the roof line and one along the ground line, we find that they meet (Figure 4-13). This converging point is called the *vanishing point*. Also, lines drawn along the windows converge at the same point.

One-Point Perspective. Drawings that use only one vanishing point are called one-point perspective drawings (Figure 4-14). The vanishing point is located on a *horizon line*, which is a horizontal line representing an imaginary horizon. Placement of the horizon line is important for how the object will appear in the drawing. Figure 4-15 illustrates how the placement of the horizon line will affect the appearance of the same object.

In one-point perspective, the vanishing point is used to develop one side of the object and either the top or the bottom. The front of the object is drawn "head-on." For this reason, the vanishing point should not be placed too far to the side of the object or the object will look distorted.

Fig. 4-13

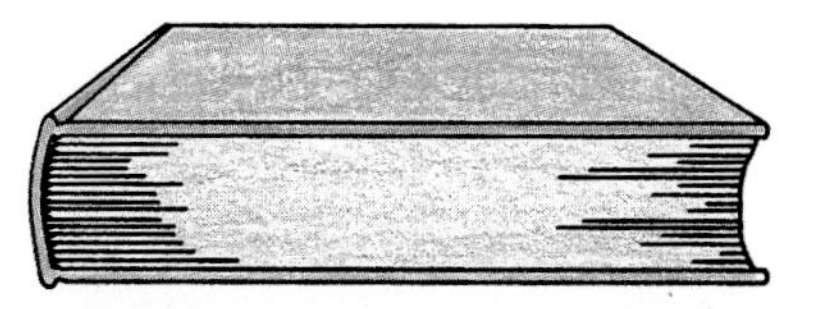

Fig. 4-14

An object in one-point perspective

One-point perspective is also very useful for drawing the interior of a room or an outdoor scene. Here, the horizon line is placed just slightly above the center, and the vanishing point is usually centered on this line (Figure 4-16 and 4-17).

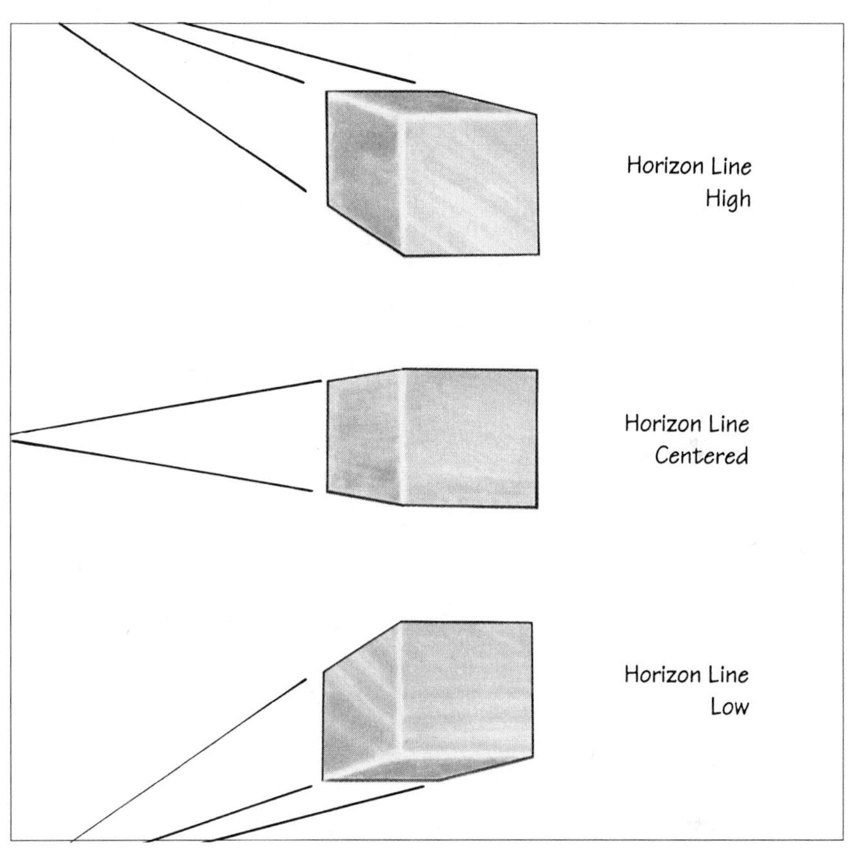

Fig. 4-15

The placement of the horizon line has an effect on the view of an object. Courtesy of David Bodmer

Fig. 4-16

One-point perspective of a kitchen interior

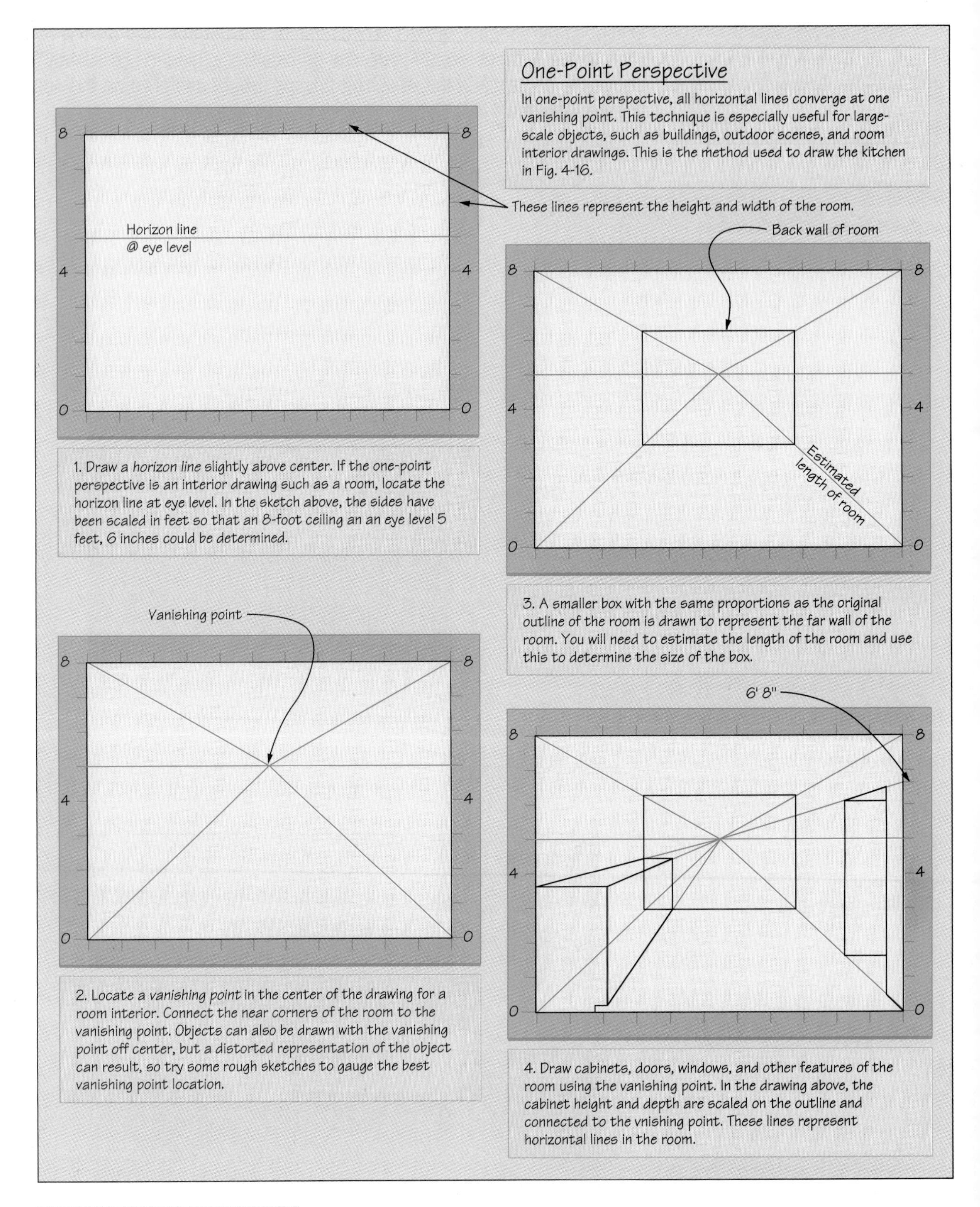
One-Point Perspective
In one-point perspective, all horizontal lines converge at one vanishing point. This technique is especially useful for large-scale objects, such as buildings, outdoor scenes, and room interior drawings. This is the method used to draw the kitchen in Fig. 4-16.
These lines represent the height and width of the room.
Horizon line
@ eye level
8
4
0
1. Draw a horizon line slightly above center. If the one-point perspective is an interior drawing such as a room, locate the horizon line at eye level. In the sketch above, the sides have been scaled in feet so that an 8-foot ceiling an an eye level 5 feet, 6 inches could be determined.
Back wall of room
Estimated length of room
3. A smaller box with the same proportions as the original outline of the room is drawn to represent the far wall of the room. You will need to estimate the length of the room and use this to determine the size of the box.
Vanishing point
2. Locate a vanishing point in the center of the drawing for a room interior. Connect the near corners of the room to the vanishing point. Objects can also be drawn with the vanishing point off center, but a distorted representation of the object can result, so try some rough sketches to gauge the best vanishing point location.
6' 8"
4. Draw cabinets, doors, windows, and other features of the room using the vanishing point. In the drawing above, the cabinet height and depth are scaled on the outline and connected to the vnishing point. These lines represent horizontal lines in the room.

Fig. 4-17

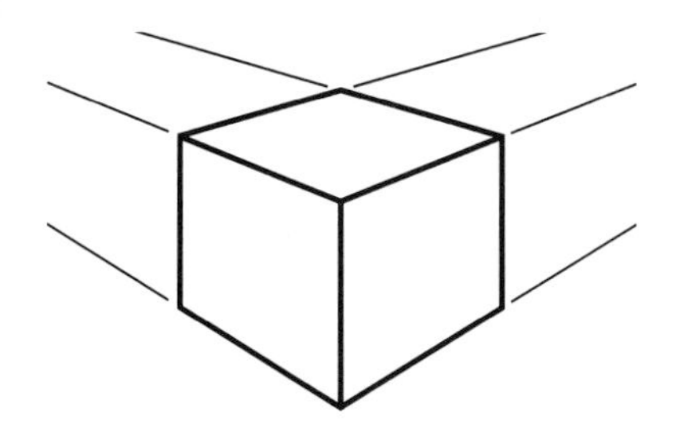

Fig. 4-18

Two-point perspective of a cube

Two-Point Perspective. Viewing an object at an angle, which is how we see most things, will not work in one-point perspective. It is necessary to provide two vanishing points, one for each of the two sides you see (Figure 4-18).

Placement of the horizon line depends on the point of view you are trying to represent in your drawing. If you wish to present a "bird's-eye view," the horizon line should be high on the page, well above the space in which you will draw your object (Figure 4-19). To view the underside of the object being drawn, place the horizon line low on the page, below your drawing space.

The placement of the two vanishing points will have an effect on the appearance of the object you are drawing. In many cases, the vanishing points will need to be off the paper on which you are drawing to give the object a realistic appearance. Placement of the vanishing points too close together distorts the appearance of the object. Figure 4-20 illustrates how placement of vanishing points can give you a number of effects.

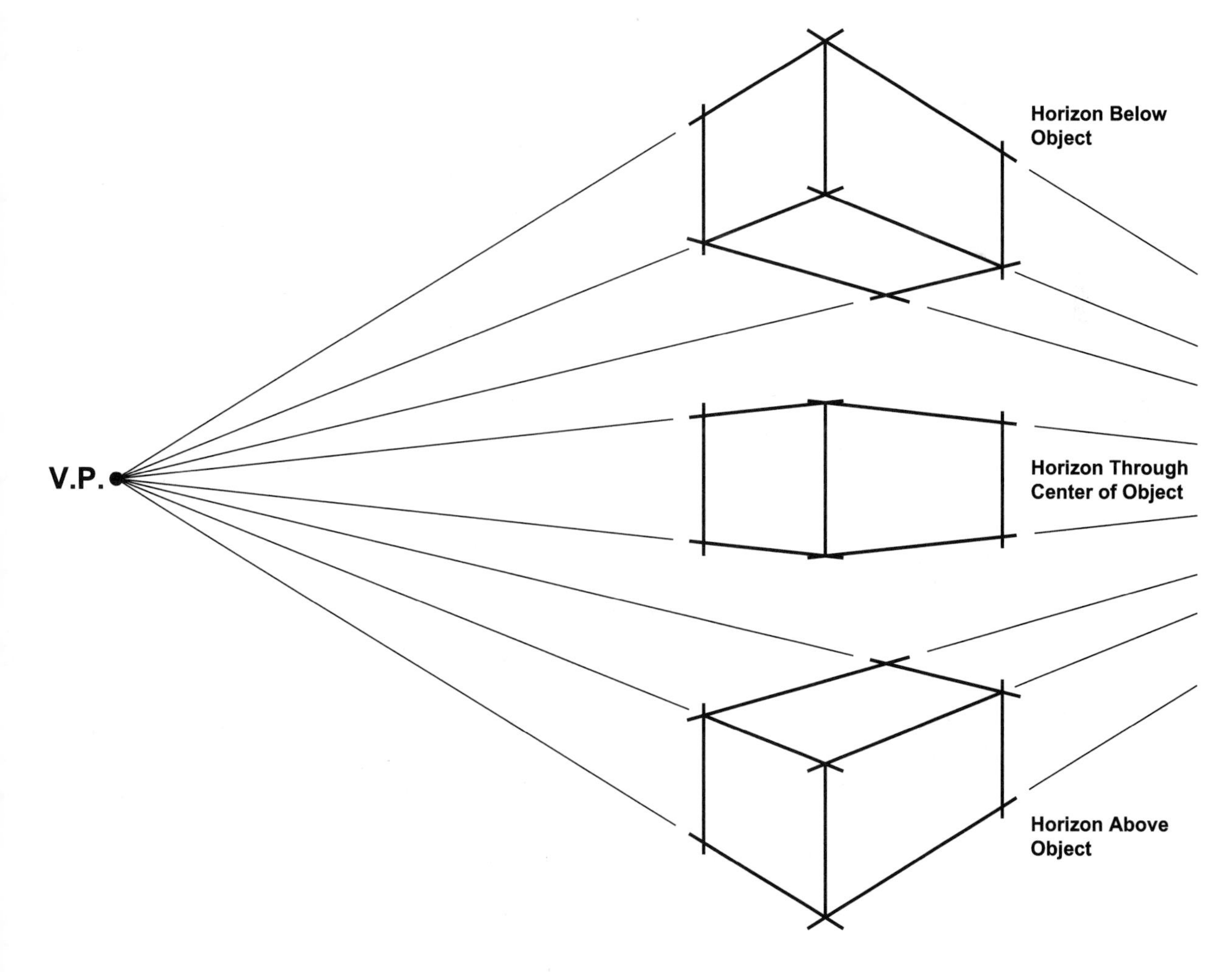

Fig. 4-19

The placement of the horizon line has an effect on the view of an object in two-point perspective.

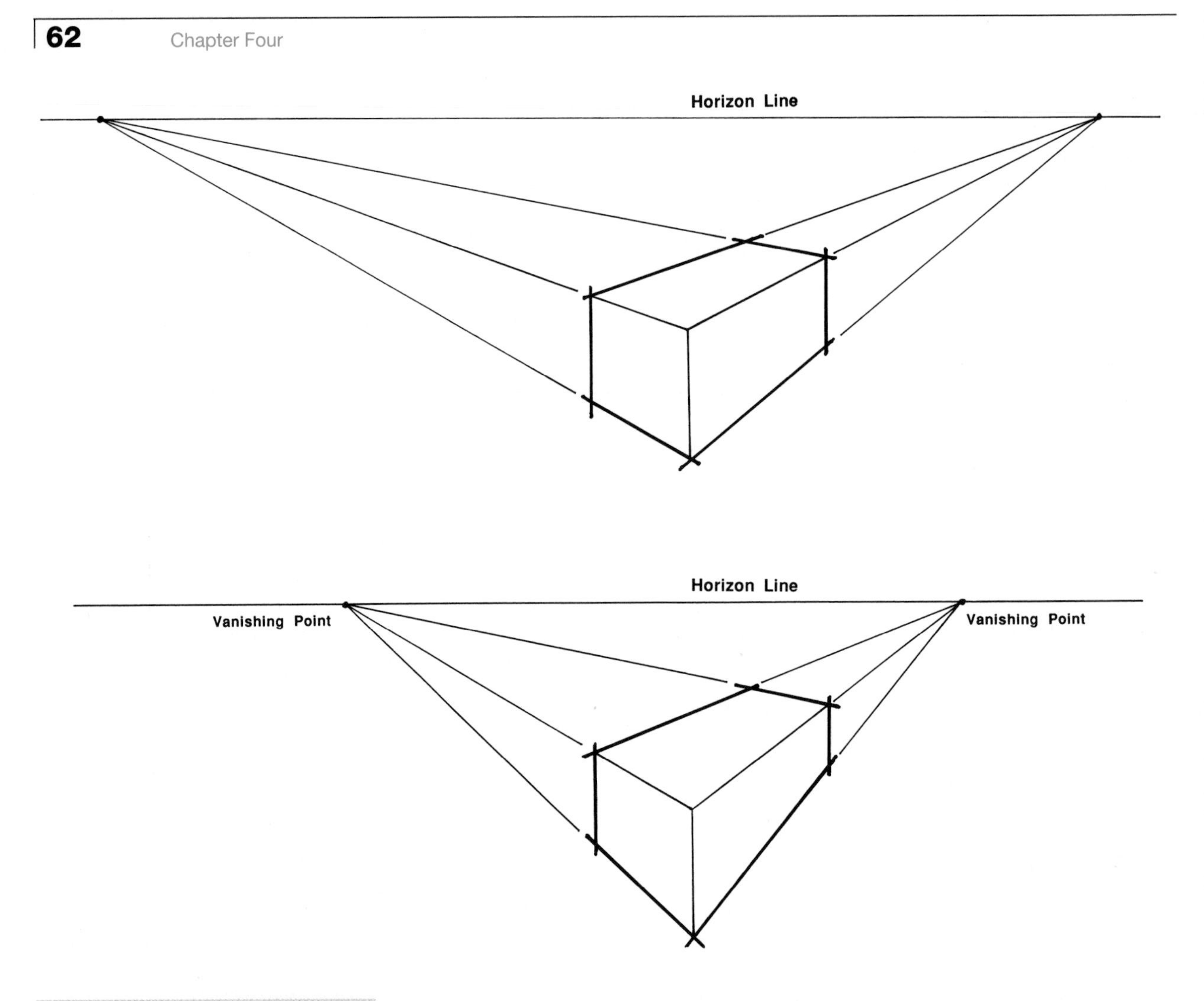

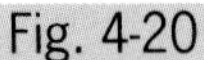
Fig. 4-20

The placement of the two vanishing points changes the shape of the object.

Crating. The cubes on the previous pages are very useful figures. Crating is the process of visualizing the object you want to draw as inside a box or a crate (Figure 4-21). Using perspective techniques, the cube is developed and the object is "found" inside the crate (Figure 4-22).

Finding Shapes Inside the Crate. The cube can be used to help you develop the other basic shapes: the *cylinder*, *cone*, and *sphere* (Figure 4-23). Diagonal lines are used to find the center of the crate side. In the cylinder, the center of each end is connected and becomes the center axis of the cylinder. The tip of the cone is at the center of the top plane of the crate, and the circumference of the cone bottom touches the center of each side of the bottom of the crate. The sphere touches one point at the center of each side of the crate. All objects can be viewed as being constructed from a combination of the basic shapes, as in Figure 4-24.

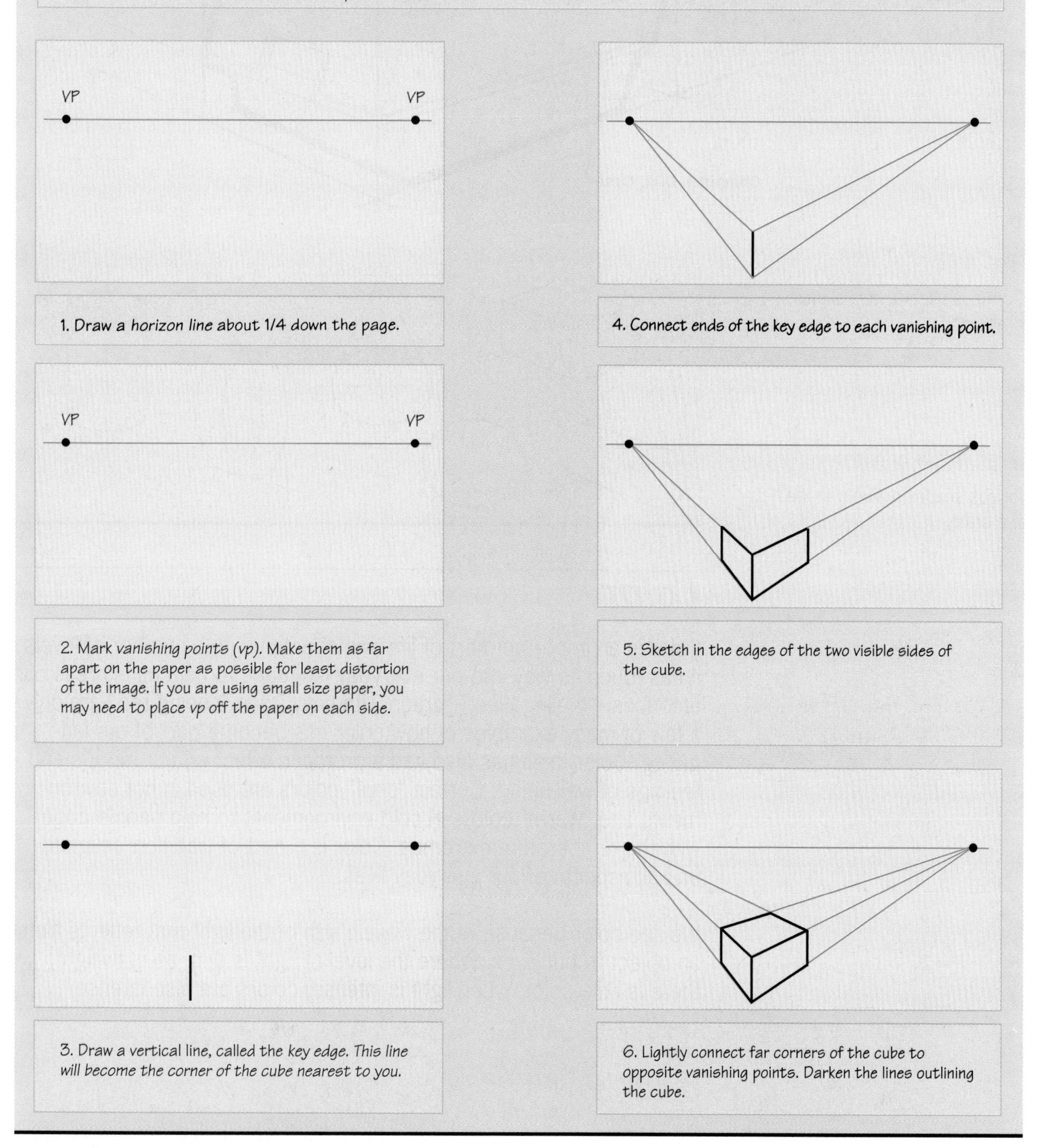
Drawing Two-Point Perspective
A perspective drawing begins with the basic shape of the cube. Once established, the shape of the object is found inside the cube "crate" (see Figs. 4–22, 4–23, 4–24). Experiment to see how the placement of the vanishing points and the distance between the key edge and the horizon line distort the shape of the cube.
VP
VP
1. Draw a horizon line about 1/4 down the page.
VP
VP
2. Mark vanishing points (vp). Make them as far apart on the paper as possible for least distortion of the image. If you are using small size paper, you may need to place vp off the paper on each side.
3. Draw a vertical line, called the key edge. This line will become the corner of the cube nearest to you.
4. Connect ends of the key edge to each vanishing point.
5. Sketch in the edges of the two visible sides of the cube.
6. Lightly connect far corners of the cube to opposite vanishing points. Darken the lines outlining the cube.

Fig. 4-21

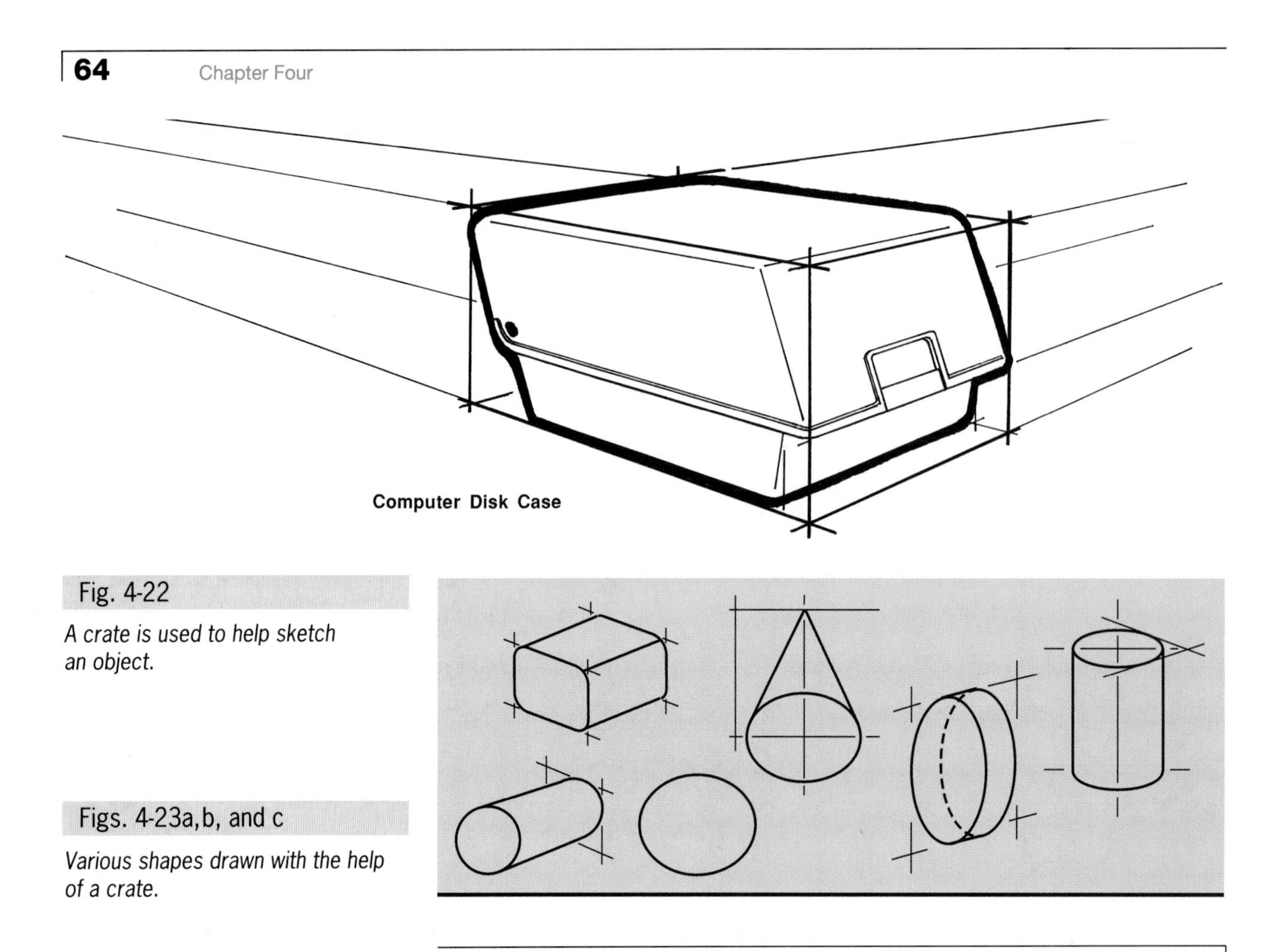

Fig. 4-22

A crate is used to help sketch an object.

Figs. 4-23a, b, and c

Various shapes drawn with the help of a crate.

Color

Color is an important part of life and affects us on a number of levels. It has found its way into our everyday language to help us express our emotions: "feeling blue," "green with envy," and "seeing red" are only a few of many examples of how color has become part of our language. Color itself has also had a profound effect on our moods and feelings of well-being. Certain "cool" colors are used in hot environments and "warm" colors in cold environments to help people cope with the temperature extremes. Color is a part of tradition, religion, and all aspects of our everyday lives.

We see color because of the wavelength of the light that reflects from an object to our eyes. Where the level of light is dim, as in twilight, there is little color. When light is intense, colors are also intense.

Primary Colors

Primary colors are the basic colors from which all other colors are made. There are three sets of primary colors. The *additive primaries*

Fig. 4-24

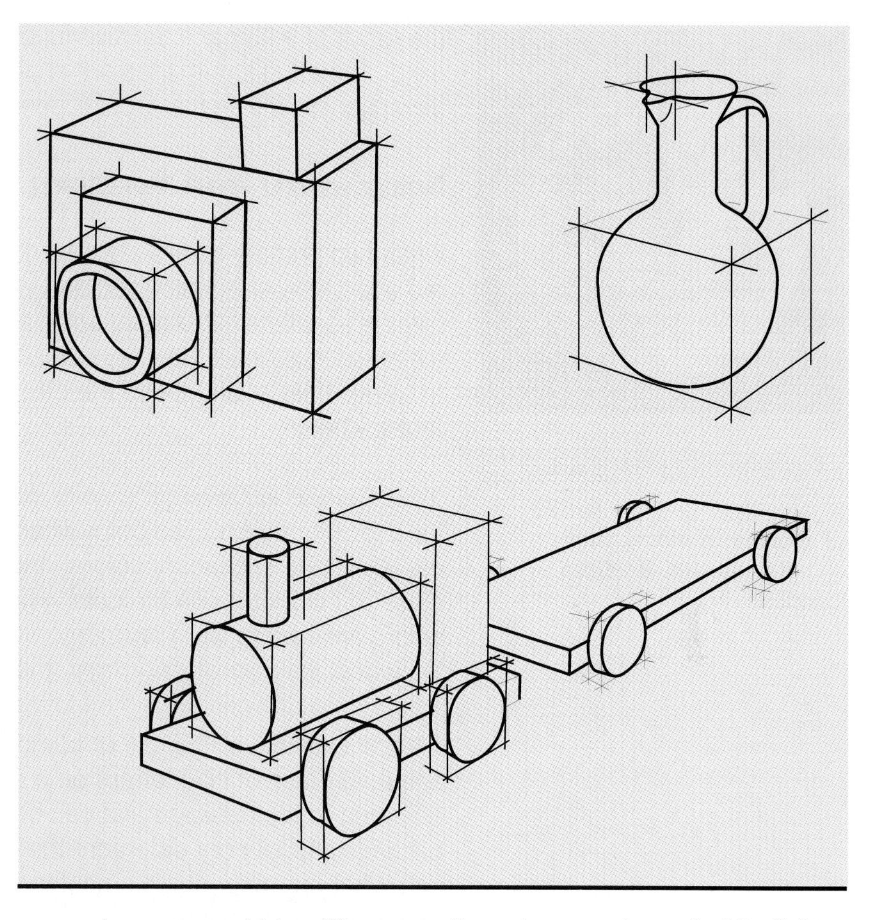

are red, green, and blue. These are the primary colors of white light. Additive color theory is based on the principle that light energy is emitted from a source, such as the sun or artificial light. Visible light consists of electromagnetic radiation between 400 and 770 nanometers (nm). Black is the absence of light. Additive theory is used to make color television images and in stage lighting. When red, green, and blue light are projected onto each other, we see white.

Subtractive primaries are yellow, magenta (process red), and cyan (process blue). Subtractive color theory is based on the principle that pigments absorb some light while reflecting others. Subtractive principles are used in producing all mass media print, such as packaging, publications, and all general commercial work. The theory is also used in making photographic color prints. When yellow, magenta, and cyan inks and photographic dyes are overlapped, we see black.

Artist's primaries are red, yellow, and blue. Because artist's primary colors are pigments, they also work by the subtractive color theory. Because printing inks are translucent and artist's pencils and paints are more opaque, the colors of the primaries are somewhat different. Talking about color as hue (color name) is awkward, because color is

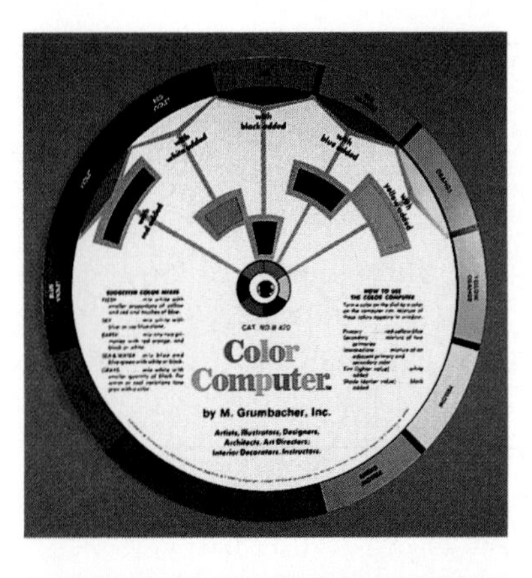

Fig. 4-25

A color wheel. Courtesy ©1972 M. Grumbacher Incorporated. All rights reserved. Used with permission.

the result of a human cognitive function; that is, we see color in our mind. The artist's primaries are important in documenting your work, because you will be using color pencils, color markers, and paints.

Secondary and Tertiary Colors

When two primary colors are mixed, a secondary color results. Mixing red and blue yields violet; red and yellow, orange; and yellow and blue, green. When a primary color and a secondary color are mixed, the result is called a "tertiary color." All the possibilities so far add up to twelve colors, and these are often illustrated in what is called a "color wheel."

The Color Wheel. The color wheel (Figure 4-25) is a handy tool for choosing colors. The color wheel contains the three primary colors, the three secondary colors, and the six tertiary colors. Colors close to each other on the color wheel work well together. These colors are said to be in *harmony*. Colors from opposite sides of the wheel are **complementary**; that is, they contrast or clash.

A good example of the use of complementary colors is found in packaging. The use of blue letters on a red background, for example, calls attention to the package and can be read from a distance. Notice the packaging of laundry detergent the next time you are in a supermarket. What are other complementary color combinations that are commonly used in packaging?

The use of color is very important. As humans, color affects the way in which we think and feel about something. The artist's primaries are important as you illustrate your design ideas. The additive primaries are important as you think about the lighting conditions that will be used to view your documentation or product. All colors will look different under sunlight, incandescent, and fluorescent light bulbs, and other lighting conditions. The poetry on the Moody Blues album *Days of Future Past* ("... Cold hearted orb which rules the night; removes the color from our sight; red is gray and yellow, white....") refer to the illusions created by moonlight.

Finally, subtractive primaries that are closely related to artist's primaries would become important if you took color photographs or had something reproduced, such as a color package, by a commercial printer.

Hue, Chroma, and Value

Hue. **Hue** describes the color we are dealing with: blue, green, red. It is described as the "pure color" on the color wheel in Figure 4-25. Hue

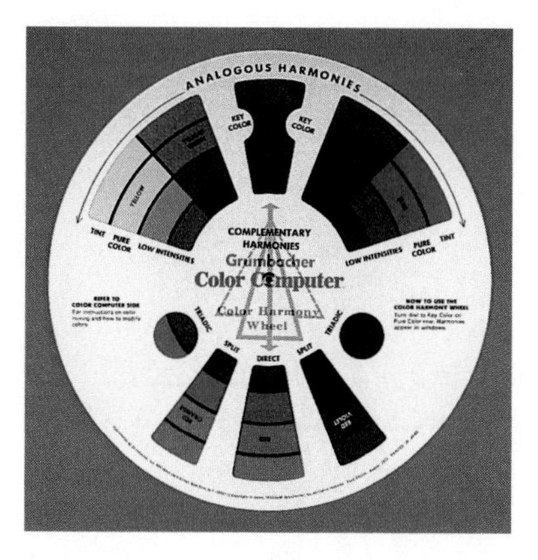

Fig. 4-26

Courtesy ©1972 M. Grumbacher Incorporated. All rights reserved. Used with permission.

refers to the actual wavelength of the light on the spectrum. Reds have the longest wavelength (lowest frequency); the other end of the spectrum contains the blue hues, which have short wavelengths (highest frequency).

Chroma. Chroma describes the intensity or brightness of a certain color. A pure hue or color is the most intense, but the color may be neutralized by mixing it with a color on the opposite side of the color wheel.

Value. Value, sometimes referred to as "tone," describes the amount of light the color actually reflects. The value of a color can be altered by adding either white or black to the hue. Adding white tends to make the color more pale or to **tint** it, washing out the hue as more white is added. This is the effect you see when you view a dark object in a window with the sun shining in. Although the object may be all the same color, the side toward the window appears to be a lighter color. Mixing black with a color tends to make a darker **shade** of the hue. Tints and shades are shown in Figure 4-26.

Sketching and Drawing Techniques

Outlining

The use of *thick lines* within a sketch or a drawing can make an object stand out. In Figure 4-27, you can see two of the sketches that are identical, except for the object outline: In one sketch, the entire object is drawn in the same line width; in the other, thicker lines have been used to give the sketch impact. The thicker lines have been added where only one surface of the object is seen; the thin lines remain where two surfaces meet and both are visible.

A second method uses a black or dark-colored marker to outline an object and make it stand out from the page (Figures 4-28a, b). Use bold, rapid strokes, and let them overlap a bit where they intersect.

Fig. 4-27

The use of thick outlines makes the features of an object stand out.

Figs. 4-28a, b

Using a marker to make an outline for an object

Highlights

Highlights and reflections are a natural part of almost everything we see. In drawing, the **highlights** and reflections are often exaggerated to give the object more impact or a look of realism. A flat, horizontal surface has vertical highlights (Figure 4-29a). These are easy to add by using pencil shading, markers, or other techniques. Curved or rounded shapes, such as spheres or cylinders, tend to distort the light hitting them (Figure 4-29b). A cylinder tends to stretch the highlight along its length, and even the inside of the cylinder, being curved, reflects light in this way. A sphere will stretch the light out in all directions but only for a short distance (Figure 4-29c). You will often see a "window" on a sphere as a highlight.

Light Source and Shading

When we see an actual object, we see a range of tone of a color, according to how much light strikes a particular surface. Because light most often comes from one main source, such as the sun or the lamp in a room, the surfaces of an object toward the light source reflect the most light. Surfaces at an angle to, or away from, the light source do not reflect much light, so they appear darker. Even when there are several sources of light, one usually dominates the object

Figs. 4-29a, b and c

Using highlights to give the appearance of a shiny surface, such as high-gloss painting

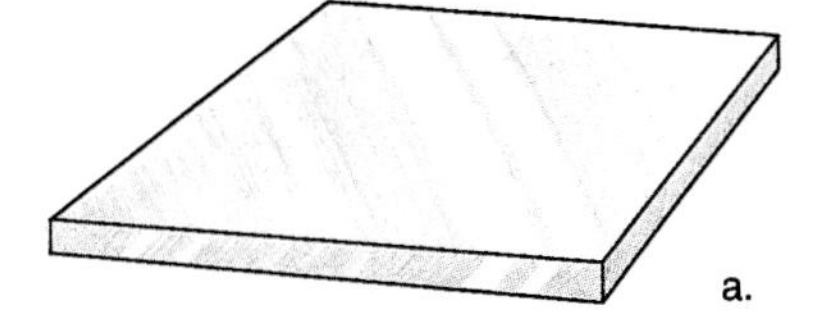

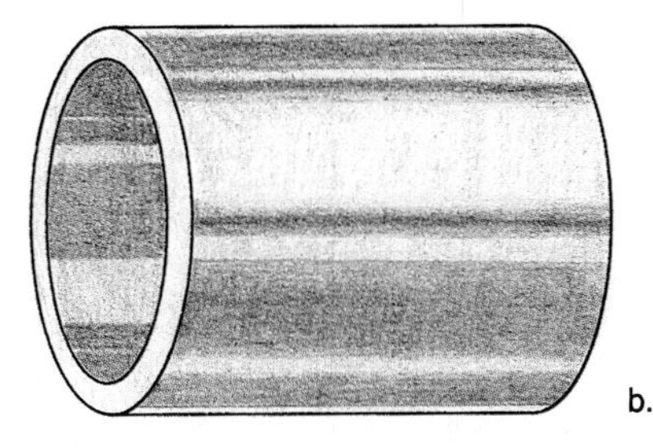

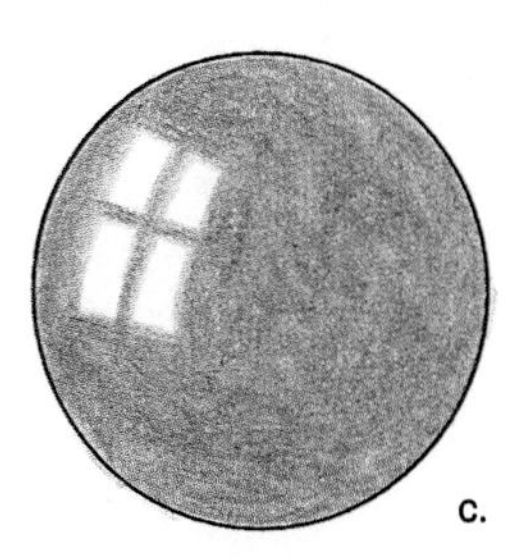

we are viewing because of the intensity of that particular light or its proximity to the object we are looking at.

The cube in Figure 4-30a is drawn by outlining the edges where two surfaces meet. This cube looks rather dull and lifeless. In Figure 4-30b, however, the same cube has been shaded with pencil to represent the amount of light reflecting on the three surfaces we can see. This second cube has depth and appears more realistic. This is because it more closely resembles the way in which we see actual objects.

For sketching purposes, it is handy to think of the light source as coming over your left shoulder (Figure 4-31). This will help you visualize which surfaces are lighter and which are darker. Hold the pencil at a low angle so that most of the length of the pencil lead contacts the paper. A light pressure on the pencil will give you a light tone; a heavier pressure can give you a very dark tone. These principles apply to both regular and color pencils.

As you move the pencil back and forth, you can blend in the tone, going back over places you have missed. Getting a smooth, even tone takes practice and patience (Figure 4-32). It is better to use a lighter pressure and go over an area several times than to use a heavier pressure and end up with a streaked, uneven effect.

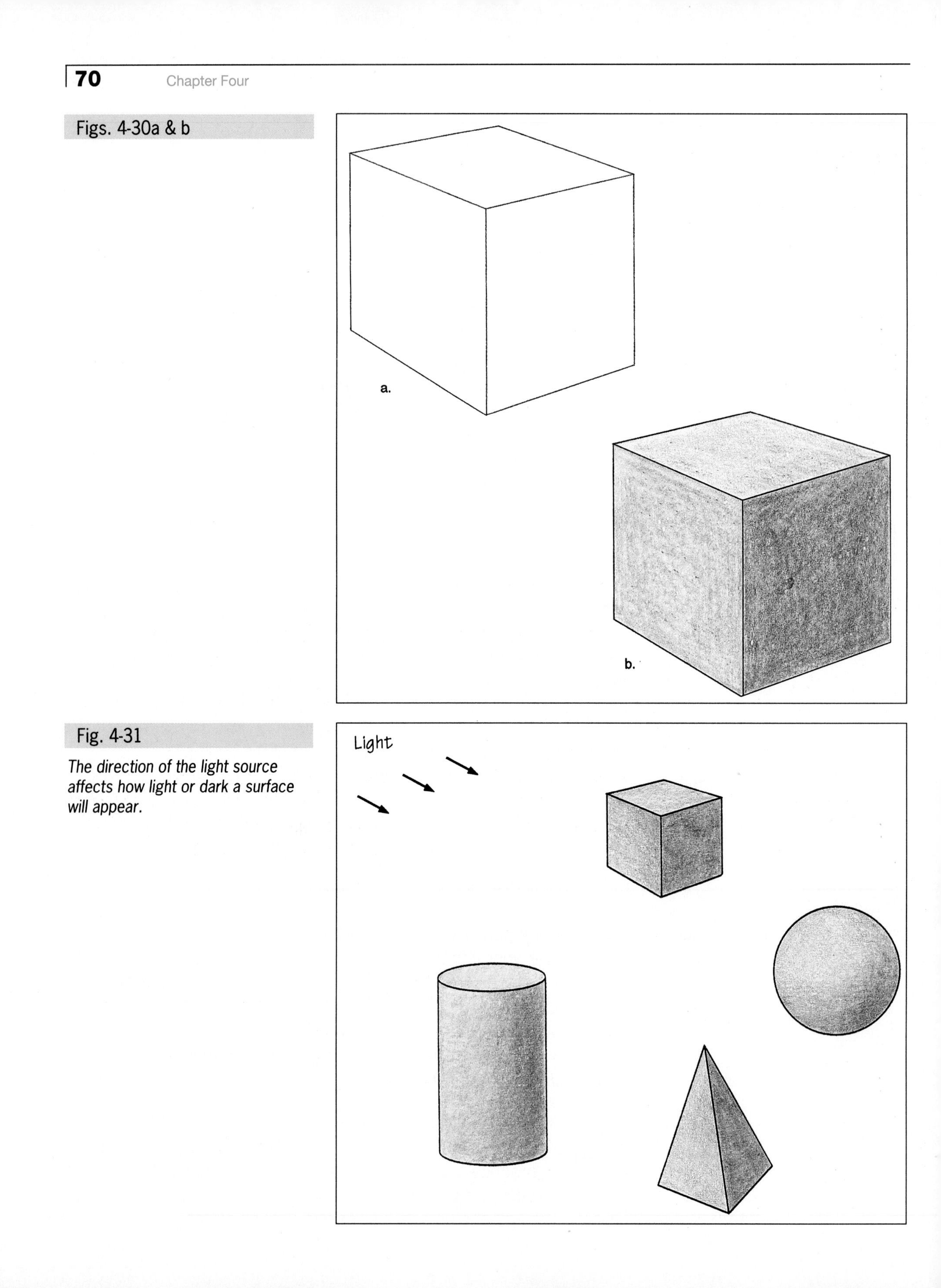

Figs. 4-30a & b

Fig. 4-31

The direction of the light source affects how light or dark a surface will appear.

Fig. 4-32

Using a pencil to get a blended shade on a cylinder

Texture and Materials

The *texture* of a material is an important feature. In products, it is useful for both functional and aesthetic purposes. It can provide a nonslip "grip" on the handle of a tool, a floor surface, or a skateboard. It can also add an interesting visual element to a product. Texture can be the result of the nature of the material itself, or it can be a result of a production process, such as molding (Figure 4-33).

Sketching the texture of a material is not difficult. Architects use a number of graphic standards that represent materials like concrete, earth, and foam insulation, (Figure 4-34). These and other techniques can add a great deal of impact to a drawing and can give the viewer a better idea of what the object you are sketching actually looks like.

Fig. 4-33

Many objects use molded lines and other features to achieve texture.

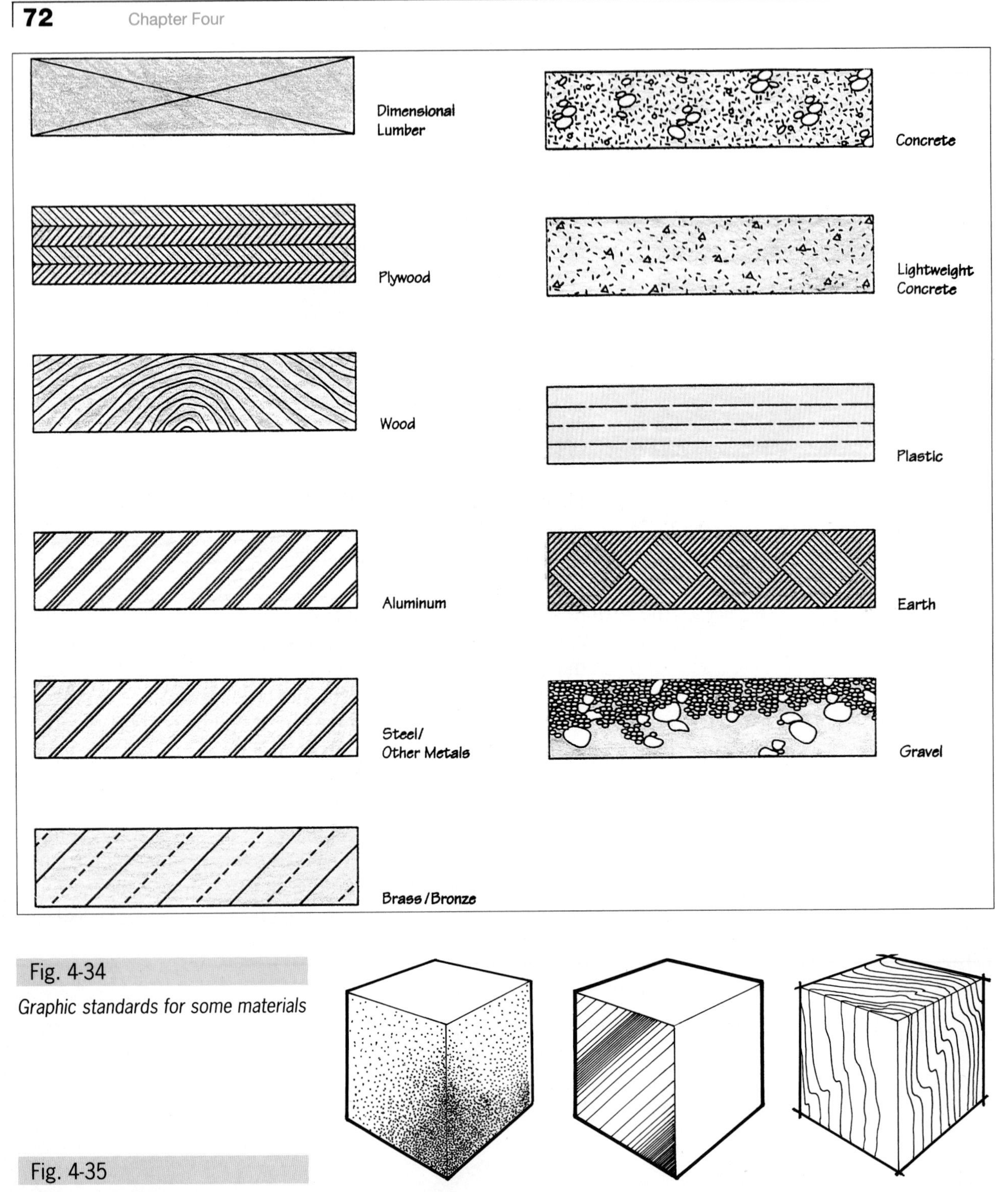

Fig. 4-34

Graphic standards for some materials

Fig. 4-35

Simple sketching techniques to achieve the appearance of texture

One of the easiest materials to sketch is wood. With a little practice, you can even sketch different kinds of wood, such as pine, mahogany, and oak. The different grain characteristics of these woods make it easy. Other textures are also easy to achieve (Figure 4-35).

Color Pencil Techniques

The use of color pencils is a relatively easy way to make sketches and drawings more interesting, effective, and informative. They may be used to make a color background and to outline, highlight, or shade.

Begin with a sharp point, and hold the pencil as you would for shading. The idea is to create a smooth blend of color by going over an area a number of times in several directions (Figure 4-36). Light pressure must be used to obtain an even color application.

Fig. 4-36

Using a color pencil to make a smooth transition in tone

Color pencils are ideal for shading, to give an object a 3-D appearance (Figure 4-37). Notice how the color on the cylinder is dark at the edges and light along the center. Carefully blend the color to get a crisp, neat drawing.

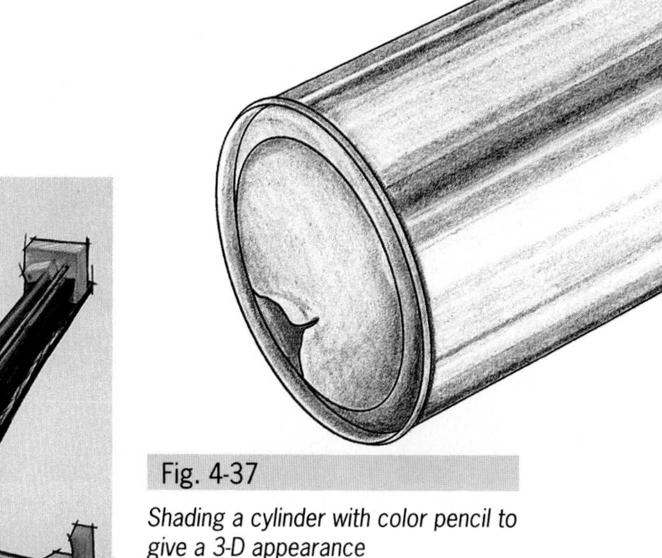

Fig. 4-37

Shading a cylinder with color pencil to give a 3-D appearance

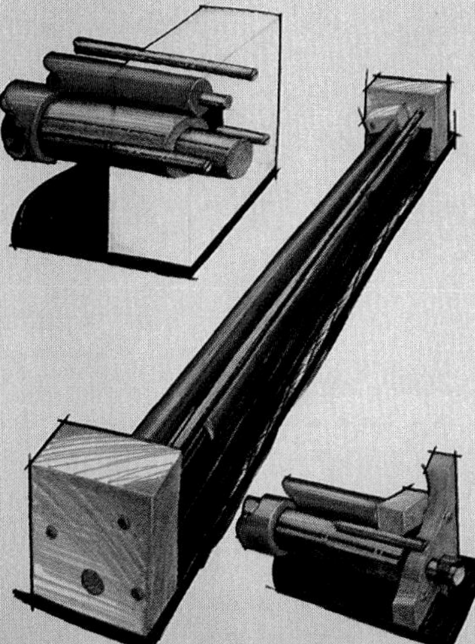

Fig. 4-38

A color marker rendering of a design idea. Courtesy of Peter Jewett

Color Marker Techniques

Color marker rendering is an effective and dramatic technique that looks more difficult than it actually is. Although it takes some practice, there are "tricks" that you can use to make your drawings look professional (Figures 4-38 and 4-39).

Marker Rendering Techniques

The techniques outlined here will help you develop a feel for the use of color markers. You will need to practice until you develop some skill and confidence in your work. When you get consistently good results, experiment with three markers of the same color but light, medium, and dark values for the three surfaces.

1. Sketch a cube in perspective or isometric. Use light pencil lines. Put two additional pieces of paper under your original drawing to prevent "bleeding" through to the work surface underneath.

2. Use two pieces of paper as a mask. Place the mask over your sketch so that only the top surface of the cube is visible. Use a medium-value marker, and work from one side of the cube's top surface to the other. Begin each stroke on the mask, and use bold, rapid, vertical strokes that end past the object lines of the cube. You are working out from the "V." Don't go back over areas you have missed.

3. Rotate the page to give yourself a comfortable position in which to use the marker. Adjust the mask to leave a thin space between the colored top surface and the left surface you will render. This space will be the corner of the cube between these surfaces. Work out from the "V," and use the same technique, but wait about a minute and give this surface a second coat of marker. This will make this side darker than the top surface.

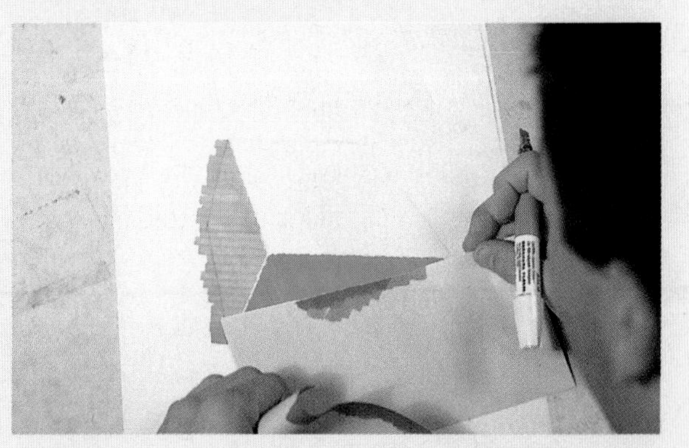

4. Again, rotate the page, and reposition the mask to expose only the right side. Be sure to position the mask pieces so that a thin, uncolored space is left between cube surfaces. Again, work out from the "V." Give this side three coats of marker, waiting a minute or so between coats. These three coats will make this surface the darkest side.

Fig. 4-39

5. To complete the rendered cube, use a black wide-tipped marker and smooth, bold strokes along the six outside object lines. This technique will make the object stand out from the page and cleanup the outside edges.

6. Overlapping the marker strokes gives the object a sharp, designer-rendered appearance. You can use this technique to outline pencil-drawn objects, too.

7. Another excellent technique for marker rendering is to cut out the cube and use stick glue to paste it on another sheet of paper. A background can be applied first with another contrasting color marker.

8. A cylinder cone and sphere are shown rendered above. Notice how the marker lines are drawn along the axis of the cylinder and how the value goes from dark along the upper edge to light, to dark along the bottom edge. A cut-out placed on the end surface to mask that area is shown. The marker lines on the cones begin at the top and radiate out as they reach the bottom.

9. Cutting out masks so that you can render each part is a time-consuming but worthwhile technique to get good results from marker rendering. Another technique is to photocopy your original drawing and render each separate part on each page. This allows you to cut out and assemble a finished drawing from the separate parts.

Working Drawings

What Are Working Drawings?

Drawings that contain the information and detail to actually produce an item are referred to as *working drawings*. These are drawings that show exact sizes, shapes, materials, and assembly instructions. Although some simple objects may be made from sketches, more complicated products require more detail to construct.

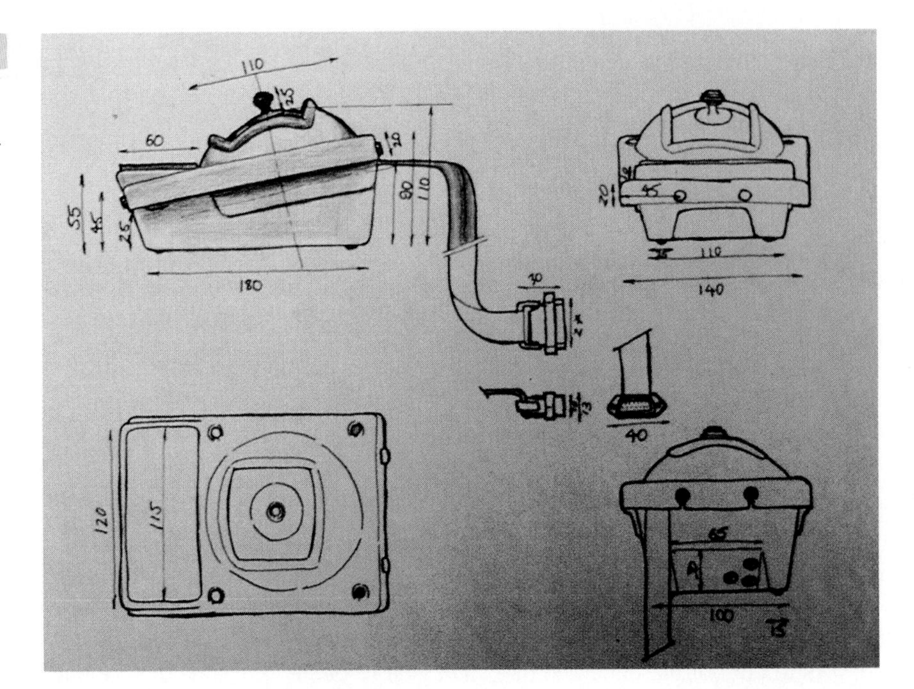

Fig. 4-40

Orthographic projection drawing of a computer mouse. A color pencil is used to bring out details. Courtesy of Peter Jewett

Orthographic Projection

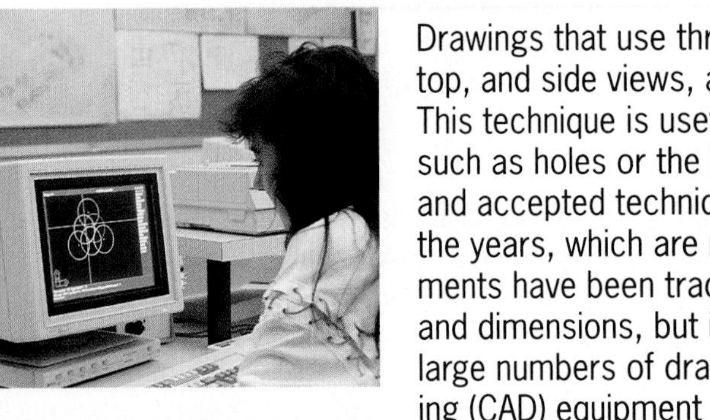

Fig. 4-41

Student doing design work using a CAD system

Drawings that use three basic views of an object, such as the front, top, and side views, are called **orthographic projection** drawings. This technique is useful for locating specific features with dimensions, such as holes or the intersection of planes (Figure 4-40). Many rules and accepted techniques have evolved in orthographic projection over the years, which are recognized as industry standards. Drafting instruments have been traditionally used to develop precise scale, lines, and dimensions, but in recent years we have seen the replacement of large numbers of draftspeople with operators of computer-aided drafting (CAD) equipment and software (Figure 4-41).

The development of working drawings for your project should include basic orthographic projection sketches or drawings where appropri-

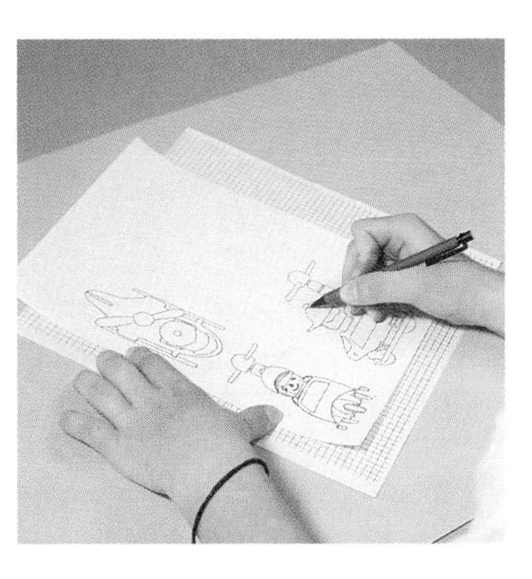

Fig. 4-42

Student using underlay grids to develop an orthographic sketch

ate, but the development of skill to product industry standard drawings would take a great deal of valuable time and practice. It is often acceptable to use an **underlay grid** beneath your drawing paper to help project features from one view to another (Figure 4-42). In this way, a three-view drawing can be sketched in minimum time. If time allows, the sketch can be traced using a straightedge to sharpen up the drawing.

Sectional, Exploded, and Cutaway Views

Sectional Views. Many objects are assembled from component parts. To better understand the relationship between the parts, several techniques may be used. The **sectional view** drawing is made of an object as if it were cut in two by a saw. (Figure 4-43).

Exploded Views. The **exploded view** drawing is especially useful in assembling an object. Instructions packaged with products that require assembly often have exploded view drawings to make it clear how separate parts are fastened to each other (Figure 4-44).

Cutaway Views. A **cutaway view** drawing is similar to a sectional view drawing except that often whole interior parts are shown intact (Figure 4-45). This kind of drawing lends itself to three-dimensional drawing, where mechanisms and interior features are shown.

Fig. 4-43

Sectional view drawing of a tape dispenser

Fig. 4-44

Exploded view drawing

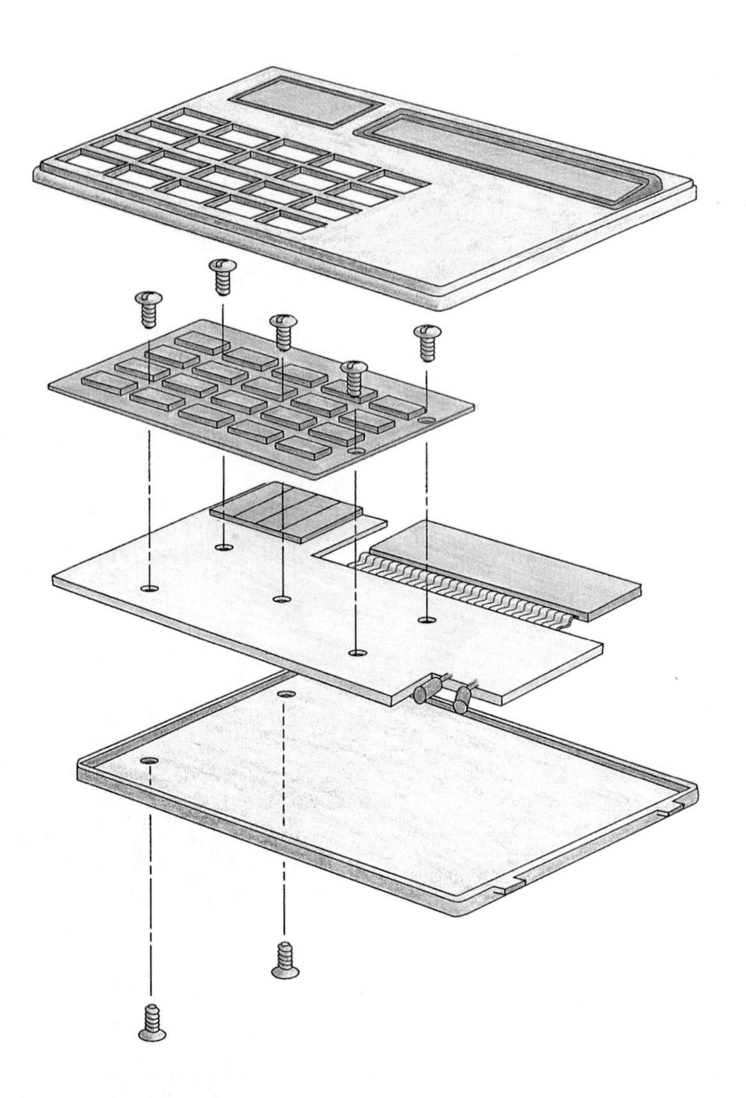

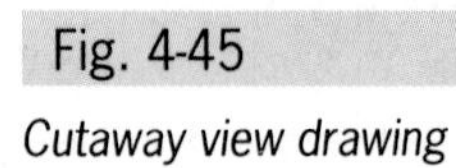

Fig. 4-45

Cutaway view drawing

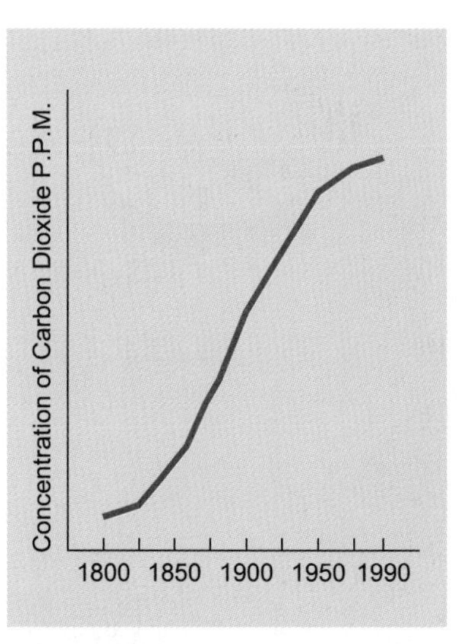

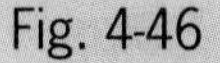
Fig. 4-46

A simple graph showing a change of a value over time

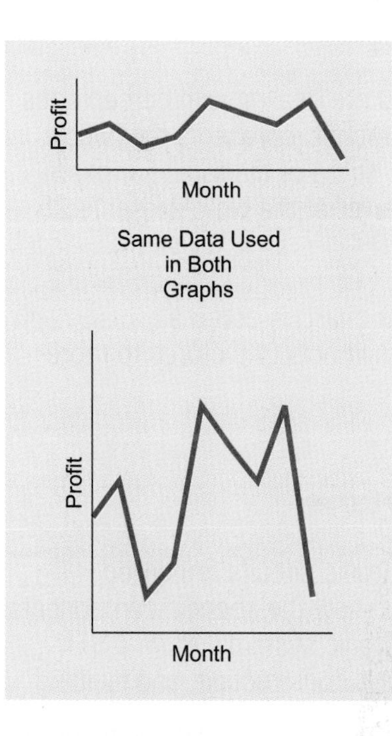

Fig. 4-47

The scales chosen for a graph must be carefully selected to represent the data accurately.

Information Graphics

Charts and Graphs

Presenting information is an important part of documentation. When information is in the form of numbers, charts and graphs are extremely useful. There are excellent examples of the use of charts and graphs in weekly newsmagazines and newspapers.

Graphs. A **graph** is a diagram that shows how some numerical quantity has changed over time (Figure 4-46). For example, a graph of the carbon dioxide found in the air in a certain city over the last ninety years would probably reveal a marked increase over that time. Graphs usually have numbers representing a quantity running up the vertical scale and a time designation along the horizontal scale. Time may be in any unit, such as years, months, hours, or nanoseconds. The scale used on graphs can be manipulated to have an impact on the viewer. Look at the two graphs in Figure 4-47. Although the same data are represented on both graphs, one graph paints a much rosier picture than the other of the change over time.

Charts. A **chart** is a diagram that shows the relationship between two or more numerical quantities. A comparison of the average carbon dioxide levels in 1991 for several cities shows certain cities have higher levels than others.

There are a number of different kinds of charts that may be used to present your data. A **bar chart** uses lines, bars, or figures to compare one quantity with another (Figure 4-48). It can be shown in either a vertical or a horizontal orientation.

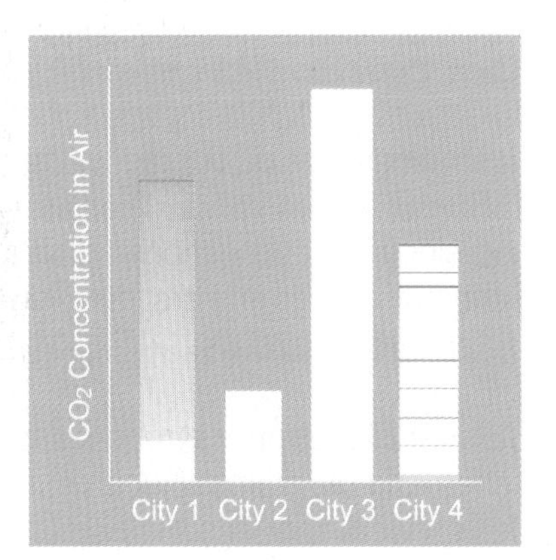

Fig. 4-48

A bar chart comparing CO_2 levels in several cities

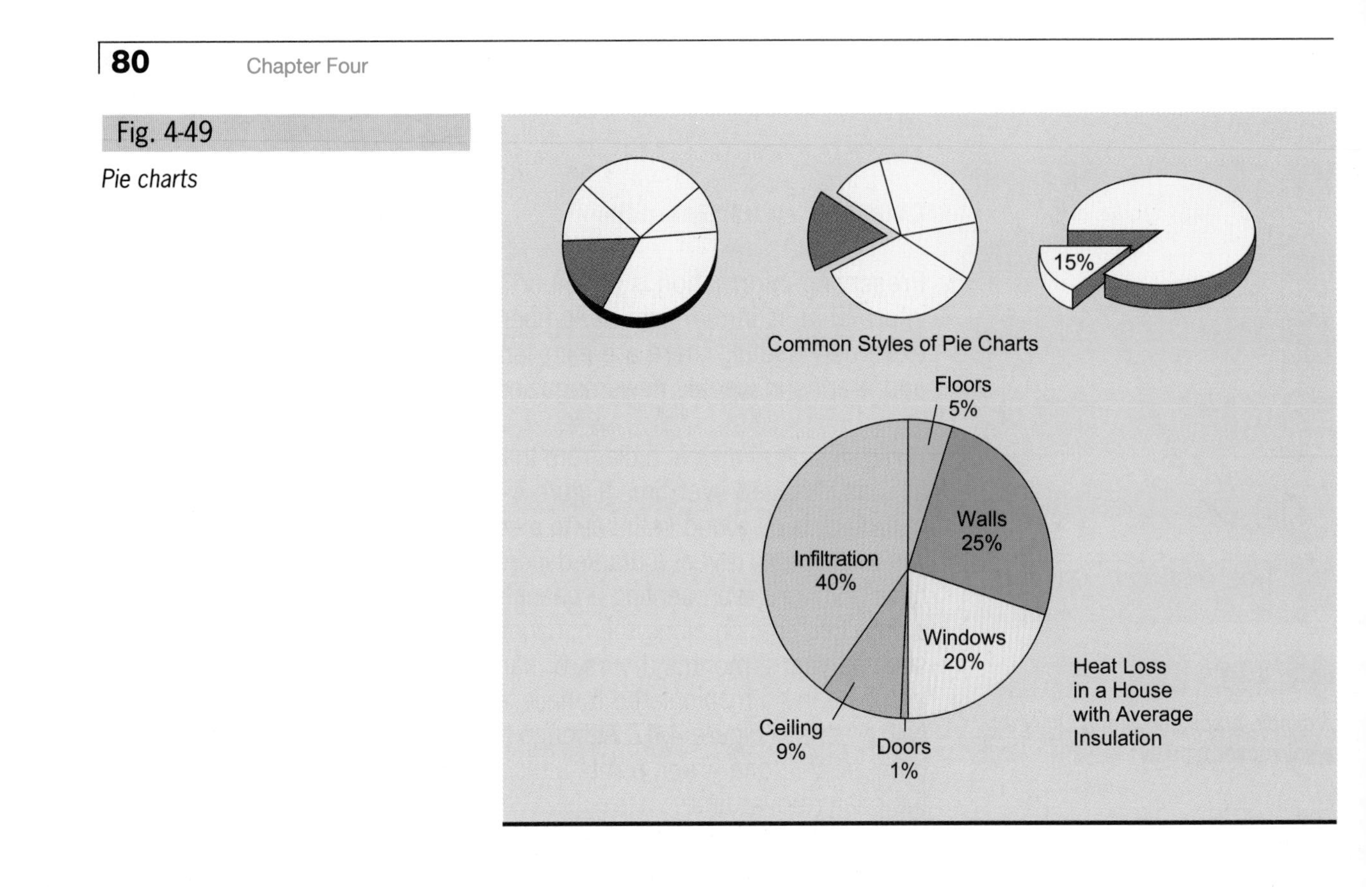

Fig. 4-49

Pie charts

A **pie chart** is especially useful when you are dealing with percentages of a whole (Figure 4-49). Because 100 percent represents the "whole pie," the size of each slice represents its share of the total: For example, half of the pie is 50 percent; one-quarter of the pie represents 25 percent; and so on.

Another kind of a chart, related to the bar chart, is called a **pictograph**. A pictograph is a chart that uses simple symbols or pictures to represent a fixed quantity (Figure 4-50).

Electronic Schematic Diagrams

Illustrations that depict the wiring of electronic circuits are called **schematic diagrams**. They are used to show the specific components and electrical connections within an electronic system (Figure 4-51). Schematic diagrams are used in designing, constructing, and faultfinding electronic circuits.

Standard electronic symbols have been developed and are used for schematic diagrams. Documenting the development of electronic design work requires the use of standard symbols.

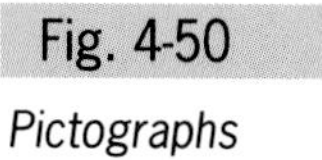

Pictographs

1986
1988
1990
1992
= 1000

= $1000

1988 1989 1990 1991 1992

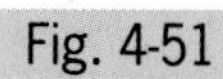

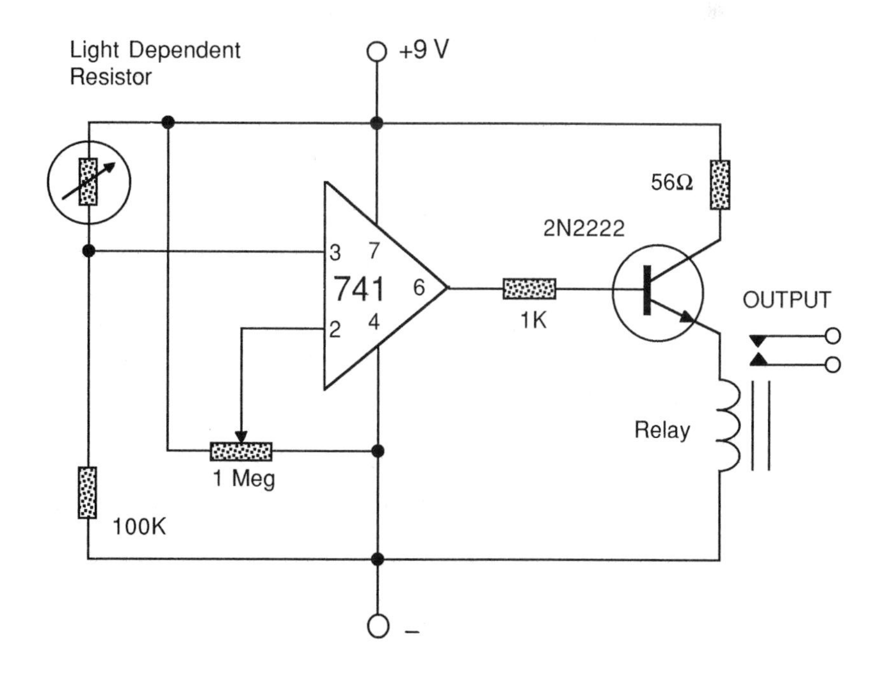

Electronic schematic of a circuit that switches a relay when light strikes a sensor

Maps

Diagrams that show the arrangement of the natural and human-made world are called *maps*. Maps are designed with a specific purpose in mind. For example, *aeronautical maps* are made to help the pilot locate a position or destination (Figure 4-52). These maps are specialized according to the flying conditions, such as “low-altitude charts” for instrument flying and “sectional charts” for visual flying. Like all maps, the critical space requirements dictate that *symbols* be used to help make objects and points of interest clear. The addition of color to a map allows more information to be put in the same space. Other special-purpose maps include nautical, geological, subway, and road maps.

Fig. 4-52

Aeronautical sectional map

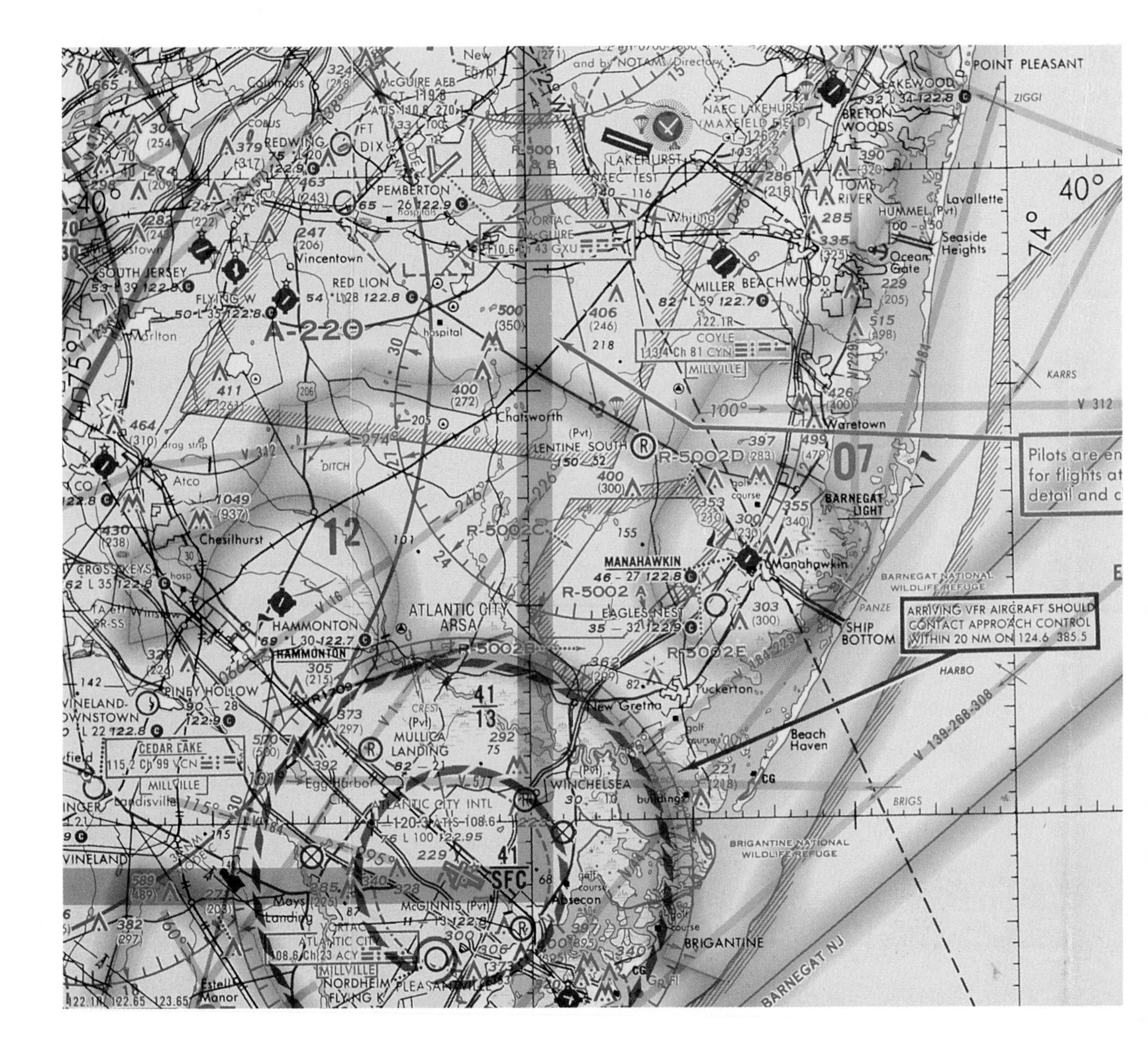

Sequence Diagrams

A series of illustrations that shows the order in which things should happen is called a **sequence diagram** (Figure 4-53). Sequence diagrams are often found in kits, products that need assembling, and instructions for making something or performing a task, such as making a cake. Sequence diagrams are an extremely effective form of visual communication.

Sequence diagrams may be developed for almost any task. Complex tasks are usually made up of a number of simpler tasks, each of which may be developed into a separate sequence. It is important that you decide exactly what it is you wish to accomplish and break down the task into a specific number of steps. Illustrate each task so that it contains the necessary information without confusing the reader. Sequence diagrams may or may not be accompanied by word descriptions.

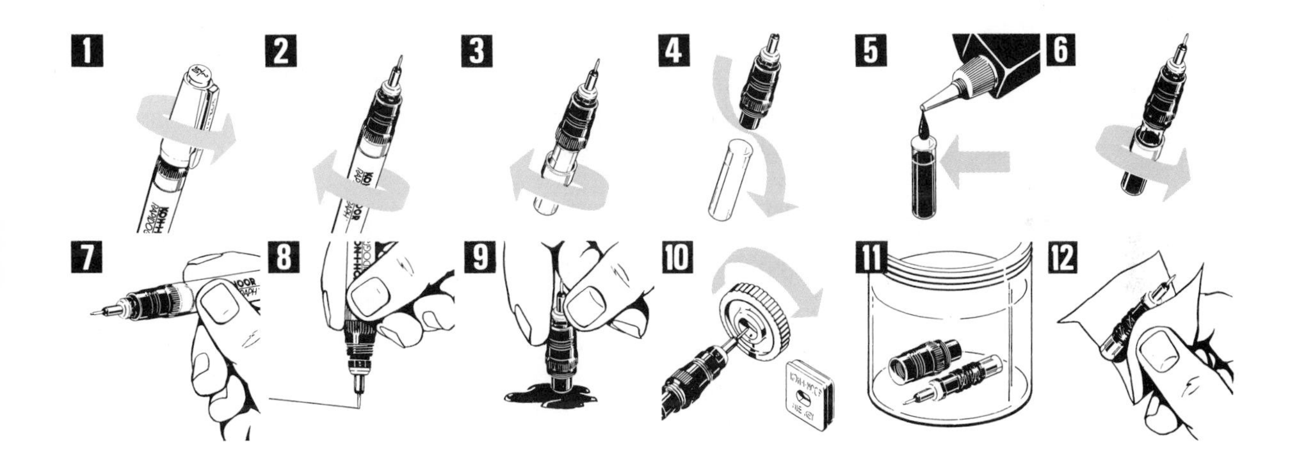

Fig. 4-53

Sequence diagram used to show a series of steps. Courtesy of Koh-i-noor

Developing a Portfolio

What kind of things should you put into a design portfolio? A design portfolio should contain a number of sections, depending on the complexity of the project. Simple design problems will have fewer sections, often combining several from the list that follows. Major projects will use all or most of the sections, or will combine the sections into another logical format. Following is a list of sections and appropriate documentation strategies.

Step	Documentation Technique
Background Problem	Description of a technological situation explaining the problem. Write, draw, annotate. Use "press type," computer-generated text, neat hand lettering, and so forth.
Design Brief and Specifications	Actual brief with specifications, conditions, and requirements. Use "press type," computer-generated text, neat hand lettering, and so forth.
Research and Investigation	Notes, sketches, letters, interview tapes, bibliography, photos. Use photocopies of catalog pages and other reference materials. Include diagrams that explain mechanical or electronic principles to be applied.
Solution Ideas	Notes, preliminary sketches, and development sketches with annotations, 2-D models. Use pencil, color pencil, marker, technical pen, and other mediums. This should be a well-documented section.
Chosen Solution	Notes, matrix comparing requirements to solutions, checklists, and so on. Presentation drawings using color pencil, marker, or other color medium.
Development of Solution	Photos of evolutionary models: wood, plastic, card/paper, foam-core™ , metal, ceramic, found objects: computer simulations, kits, clay, VHS, and so on. Working drawings, such as orthographic projection. Color rendering of final appearance of solution.
Modeling or Prototyping	Photos of various stages of completion; descriptions of necessary adjustments and changes, and so on.
Testing	Data checklists, graphs, charts, photos, slides, VHS.
Evaluation (Solution Evaluation and Self-evalution)	Description of test results and self-criticism: "press type," computer-generated text, neat hand lettering, and so forth.

In addition, a title page is always appropriate, and, if the portfolio is lengthy, a table of contents may help the organization. The portfolio sections correspond with the design steps outlined in earlier chapters.

Portfolio Pages

Page Orientation. A horizontal or *landscape* orientation of the paper is well suited for portfolio work (Figure 4-54). Choose a paper size that is large enough to allow for a border and a title block and still give you plenty of room to draw. A standard 11" × 17" page is large enough without being too cumbersome to carry around, but other sizes are acceptable. You may also need to develop a cover or case to keep pages from tearing and to protect your work from wind, rain, and other hazards.

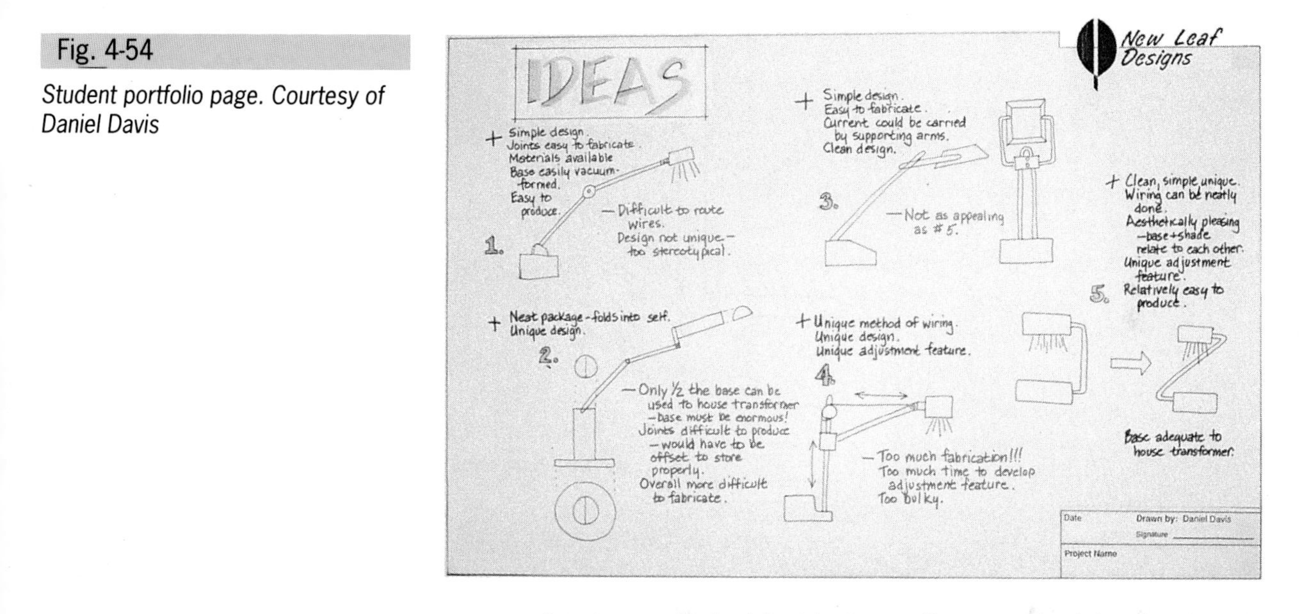

Fig. 4-54

Student portfolio page. Courtesy of Daniel Davis

INTERNATIONAL TECHNOLOGY EDUCATION ASSOCIATION

Fig. 4-55

An example of logo designs. Courtesy of ITEA

Logo. A company that wishes to be easily recognized develops a symbol for use on its products, correspondence, and advertising called a **logo**. You may develop your own logo for your design work and incorporate it into both your portfolio and your final design solutions. In the portfolio, the logo may appear on each page (Figure 4-55).

Binding. Pages may be bound on either the left side or the top, using plastic bindings, a modified three-ring binder, or one of a variety of other methods.

Portfolio Page Content

It is important to strike a balance between putting too little and too much on a page. A typical page in a portfolio might have a number of drawings (Figures 4-56a,b,c). There is no reason, however, that all of these drawings must originally have been drawn on the same piece of paper. You can redraw from original sketches onto one page or cut out a number of different drawings and glue them on a portfolio page. These techniques will allow you to present ideas in an interesting and concise format.

Figs. 4-56a, b and c

Several pages of a design portfolio. Courtesy of C. Smith

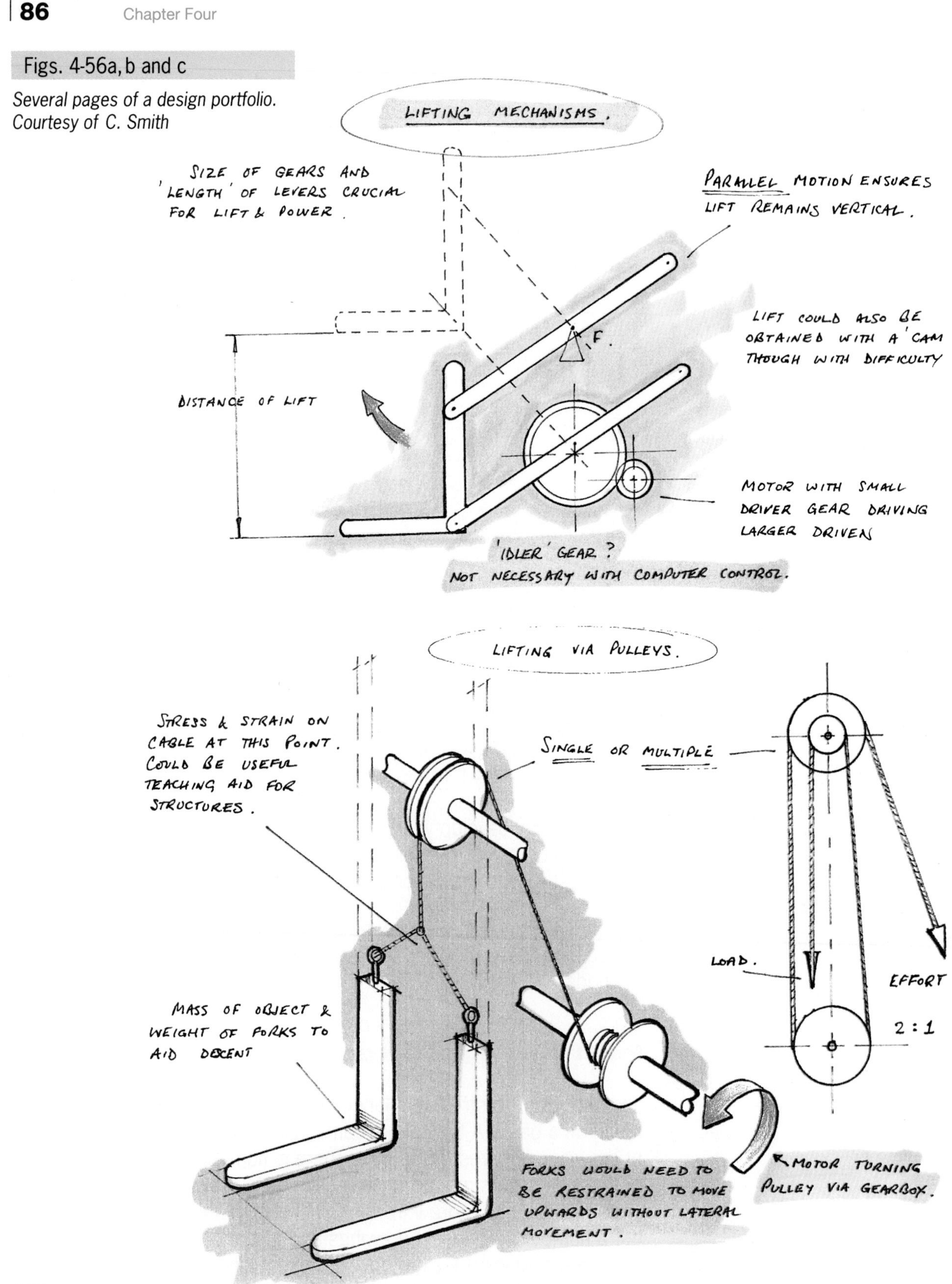

4-56a

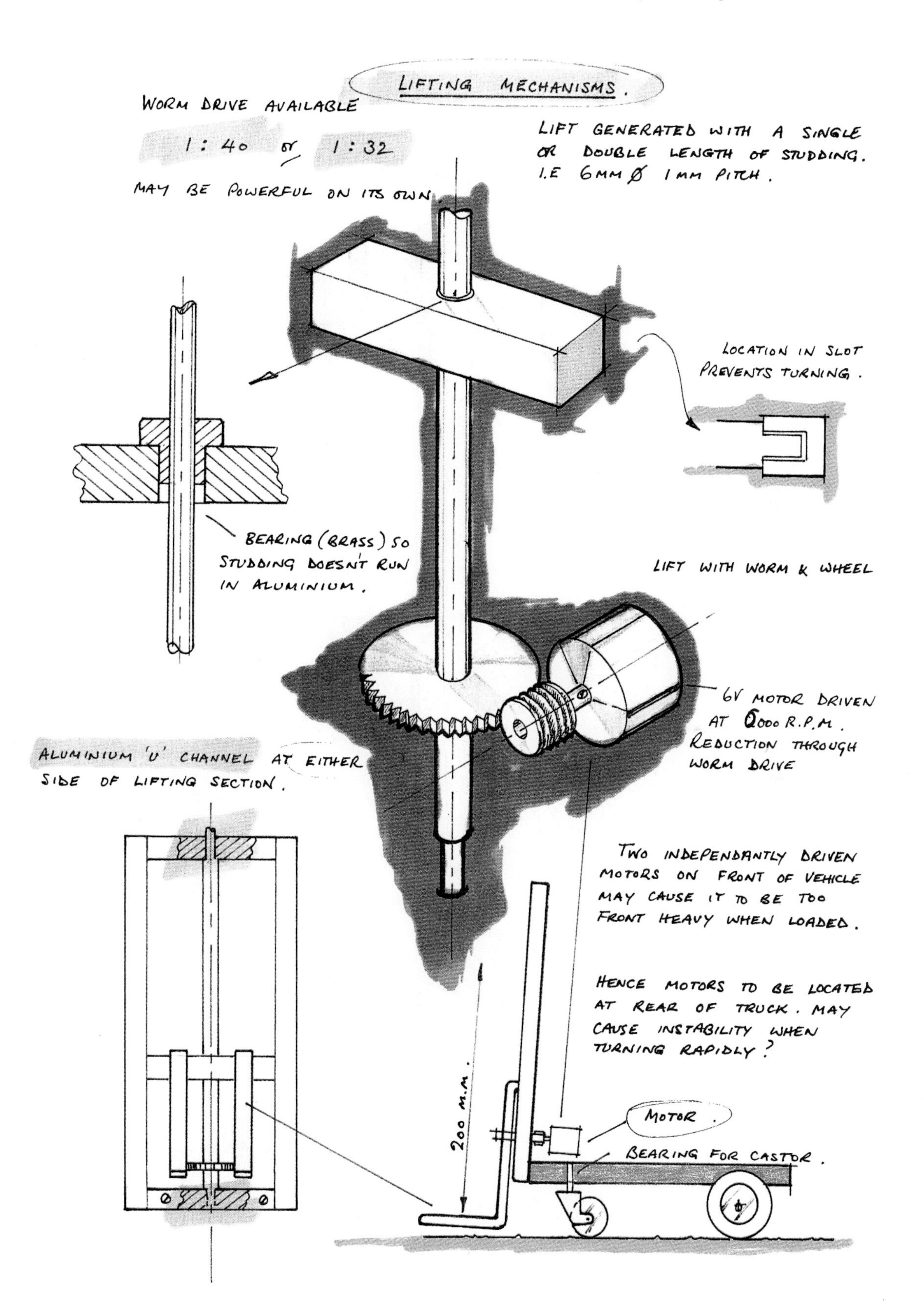
LIFTING MECHANISMS.
WORM DRIVE AVAILABLE
1 : 40 or 1 : 32
MAY BE POWERFUL ON ITS OWN.
LIFT GENERATED WITH A SINGLE OR DOUBLE LENGTH OF STUDDING. I.E 6MM Ø 1MM PITCH.
LOCATION IN SLOT PREVENTS TURNING.
BEARING (BRASS) SO STUDDING DOESN'T RUN IN ALUMINIUM.
LIFT WITH WORM & WHEEL
6V MOTOR DRIVEN AT 8000 R.P.M. REDUCTION THROUGH WORM DRIVE
ALUMINIUM 'U' CHANNEL AT EITHER SIDE OF LIFTING SECTION.
TWO INDEPENDANTLY DRIVEN MOTORS ON FRONT OF VEHICLE MAY CAUSE IT TO BE TOO FRONT HEAVY WHEN LOADED.
HENCE MOTORS TO BE LOCATED AT REAR OF TRUCK. MAY CAUSE INSTABILITY WHEN TURNING RAPIDLY?
200 M.M.
MOTOR.
BEARING FOR CASTOR.

4-56b

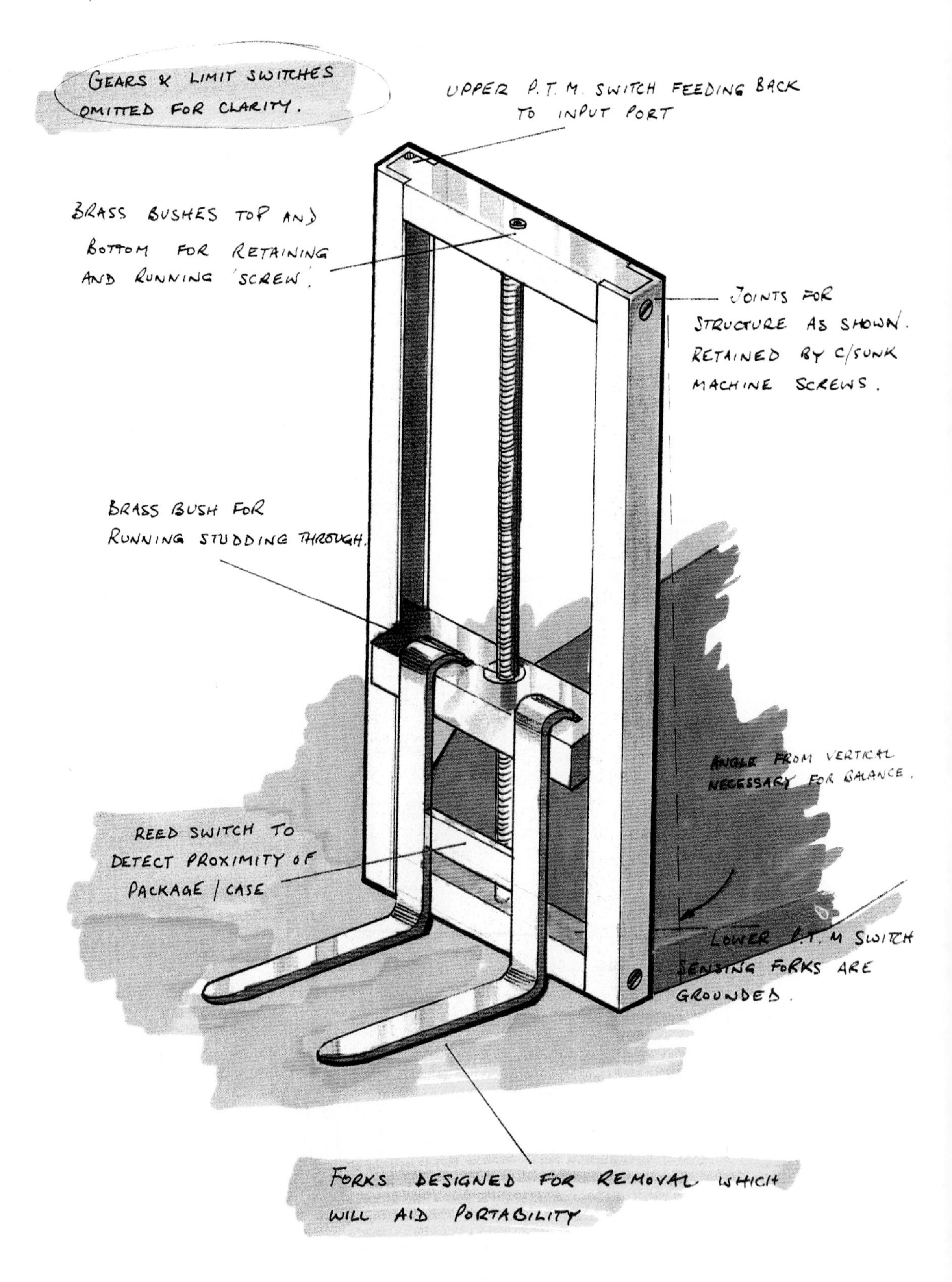
GEARS & LIMIT SWITCHES OMITTED FOR CLARITY.
UPPER P.T.M. SWITCH FEEDING BACK TO INPUT PORT
BRASS BUSHES TOP AND BOTTOM FOR RETAINING AND RUNNING 'SCREW'.
JOINTS FOR STRUCTURE AS SHOWN. RETAINED BY C/SUNK MACHINE SCREWS.
BRASS BUSH FOR RUNNING STUDDING THROUGH.
ANGLE FROM VERTICAL NECESSARY FOR BALANCE.
REED SWITCH TO DETECT PROXIMITY OF PACKAGE / CASE
LOWER P.T.M SWITCH SENSING FORKS ARE GROUNDED.
FORKS DESIGNED FOR REMOVAL WHICH WILL AID PORTABILITY

4-56c

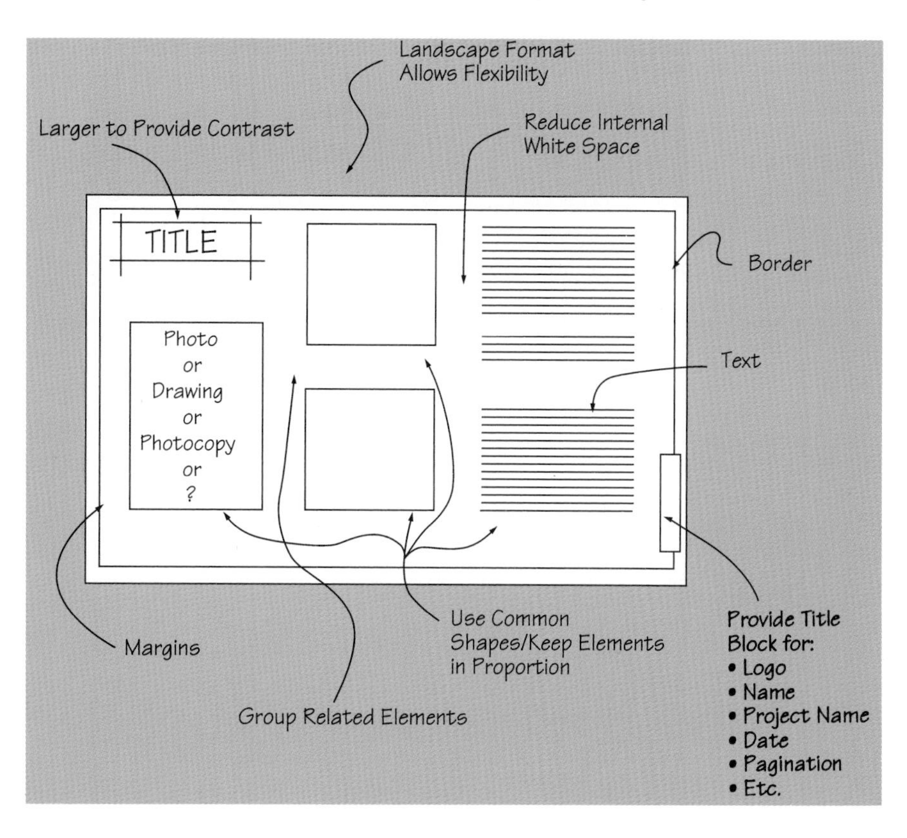

Fig. 4-57

Items to be considered in designing a portfolio page

Portfolio Page Layout

Organizing the appearance of the page is an important part of portfolio presentation. Figure 4-57 illustrates a number of strategies you can use to make a page look interesting.

Observation/Analysis/Synthesis

1. Using an HB pencil, create a one-point perspective sketch of one of the following: VHS cassette tape, audio cassette tape, "Walkman" type personal stereo.
2. Using color pencils, make a cube stand out by shading around it.
3. Using color pencils, give a cube a 3-D appearance by shading with three values of one color.
4. Using color markers, give a cube a 3-D appearance by shading with three values of one color.
5. Using a scale and calipers, sketch a full-size or half-scale orthographic projection drawing with a front, top, and side view of one of the following: iron, hair dryer, electric drill, personal stereo, cordless telephone.

6. Research and reproduce a number of symbols used for a special-purpose map, such as a geological survey, aeronautic sectional, and so on.
7. Develop a sequence diagram without word descriptions for one of the following: setting the time on a VCR, making a dessert, removing a broken bolt from a metal casting, safely starting a model airplane gasoline engine, changing a tire on a bicycle.
8. Develop a special-purpose map for one of the following: school bus route, paper route, marathon run.
9. Design and produce a master portfolio page, including an original logo, company name, and a block for project name, date, and other necessary information.
10. Use the techniques in this chapter to fully document a design problem.

CHAPTER 5 Investigation and Research

Investigation Strategies

The purpose of research is to collect data and information that will assist you in solving a problem. The key to doing this research is asking the right questions and consulting the right sources.

Thinking about problems and opportunities

Jennifer has been thinking about the problem she has with the lighting in her room. Dave's family travels a lot and enjoys riding bikes, but they always have trouble finding space for their bicycles when they pack for their trips. José enjoys sports and would like to improve his soccer shot. Natala's mother is opening a new store and needs a sign that will hang above the front door. These are all problems or opportunities that raise many new questions.

Each of these students will need to answer many obvious questions, but, more important, they need to ask additional questions. Asking the right question is probably the most important part of doing investigation and research. Effective designers and engineers know how to ask the right questions and are skilled at using resources to get the right answers.

Asking Questions

For most research for writing school reports that you have done in the past, you probably asked only *closed questions*, which generated specific answers. For example, reports entitled "What are the causes of the Civil War?" or "How did Thomas Edison invent the light bulb?" would require looking up answers to these closed questions in a variety of library sources. The typical report-writing process usually involves taking careful notes on the specific topic and preparing a well-organized and thoughtfully written report.

Research for technological problem solving is different. At the heart of finding appropriate solutions to problems is asking the right questions. An *open-ended question* can have more than one answer and often leads to more questions—for example, "How can I design a bicycle that can be transported in the trunk of a car?" Answering this question requires research on the design of bicycles and car trunks. Answers will vary, and each answer will lead to more questions about materials, structures, and design aesthetics. Being able to ask the right questions is crucial to all research.

Fig. 5-1

Student beginning his research using a library catalog system

Collecting Published Information

Research in a library can be an overwhelming challenge (Figure 5-1). There is so much information stored in one place, and it often seems like the elusive librarian who put it there is the only one who understands how it's organized. The books in all libraries are organized by means of the *Dewey Decimal Classification System* or the *Library of Congress Classification System*, and they can be accessed by means of the card catalog or a computerized catalog. These catalogs are an index to the library. The key to every resource in the library is understanding the library indexes.

Finding Key Words

The subject headings in the card catalog and the topics in periodical indexes, book indexes, and data bases all are made up of similar key

Fig. 5-2

Library of Congress Subject Headings book

words or subjects. Before you can research your topic, you need to think about all of the *key words* under which your subject might be found. For example, the topic "engineering graphics" might be found under the key words *mechanical drawing* or *technical illustration.* The two-volume *Library of Congress Subject Headings* is a useful resource for finding the exact key words used by most indexing systems (Figure 5-2). List the key words under which your subject might be found. An encyclopedia article on your subject can help you find these key words.

Read the Encyclopedias on Your Topic

The first step in research is finding a good overview of your subject. Often the best source of this background information is an encyclopedia. The encyclopedia will help you find key words or subtopics to assist you in generating more questions in order to form a framework or an outline for your research. There are many specialized science and technology encyclopedias available in school libraries. Use the index to get a complete overview of your topic. The index will keep you from missing good information in another volume that might be listed under a subtopic that you did not think of.

Using the Card Catalog

Use the subject cards (the subject is usually printed in capital letters) in the card catalog to locate books on your topic (Figure 5-3). The notations at the bottom of the card can lead you to other related topics. The **call number,** located in the upper-left corner of the card,

Fig. 5-3

Subject card

TRANSPORTATION.

380 Moolman, Valerie.
M The future world of transportation /
by Valerie Moolman and the editors of Grolier.
— New York: Watts, 1984.
112 p. : ill. (some col.) ; 29 cm. —
(Walt Disney World EPCOT Center Book)
Includes index.
Summary: Traces the history and discusses the future of transportation with emphasis on the development of different types of vehicles and modes of transport. Based on the "World of Motion" exhibit at Walt Disney's EPCOT center.
ISBN 0-531-04882-9 (lib. bdg.)
1. Transportation.

Fig. 5-4

Computer catalog screen

COMPUTERIZED CARD CATALOG

Search Request:	a = Transportation b = Bicycles

Author:	Moolman, Valerie
Title:	Transportation
Publisher:	New York: Watts, 1984
Description:	112p, traces the history and future of transportation, with emphasis on the development of different types of vehicles. Based on the "World of Motion" exhibit at EPCOT Center, Walt Disney's World.

Commands:	N = Next Screen	P = Previous Screen
	I = Index	H = Help
Location:	Main Library	
Call Number:	C 380 M	

RADIO
History
A less-than-civil war [K. Burns' PBS documentary Empire of the air] H. F. Waters. il por *Newsweek* 119:61 Ja 27 '92
RADIO ADVERTISING
See also
Books—Advertising
RADIO ASTRONOMY
Runaway pulsar [PSR 1757-24; research by Dale A. Frail and Shrinivas R. Kulkarni] il *Sky and Telescope* 83:9-10 Ja '92
RADIO BROADCASTING
See also
Radio stations
Censorship
Constitutionally correct [E. J. Kemp fighting effort to close KFI radio station for having aired talk show critical of handicapped anchor B. Walker's decision to have children] *National Review* 44:16 Ja 20 '92
Conversation programs
See Radio broadcasting—Talk shows
Rock music
Rules to sing about [B. Adams' album Waking up the neighbours runs afoul of Canadian content regulations as related to radio airtime] K. Doyle. il *Maclean's* 105:2 Ja 27 '92
Talk shows
Constitutionally correct [E. J. Kemp fighting effort to close KFI radio station for having aired talk show critical of handicapped anchor B. Walker's decision to have children] *National Review* 44:16 Ja 20 '92
Canada
See also
Canadian Broadcasting Corporation
Canadian Radio-Television and Telecommunications Commission
RADIO-CANADA *See* Canadian Broadcasting Corporation
RADIO CENSORSHIP *See* Radio broadcasting—Censorship
RADIO IN ASTRONOMY *See* Radio astronomy
RADIO IN NAVIGATION
See also
Loran
RADIO INDUSTRY
See also
American Broadcasting Companies, Inc.
National Broadcasting Co., Inc.
Radio stations
RADIO LAWS AND REGULATIONS
See also
United States. Federal Communications Commission
Canada
See also
Canadian Radio-Television and Telecommunications Commission
RADIO PROGRAMS *See* Radio broadcasting
RADIO SOURCES (ASTRONOMY)
See also
Pulsars
Quasars
RADIO STATIONS
Constitutionally correct [E. J. Kemp fighting effort to close KFI radio station for having aired talk show critical of handicapped anchor B. Walker's decision to have children] *National Review* 44:16 Ja 20 '92

Fig. 5-5a

Page from Reader's Guide to Periodical Literature. *Courtesy of Reader's Guide to Periodical Literature, 1990, sample subject entry and sample entry. Copyright ©1990 by the H.W. Wilson Company. Material reproduced with permission of the publisher.*

should be written down while you are at the catalog. This call number specifies where the book is located on the library shelves. The copyright date of the book is especially significant if your topic is one for which only current information is acceptable.

Many libraries now have computerized catalogs that must be accessed through a terminal (Figure 5-4). Using key words, the user types in the subject and the computer gives a list of available books on the topic. Be careful, as many computerized catalogs contain books available at many branches of a library system. By following the screen instructions carefully, you can discover the location and availability of each book. Libraries with this type of system can get you materials from other libraries through interlibrary loan.

Using Periodical Indexes

Periodicals or magazines are often the best source for topics requiring current information. Every library has H.W. Wilson's *Reader's Guide to Periodical Literature* index in book or computerized CD-ROM form (Figure 5-5a, b). The *Applied Science and Technology Index* is published by the same company and indexes articles in technology, engineering, and science. This index works like others, listing magazine articles by subject headings. Subheadings listed under most headings give articles in more specific areas. Even if your library does not have the magazines listed in the periodical index, you should write down all articles that seem to be what you need. The magazines available in your library should be posted on a master list near the periodical

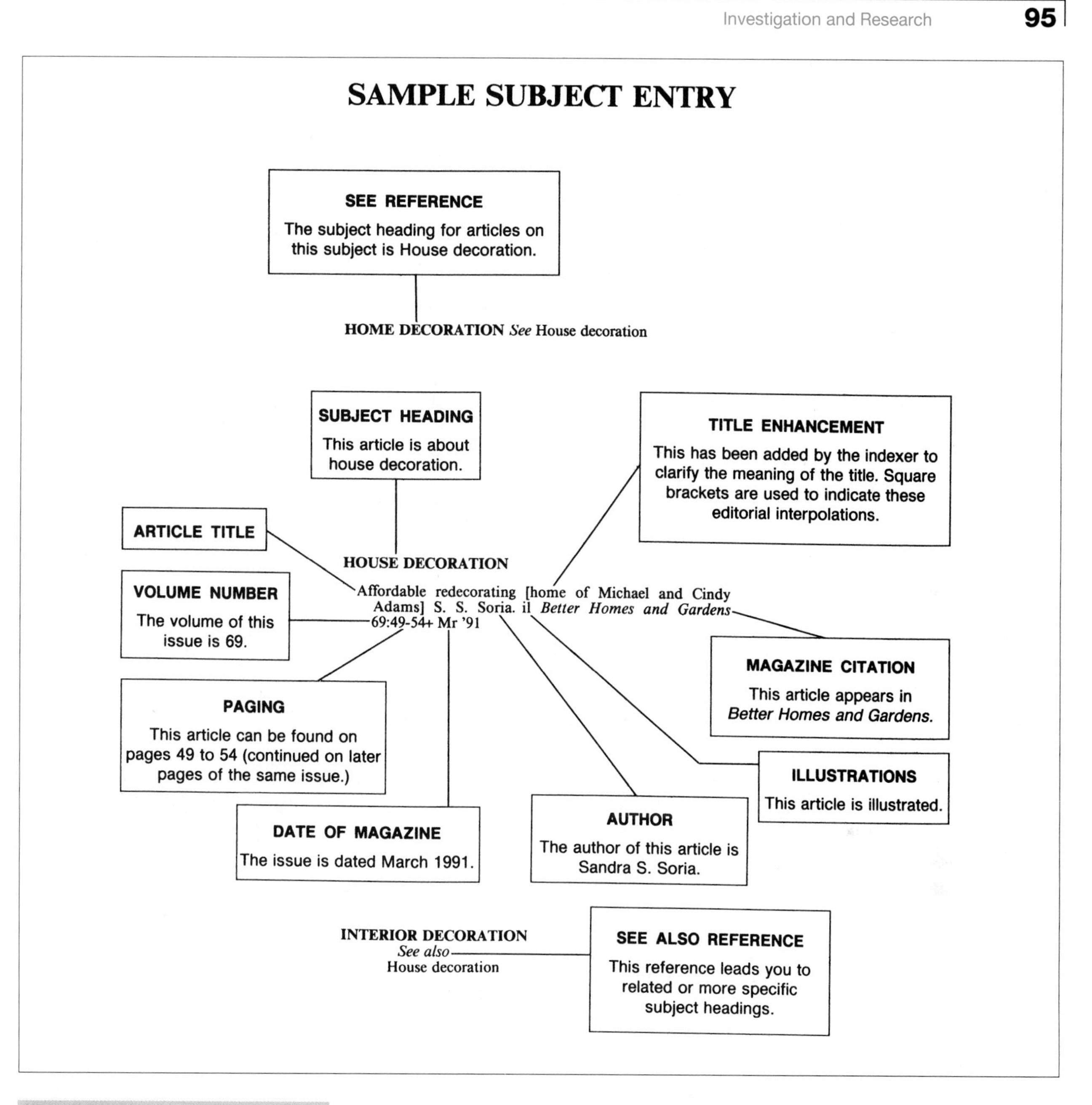

Fig. 5-5b

Page from Reader's Guide to Periodical Literature. *Courtesy of Reader's Guide to Periodical Literature, 1990, sample subject entry and sample entry. Copyright ©1990 by the H.W. Wilson Company. Material reproduced with permission of the publisher.*

index. Whatever is not available in your library can usually be acquired through interlibrary loan from a local public or college library.

The computerized periodical index is a bibliographic data base that can be accessed from CD-ROM by typing in key words (descriptors). First, type in the general descriptor for your subject. If there are too many articles listed, go back to the search screen and narrow your topic by listing one or two other key words. The articles will appear in chronological order, with the most current article on the subject listed first. A printer is usually available so that you can print out a list of the articles to take to the circulation desk (Figure 5-6).

Fig. 5-6

Printout of computerized periodical index

```
SUBJECTS COVERED:
Bicycles/Design
Trimble, James

3 RDG
Kukoda, John
Birth of a bike (Schwinn Paramount OS)
Bicycling v31 p160 -5 April '90
il

SUBJECTS COVERED:
Bicycles/Design
Schwinn Bicycle Co.

4 RDG
Transport (designers G. Bliss and J. Van der Tuin)
The New Yorker v65 p50 -1 September 25 '89

SUBJECTS COVERED:
Bicycles/Design
Van der Tuin, Jan
Bliss, George

5 RDG
Weber, Bruce
Bike to the future (students at Art Center College of Design
create innovative bicycles)
The New York Times Magazine p82 February 12 '89
il

SUBJECTS COVERED:
Bicycles/Design
Art Center College of Design

1 RDG
Kukoda, John
The ties that bind (comparison of construction methods)
Bicycling v31 p66-72 November '90
il

SUBJECTS COVERED:
Bicycles/Frames

5 RDG
Boorstin, Daniel J.
The Edison effect (difference between discovery and invention)
U.S. News & World Report v105 p61 September 26 '88
il

SUBJECTS COVERED:
Science
Inventions

9 RDG
Andrews, Peter
American inventions that changed our lives
Good Housekeeping v203 p212+ July '86
il
```

Recording Information

As you find reference books, books in the card catalog, and magazine articles in periodical indexes, it is crucial that you take down all bibliographic information that you will use later for the bibliography that must accompany any research. You should record all information needed in the bibliographic form that you have been told to use. Figure 5-7 shows the *Modern Language Association* (MLA) bibliographic form that is commonly used in schools.

Fig. 5-7
Bibliography format

Book

Author's last name, first name. *Title of book*. Place of publication: Publisher, Date.

Macaulay, David. *The Way Things Work*. Boston: Houghton Mifflin Company, 1988.

Periodical

Author's last name, first name. "Title of article." *Title of publication* Date: pages.

Sias, Jim. "Packaging for Posterity." *Technology, Innovation & Entrepreneurship for Students* September–October 1990: 33–40.

Encyclopedia

"Title of article." *Name of Encyclopedia*. Date ed.
"Bicycles." *Encyclopedia Americana*. 1991 ed.

Take good notes using *direct quotes* (the author's exact words) or *paraphrasing* (conveying the meaning in your own words). Quotes should be used for definitions, important statistics, and especially meaningful points made by the writer. When you paraphrase, do not just change a few words and call it your own, but read the text and then look away from it and rewrite it in your own words. In both cases, it is usually appropriate to use a footnote to credit the source of the information.

Collecting Information From People

Although a library is the place to begin this type of research, published information may only provide part of what you need. The information found in specialized encyclopedias, books, periodicals, and other printed sources may lead to more questions that only can be answered by interviewing people who have a body of knowledge about your topic. For example, in answering the question "What device could I develop to assist Mr. Smith, who is physically challenged, to reach tall shelves in the supermarket?" it would be necessary to interview Mr. Smith. Since Mr. Smith probably has experienced this problem, he would have important information about the requirements of such a device. Perhaps you have heard of a company that manufactures devices for the physically challenged. You could contact the company by phone.

There is a directory of corporations that lists addresses and phone numbers available at the public library. If you plan to get information through the mail, it is best to make an informal contact by phone first before you write a follow-up letter. *Blind letters,* addressed to the public relations office, head designer, or some similar title are often ignored or don't get to the person who has the right answer. When writing a letter, use some official stationery and an appropriate business letter format.

Many companies even have toll free 800 numbers to make contacting them easier. If you call, be sure to prepare specific questions about the product ahead of time. Record their answers to your questions in an organized manner. Be sure to get the name and title of the person with whom you spoke. This information is needed for the bibliography and should follow the designated format for an interview reference.

Investigating strategies include asking the right questions and collecting information from published sources and knowledgeable people. Use appropriate resources, including your teacher and school or community librarian. Plan your research. Finding the right answers and asking the right questions is an important first step in technological design and problem solving.

Human Factors Engineering

Fig. 5-8

Products must be designed for the intended user.

One question that will require some research for nearly every problem situation involves the question concerning the human user; for example, "Did you ever try to ride a tricycle as a teenager?" We expect to live and work in environments and to use products that "fit" us as human users. The degree to which the things in our environments are comfortable is a measure of good planning and the application of human factors engineering principles (Figure 5-8). **Human factors engineering**, or **ergonomics**, the term more commonly used in Europe, is the design of products, as well as of working and living space, to fit the needs of humans. Said another way, it is the application of the knowledge about people and their environments. In fact, human factors engineering is the synthesis of knowledge from psychology, anthropology, physiology, biology, and engineering. The more we know about the human user, the better we can plan the environments we create.

The Myth of the Average Person

Most people believe that things are designed for the average person. This isn't always true. For example, Greg LeMond has a bicycle that

Fig. 5-9

Sally Ride's spacesuit was custom designed. Courtesy of NASA

was designed specifically for him, and Sally Ride has a spacesuit that was made specifically for her (Figure 5-9). Should a door, which is normally 6 feet, 8 inches tall, be 5 feet, 5.5 inches for the average height of all adult men and women? Obviously not! The door is designed to allow nearly all adults to pass through without bending over.

Human Scale

Because almost all problems or opportunities in technology involve the human user in some way, the designer needs to be able to get information about human physical characteristics. **Anthropometrics** is the branch of science that deals with human measurement by describing human size, shape, and other characteristics. Anthropometric data is a form of **descriptive statistics.** It is based on **probability,** or the likelihood of some event happening. Suppose you were to observe people entering your school. It would be more probable that you would see a higher frequency of average-height people than very short or tall people. If you made a graph of the height of the people you observed, it would most likely look like a **Gaussian,** or normal, **distribution** curve (Figure 5-10). This is because the *frequency distribution* of the people entering the school is like the normal population. Human factors engineers use anthropometric data to design environments and products.

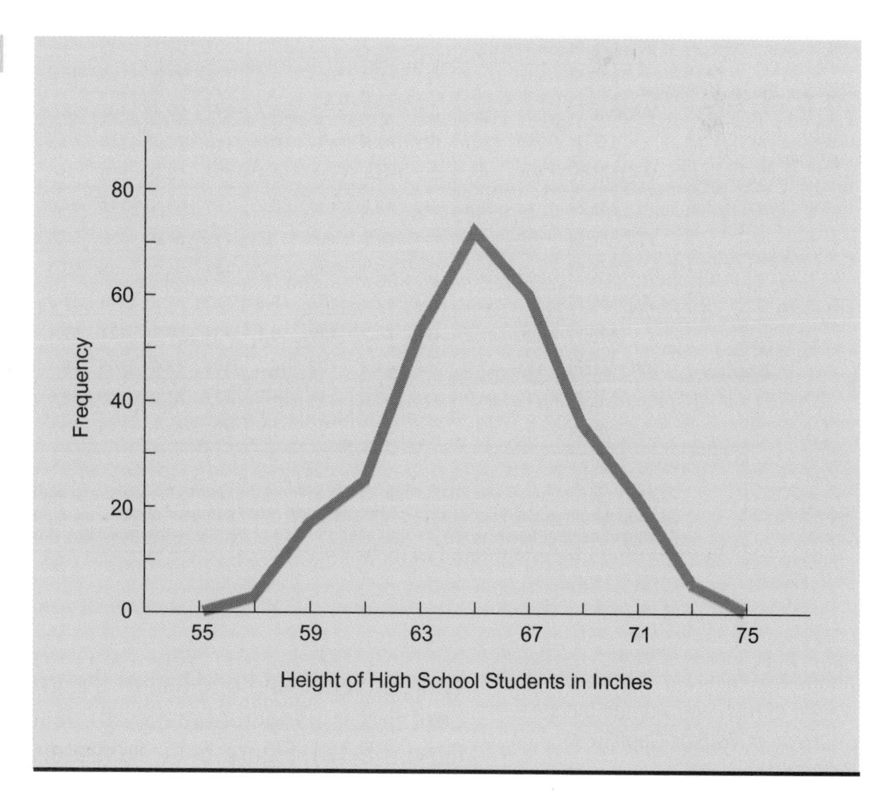

Fig. 5-10

Data representing a population takes on a "bell-shaped" form, called a Gaussian curve.

Picking the Right Numbers

Most students understand basic descriptive statistics. When test grades are returned, you hope to be above the *average* or **mean** score for the test! At this point, approximately one-half of the group

WEIGHT

Adult Male and Female Weight* in Pounds and Kilograms by Age, Sex, and Selected Percentiles†

Percentile		18 to 79 (Total)		18 to 24 Years		25 to 34 Years		35 to 44 Years		45 to 54 Years		55 to 64 Years		65 to 74 Years		75 to 79 Years	
		lb	kg	lb	kg	lb	kg	lb	kg	lb	kg	lb	kg	lb	kg	lb	kg
99	MEN	241	109.3	231	104.8	248	112.5	244	110.7	241	109.3	230	104.3	225	102.0	212	96.2
	WOMEN	236	107.0	218	98.9	239	108.4	238	108.0	240	108.9	244	110.7	214	97.1	205	93.0
95	**MEN**	**212**	**96.2**	**214**	**97.1**	**223**	**101.2**	**219**	**99.3**	**219**	**99.3**	**213**	**96.6**	**207**	**93.9**	**198**	**89.8**
	WOMEN	**199**	**90.3**	**170**	**77.1**	**191**	**86.6**	**204**	**92.5**	**205**	**93.0**	**211**	**95.7**	**196**	**88.9**	**193**	**87.5**
90	MEN	205	93.0	193	87.5	208	94.3	207	93.9	209	94.8	203	92.1	198	89.8	191	86.6
	WOMEN	182	82.6	157	71.2	173	78.5	184	83.5	190	86.2	195	88.5	183	83.0	178	80.7
80	MEN	190	86.2	180	81.6	195	88.5	193	87.5	194	88.0	190	86.2	183	83.0	170	77.1
	WOMEN	164	74.4	145	65.8	152	68.9	165	74.8	171	77.6	176	79.8	169	76.7	162	73.5
70	MEN	181	82.1	171	77.6	185	83.9	184	83.5	185	83.9	180	81.6	172	78.0	161	73.0
	WOMEN	152	68.9	137	62.1	143	64.9	153	69.4	158	71.7	165	74.8	160	72.6	155	70.3
60	MEN	173	78.5	164	74.4	177	80.3	177	80.3	178	80.7	172	78.0	166	75.3	150	68.0
	WOMEN	144	65.3	131	59.4	136	61.7	144	65.3	149	67.6	154	69.9	151	68.5	147	66.7
50	MEN	166	75.3	157	71.2	169	76.7	171	77.6	171	77.6	165	74.8	161	73.0	146	66.2
	WOMEN	137	62.1	126	57.2	130	59.0	137	62.1	143	64.9	146	66.2	145	65.8	137	62.1
40	MEN	159	72.1	151	68.5	162	73.5	164	74.4	163	73.9	158	71.7	153	69.4	141	64.0
	WOMEN	131	59.4	122	55.3	125	56.7	131	59.4	137	62.1	140	63.5	138	62.6	127	57.6
30	MEN	152	68.9	145	65.8	154	69.9	158	71.7	156	70.8	151	68.5	146	66.2	137	62.1
	WOMEN	125	56.7	117	53.1	120	54.4	125	56.7	130	59.0	134	60.8	132	59.9	119	54.0
20	MEN	144	65.3	140	63.5	146	66.2	151	68.5	149	67.6	143	64.9	138	62.6	132	59.9
	WOMEN	118	53.5	111	50.3	114	51.7	119	54.0	122	55.3	129	58.5	125	56.7	113	51.3
10	MEN	134	60.8	131	59.4	136	61.7	141	64.0	139	63.0	131	59.4	126	57.2	120	54.4
	WOMEN	111	50.3	104	47.2	107	48.5	113	51.3	113	51.3	120	54.4	114	51.7	105	47.6
5	**MEN**	**126**	**57.2**	**124**	**56.2**	**129**	**58.5**	**134**	**60.8**	**131**	**59.4**	**123**	**55.8**	**117**	**53.1**	**107**	**48.5**
	WOMEN	**104**	**47.2**	**99**	**44.9**	**102**	**46.3**	**109**	**49.4**	**106**	**48.1**	**112**	**50.8**	**106**	**48.1**	**95**	**43.1**
1	MEN	112	50.8	115	52.2	114	51.7	121	54.9	116	52.6	112	50.8	99	44.9	99	44.9
	WOMEN	93	42.2	91	41.3	92	41.7	100	45.4	95	43.1	95	43.1	92	41.7	74	33.6

*All measurements were made with the examinee stripped to the waist and without shoes, but wearing paper slippers and a lightweight, knee-length examining gown. Men's trouser pockets were emptied.
†Measurement below which the indicated percent of people in the given age group fall.

Fig. 5-11

Weight of men and women by age. Courtesy of Watson-Guptill Publications.

scored higher and one-half of the group scored lower. Another concept that is used in statistics to describe data is **percentiles** (Figure 5-11). Figure 5-11 illustrates the frequency distribution by weight for all adults (ages 18–79) in the United States. As can be observed, the average or 50th percentile weight is 166 pounds for men and 137 pounds for women. Each percentile describes a point on the frequency distribution. For example, the 5th percentile indicates that 5 percent of the population is below this point and 95 percent of the population is above this point. The 5th percentile weight of this population is 115 pounds.

Conversely, the 95th percentile indicates that 5 percent of the population is above this point and 95 percent of the population is below this point. The 95th percentile weight for this population is 212 pounds for men and 199 pounds for women. Looking at the Gaussian curve, you will observe that the highest frequency of people is around the 50th percentile. It is important to observe that fewer people fall at the two extreme ends of the curve.

Fig. 5-12

An adjustable chair is a solution that fits a larger population. Courtesy of Steelcase, Incorporated

Choosing the right numbers becomes very important for appropriate design solutions. If the product only fits a few people, it may be easier to design but the potential market will be very small. How many people could fit into a pair of basketball shoes designed for Michael Jordan? A product designed for a larger population is much more difficult to design and produce but has a larger potential market. An adjustable chair illustrates how a well-planned design can lead to a product that fits a larger population (Figure 5-12). The cost-benefit of designing for a larger population, however, diminishes near the extremes of the normal distribution curve. For this reason, most designers and engineers plan to use measurements for the middle 90 percent of the population.

Not All Measures Are Equal

The amount of difference in the population is also important to know. The **standard deviation (SD)** is a measure of the degree of variation from the mean. Said another way, SD describes how far the population is stretched out along the base of the normal distribution curve. A measure of 1 SD above and 1 SD below the mean (±1 SD) includes 68 percent of the population, while ±2 SD includes 95 percent of the population. The 5th to 95th percentiles are nearly 2 standard deviations on either side of the mean. These differences occur in a population for a variety of reasons. The greatest differences in populations occur between group age and gender. Other factors causing differences include ethnic, economic, occupational, and educational backgrounds.

Differences are important to know about when you are planning a design solution, and they help to define your target population. When you are using anthropometric data, it is important to know your target

STATURE

Adult Male and Female Stature* in Inches and Centimeters by Age, Sex and Selected Percentiles†

Percentile		18 to 79 (Total)		18 to 24 Years		25 to 34 Years		35 to 44 Years		45 to 54 Years		55 to 64 Years		65 to 74 Years		75 to 79 Years	
		in	cm	in	cm	in	cm	in	cm	in	cm	in	cm	in	cm	in	cm
99	MEN	74.6	189.5	74.8	190.0	76.0	193.0	74.1	188.2	74.0	188.0	73.5	186.7	72.0	182.9	72.6	184.4
99	WOMEN	68.8	174.8	69.3	176.0	69.0	175.3	69.0	175.3	68.7	174.5	68.7	174.5	67.0	170.2	68.2	173.2
95	**MEN**	**72.8**	**184.9**	**73.1**	**185.7**	**73.8**	**187.5**	**72.5**	**184.2**	**72.7**	**184.7**	**72.2**	**183.4**	**70.9**	**180.1**	**70.5**	**179.1**
95	**WOMEN**	**67.1**	**170.4**	**67.9**	**172.5**	**67.3**	**170.9**	**67.2**	**170.7**	**67.2**	**170.7**	**66.6**	**169.2**	**65.5**	**166.4**	**64.9**	**164.8**
90	MEN	71.8	182.4	72.4	183.9	72.7	184.7	71.7	182.1	71.7	182.1	71.0	180.3	70.2	178.3	69.5	176.5
90	WOMEN	66.4	168.7	66.8	169.7	66.6	169.2	66.6	169.2	66.1	167.9	65.6	166.6	64.7	164.3	64.5	163.8
80	MEN	70.6	179.3	70.9	180.1	71.4	181.4	70.7	179.6	70.5	179.1	69.8	177.3	68.9	175.0	68.1	173.0
80	WOMEN	65.1	165.4	65.9	167.4	65.7	166.9	65.5	166.4	64.8	164.6	64.3	163.3	63.7	161.8	63.6	161.5
70	MEN	69.7	177.0	70.1	178.1	70.5	179.1	70.0	177.8	69.5	176.5	68.8	174.8	68.3	173.5	67.0	170.2
70	WOMEN	64.4	163.6	65.0	165.1	64.9	164.8	64.7	164.3	64.1	162.8	63.6	161.5	62.8	159.5	62.8	159.5
60	MEN	68.8	174.8	69.3	176.0	69.8	177.3	69.2	175.8	68.8	174.8	68.3	173.5	67.5	171.5	66.6	169.2
60	WOMEN	63.7	161.8	64.5	163.8	64.4	163.6	64.1	162.8	63.4	161.0	62.9	159.8	62.1	157.7	62.3	158.2
50	MEN	68.3	173.5	68.6	174.2	69.0	175.3	68.6	174.2	68.3	173.5	67.6	171.7	66.8	169.7	66.2	168.1
50	WOMEN	62.9	159.8	63.9	162.3	63.7	161.8	63.4	161.0	62.8	159.5	62.3	158.2	61.6	156.5	61.8	157.0
40	MEN	67.6	171.7	67.9	172.5	68.4	173.7	68.1	173.0	67.7	172.0	66.8	169.7	66.2	168.1	65.0	165.1
40	WOMEN	62.4	158.5	63.0	160.0	62.9	159.8	62.8	159.5	62.3	158.2	61.8	157.0	61.1	155.2	61.3	155.7
30	MEN	66.8	169.7	67.1	170.4	67.7	172.0	67.3	170.9	66.9	169.9	66.0	167.6	65.5	166.4	64.2	163.1
30	WOMEN	61.8	157.0	62.3	158.2	62.4	158.5	62.2	158.0	61.7	156.7	61.3	155.7	60.2	152.9	60.1	152.7
20	MEN	66.0	167.6	66.5	168.9	66.8	169.7	66.4	168.7	66.1	167.9	64.7	164.3	64.8	164.6	63.3	160.8
20	WOMEN	61.1	155.2	61.6	156.5	61.8	157.0	61.4	156.0	60.9	154.7	60.6	153.9	59.5	151.1	59.0	149.9
10	MEN	64.5	163.8	65.4	166.1	65.5	166.4	65.2	165.6	64.8	164.6	63.7	161.8	64.1	162.8	62.0	157.5
10	WOMEN	59.8	151.9	60.7	154.2	60.6	153.9	60.4	153.4	59.8	151.9	59.4	150.9	58.3	148.1	57.3	145.5
5	**MEN**	**63.6**	**161.5**	**64.3**	**163.3**	**64.4**	**163.6**	**64.2**	**163.1**	**64.0**	**162.6**	**62.9**	**159.8**	**62.7**	**159.3**	**61.3**	**155.7**
5	**WOMEN**	**59.0**	**149.9**	**60.0**	**152.4**	**59.7**	**151.6**	**59.6**	**151.4**	**59.1**	**150.1**	**58.4**	**148.3**	**57.5**	**146.1**	**55.3**	**140.5**
1	MEN	61.7	156.7	62.6	159.0	62.6	159.0	62.3	158.2	62.3	158.2	61.2	155.4	60.8	154.4	57.7	146.6
1	WOMEN	57.1	145.0	58.4	148.3	58.1	147.6	57.6	146.3	57.3	145.5	56.0	142.2	55.8	141.7	46.8	118.9

*Height, without shoes. See Table 1A for definition of stature.
†Measurement below which the indicated percent of people in the given age group fall.

Fig. 5-13

Height of men and women by age. Courtesy of Watson-Guptill Publications.

population. The chart in Figure 5-13 shows the differences in height between the 1st and 99th percentiles of men and women in various age categories.

Using Anthropometric Data

We have already learned that many measures of human beings are important. In the case of **custom production**, for someone such as a cyclist or an astronaut, anthropometric measurements would need

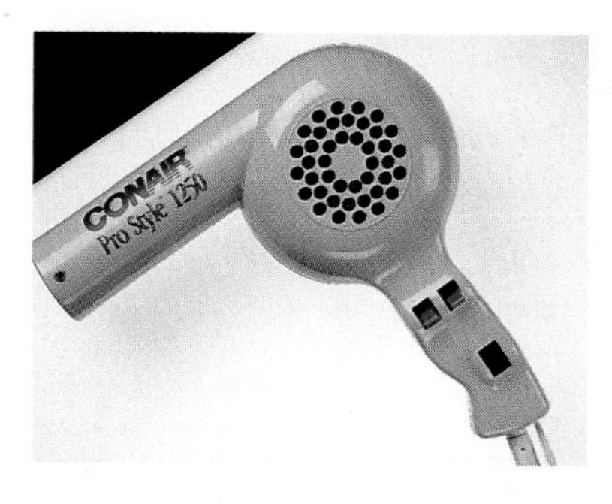

Fig. 5-14

A hair dryer handle and controls are designed to fit the human user. Courtesy of Conair

to be made for only that individual. Since most products and environments are designed for **mass production**, the best or *optimal match* must be planned. Typically, the measurement for the 5th–95th percentiles will be used.

Elderly and Physically Challenged. An important group that must not be overlooked are the physically challenged and elderly. The design for each product or environment requires different anthropometric measurements; for example, in designing a chair, you would be concerned with sitting posture, lumbar or lower-back support, and similar measures. While designing a hair dryer, anthropometric measurements for hand grip and control positions would be important (Figure 5-14). Anthropometric measures typically describe reach, clearance, and posture, or body position.

Reach, Clearance, and Posture

Reach and clearance are two measures that are selected from different parts of the population (Figure 5-15). The proper measurement for *reach* is based on the smallest member of the population and is expressed as a maximum dimension. It is logical that, if the smallest person can reach the object in question, such as a control switch on an appliance, then all other (larger) members of the population will also be able to reach the same control switch. The *clearance* dimension specifies head room, hip room, and shoulder room needed in a given situation, such as the seat of a chair. Unlike reach, the clearance dimension must accommodate the largest member of the population. For this reason, clearance is specified as a minimum dimension. It is important that the knee clearance under a table be enough (minimum) for the largest member of the population. If the largest person fits the clearance, then the other (smaller) members of the population will fit. The final measure is **posture**. In considering posture, we need to know various measurements when a person is sitting, bending, standing, reaching, gripping, and kneeling, to name a few. Posture is important when you are designing chairs, storage cabinets, control panels, and all other products and environments in which the body position matters.

Designing with Anthropometric Data

Most designs involve one of the basic measurements illustrated in Figure 5-16, page 105. The designer and engineer must select the appropriate measurement. For example, stature would be selected to plan a door opening or a clearance for an overhead obstruction. Eye height would be important to determine the placement of an instru-

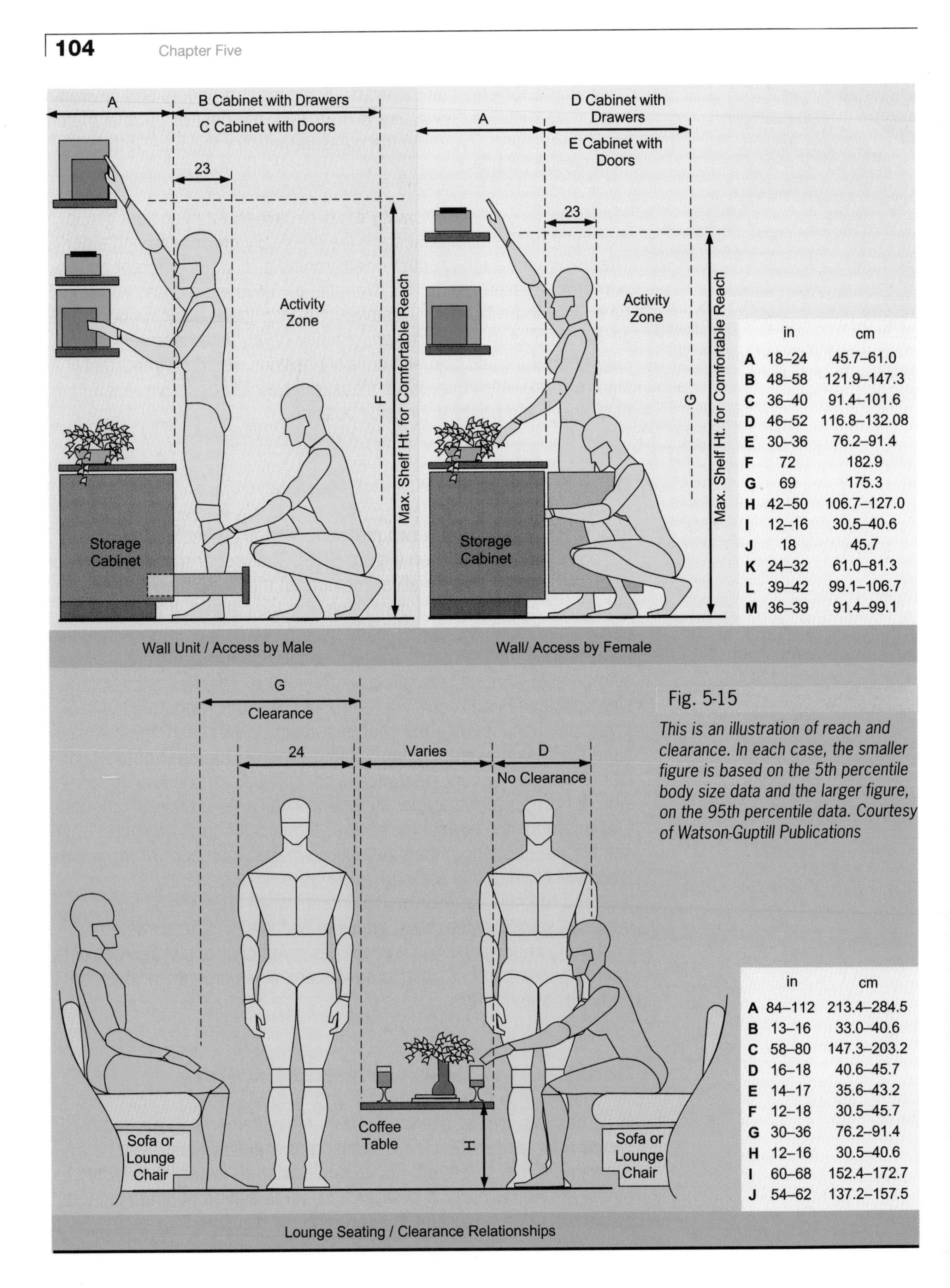

	in	cm
A	18–24	45.7–61.0
B	48–58	121.9–147.3
C	36–40	91.4–101.6
D	46–52	116.8–132.08
E	30–36	76.2–91.4
F	72	182.9
G	69	175.3
H	42–50	106.7–127.0
I	12–16	30.5–40.6
J	18	45.7
K	24–32	61.0–81.3
L	39–42	99.1–106.7
M	36–39	91.4–99.1

	in	cm
A	84–112	213.4–284.5
B	13–16	33.0–40.6
C	58–80	147.3–203.2
D	16–18	40.6–45.7
E	14–17	35.6–43.2
F	12–18	30.5–45.7
G	30–36	76.2–91.4
H	12–16	30.5–40.6
I	60–68	152.4–172.7
J	54–62	137.2–157.5

Fig. 5-15

This is an illustration of reach and clearance. In each case, the smaller figure is based on the 5th percentile body size data and the larger figure, on the 95th percentile data. Courtesy of Watson-Guptill Publications

Fig. 5-16

Anthropometric body measurements that are most useful to the designer of interior space. Courtesy of Watson-Guptill Publications

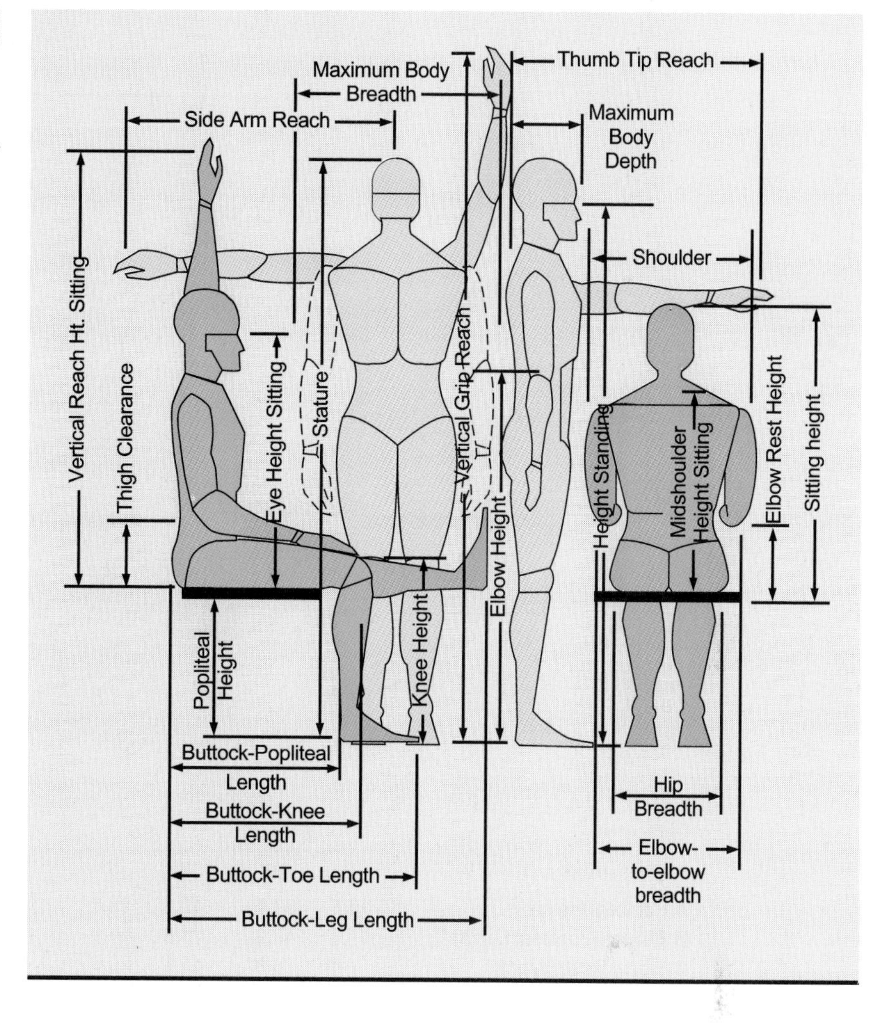

ment panel or a privacy screen height in an office complex. Each product or environment must be designed to fit the human user!

An example of using anthropometric data can be illustrated in designing a chair (Figure 5-17). Let us assume that the chair will be a general-purpose chair to be used by any adult, primarily in the North American market. For this population, the 5th–95th percentiles will be used, making the chair reasonably comfortable for 90 percent of the population. All elements of the chair must be planned. The seat height is determined by the popliteal height, measurement "A" in Figure 5-18 page 107. The **popliteal height** is the measure from the floor to the back part of the leg behind the knee joint while seated. As can be seen from the chart, the popliteal height for the identified population is between 14 inches (5th percentile, females) and 19.3 inches (95th percentile males). For this reason, the standard height of a chair seat is 16 inches. However, if the chair seat we are designing were set at 16 inches, not all people in our population would find

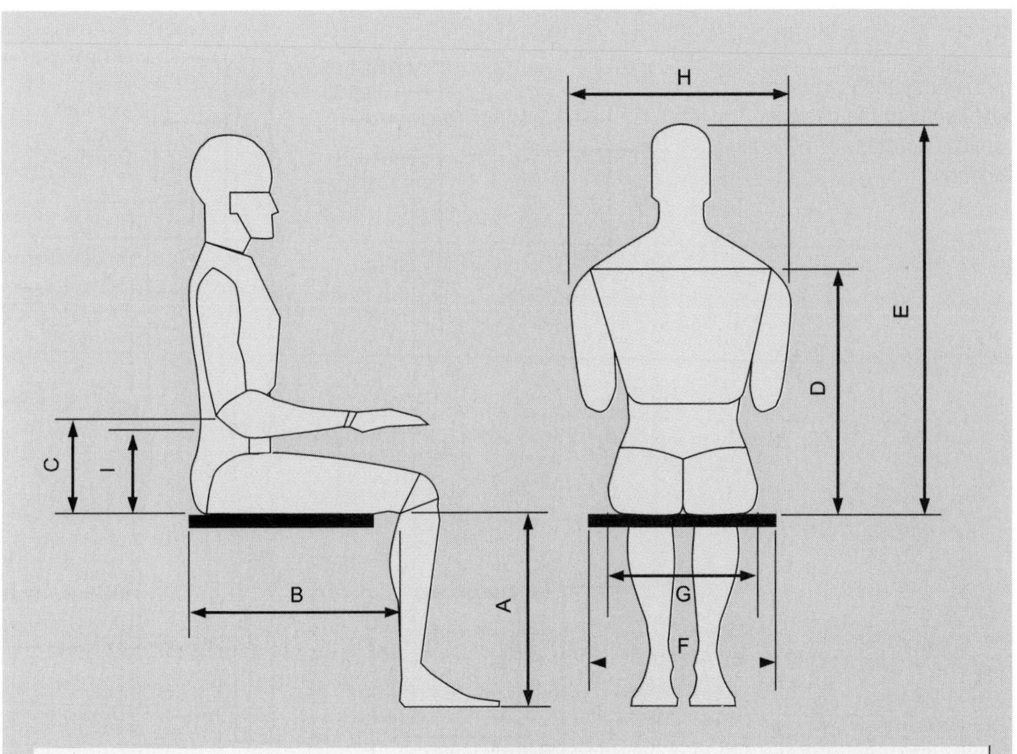

Measurement	MEN Percentile 5 in	cm	95 in	cm	Women Percentile 5 in	cm	95 in	cm
A Popliteal Height	15.5	39.4	19.3	49.0	14.0	35.6	17.5	44.5
B Buttock-Popliteal Length	17.3	43.9	21.6	54.9	17.0	43.2	21.0	53.3
C Elbow Rest Height	7.4	18.8	11.6	29.5	7.1	18.0	11.0	27.9
D Shoulder Height	21.0	53.3	25.0	63.5	18.0	45.7	25.0	63.5
E Sitting Height Normal	31.6	80.3	36.6	93.0	29.6	75.2	34.7	88.1
F Elbow-to-Elbow Breadth	13.7	34.8	19.9	50.5	12.3	31.2	19.3	49.0
G Hip Breadth	12.2	31.0	15.9	40.4	12.3	31.2	17.1	43.4
H Shoulder Breadth	17.0	43.2	19.0	48.3	13.0	33.0	19.0	48.3
I Lumbar Height	See note.							

Fig. 5-17

Key anthropometric dimensions required for a chair design. Courtesy of Watson-Guptill Publications

the chair comfortable. If the popliteal height is too high, then additional pressure is put on the underside of the knee joint, cutting off lower-leg circulation. On the other hand, if the seat is too low, the legs must extend forward or backward, reducing good posture (Figure 5-18). By far, a seat that is too low is better than a seat that is too high. In addition to the popli-teal-height information, adjustable-seat mechanisms for up to 3.5 inches are readily available at a reasonable cost. Given this information, it would be appropriate to design a chair seat with a seat height that is adjustable between 14.5 inches and 18 inches.

Fig. 5-18

Effect of chair seat height on posture

The remainder of the chair design would take into consideration other measures identified in Figure 5-17. Each additional element of the chair, such as the back or the arm rest, adds additional design complexity and production cost. The concepts of form and function, also important to design, will be covered in Chapter 10.

Designing for the Elderly and Physically Challenged

More data are becoming available for the elderly and physically challenged (Figure 5-19). Today the elderly represent over 1.5 million Americans, and that number is increasing annually. In addition to the elderly, an even larger number (3 million people) are classified as physically challenged. Worldwide, these numbers are considerably larger. These groups of people have great difficulty coping with their physical environment. This difficulty is caused primarily because this population has diminished reach and mobility capabilities. Fortunately, in the United States and other developed countries, new construction for public buildings must comply with handicap access regulations. In addition, because of the general affluence of our society, many products are available to help extend the human capability of the physically challenged and elderly. These products include motorized wheelchairs, braille readers, prosthetic devices, and hand-operated vehicles. Nevertheless, many needs and opportunities still exist for solving problems for the elderly and physically challenged populations.

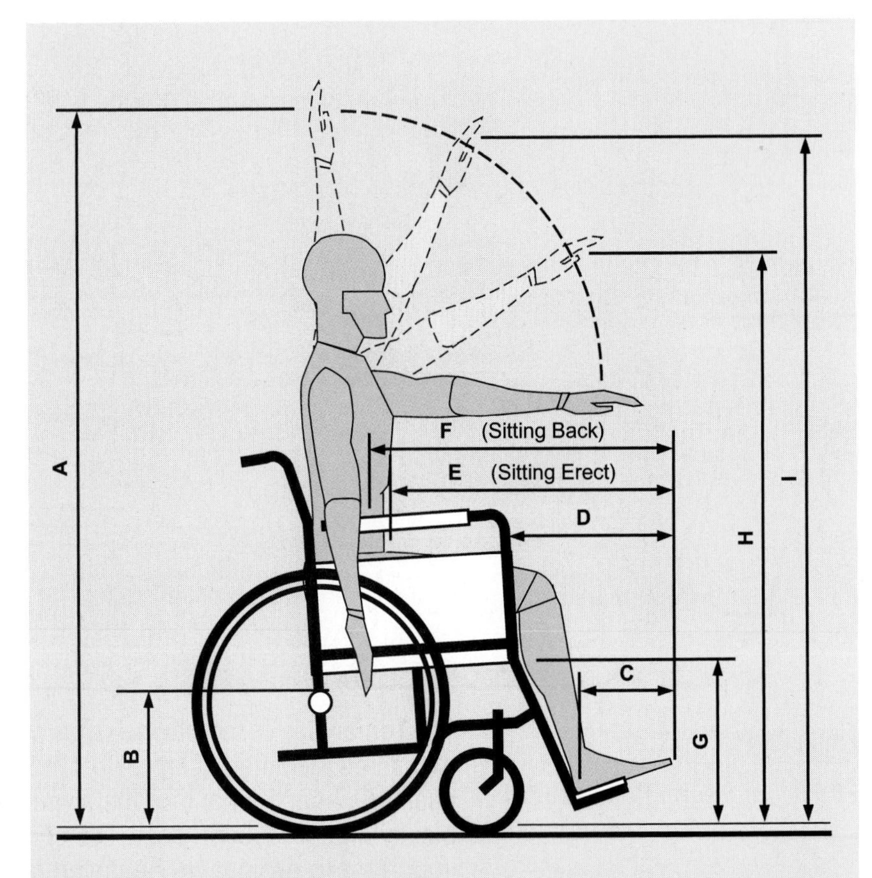

	Men		Women	
	in	cm	in	cm
A	62.25	158.1	56.75	144.1
B	16.25	41.3	17.5	44.5
C	8.75	22.2	7.0	17.8
D	18.5	47.0	16.5	41.9
E	25.75	65.4	23.0	58.4
F	28.75	73.0	26.0	66.0
G	19.0	48.3	19.0	48.3
H	51.5	130.8	47.0	119.4
I	58.25	148.0	53.24	135.2

Fig. 5-19

Anthropometric measurements for chair-bound people. Courtesy of Watson-Guptill Publications

Summary

Technology helps to extend human capability by creating new products for living and working space environments. Research and investigation begins with the process of asking many closed- and open-ended questions. Answers to these important questions can be found in libraries and businesses, and from individuals. Note taking and organizational skills need to be utilized. Nearly every problem solution involves the human user in one way or another. In order to ensure that these products or environments fit the human user, human factors engineering practices are utilized. Knowledge about physical measurements of humans is available as descriptive anthropometric data. Differences in the population, such as age and gender, must be considered. In some instances, special populations, such as the physically challenged or elderly, must receive special attention.

Observation/Analysis/Synthesis

1. Visit your school library, and have the librarian review the specific books, catalogues, and indexes available for your research.
2. Become familiar with a CD-ROM data base by looking up information related to our solid waste problem. How can you limit the topic?
3. Collect anthropometric data for height and weight for each member of your technology class. Prepare an anthropometric table; calculate the mean, and the 5th and 95th percentiles (see Appendix A for formulas).
4. Determine the standard deviation for your population (see Appendix A for formulas). How do you think your class compares with the population of all students in your school?
5. Using the data available from one of the reference books found in Appendix B, make a two-dimensional anthropometric model. The model should be made of stiff cardboard or other appropriate material with movable joints. The model can be used to test the human scale of products or environments.
6. Using the data available from one of the reference books found in Appendix B, analyze the appropriateness of the human scale of a chair, table, or other artifact available in the classroom. Present the data using tables and illustrations, as outlined in Chapter 4.
7. Identify a problem, and write a design brief for a product or a change in the personal environment of a physically challenged or elderly person.

CHAPTER 6 Designing Structural Systems

Introduction

A **structure** is a body that will resist external forces without changing its shape, except for that due to the elasticity of the material. In a technological (human-made) structure, deformation will occur only if the joints or the material itself fails.

Figure 6-1a illustrates how four bars are pinned to form a rectangle. If a force is applied, as in Figure 6-1b, the bars move and the shape becomes deformed. Even if we replaced the pins with strong, tightened bolts and nuts, the ability to resist the force would only be in the friction caused by the tightness of the bolts, not in the shape of the object.

If a fifth member were added diagonally to the object, as in Figure 6-1c, then its shape now would have the ability to resist the applied

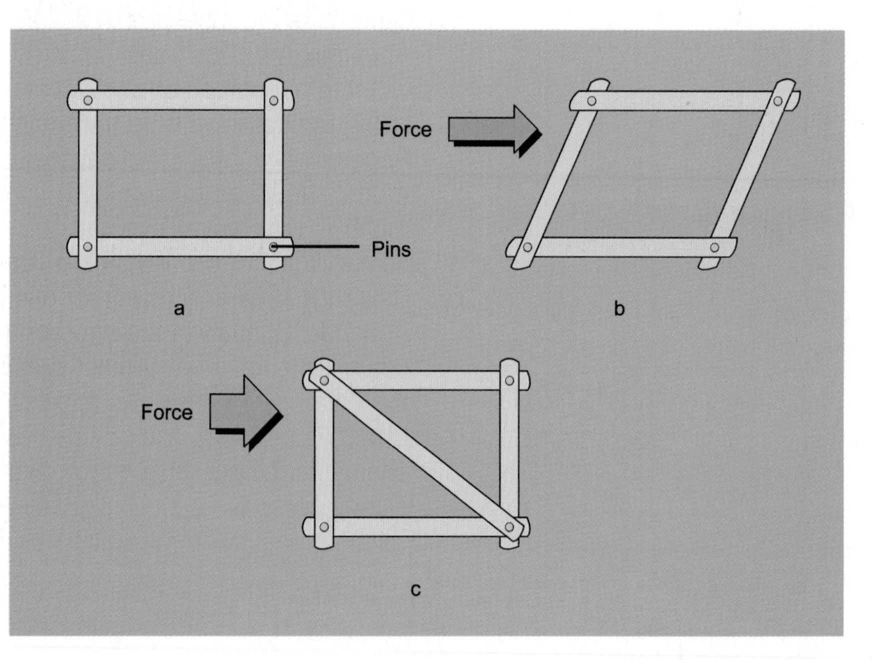

Figs. 6-1a,b and c

What makes something a structure?

force. The shape now would be a structure, because the material of the bars or the joints would have to fail before it could deform. The *triangle* shape is a key element in structural design.

Structural Systems

Structures occur in both the natural and the technological world (Figure 6-2). Structures must support a load and hold each structural

Fig. 6-2

Natural and technological structures

Fig. 6-3

The human body withstands external loads and acts as a structure.

element in a relative position to other parts. Most important, structures should not fail!

Much can be learned from natural structures. The natural structure we are most familiar with is the human endoskeleton, which gives our body form. It must be capable of supporting both internal loads created by the body itself (such as muscles and organs) and external loads, such as those created by standing, sitting, or running into the occasional door.

Skeletons consist of components that include joints and muscles. Joints are capable of certain movement and are held together by igaments and cartilage. Muscles provide the power to move body parts and are connected to the bones by tendons. The body is also capable of sensing internal and external stimuli, processing information, and producing responses. The human body is an extremely complex and marvelous creation. By looking at human capability and systems, much can be learned and applied to solving technological problems.

Technological Structures

As there are many natural structures, many types of technological structures have been invented. One way to look at structures is by the purpose they serve. Structures support, house, contain, protect, and transport, as well as serve other purposes (Figure 6-4). All structures, regardless of their purpose, must support internal and external loads, and must hold parts in relative positions. Most important, they should not fail prematurely.

System Failures

It should be assumed that all technological structures will fail at some point. **Planned obsolescence** is the name given to the concept of planning the failure of a technological product after a certain amount of use. Most cars don't last past 100,000 miles. The U.S. Depart-ment of Commerce uses the term **durable goods** to describe products that are intended to last more than three years. **Nondurable goods** are designed to fail in under three years. Technological systems that fail prematurely do so because they were poorly designed. Sometimes this is because there is insufficient knowledge about the system.

The Tacoma Narrows Bridge failed because of resonant oscillation, or vibrations created by naturally occurring wind passing by the bridge deck (Figure 6-5). Today, these oscillations are better understood and are controlled by different bridge deck designs. The failure of the Tacoma Narrows Bridge represents a Type IV technological impact: It was unexpected and undesired! Some systems fail because the design incorporates materials, components, and processes incorrectly. (Have you ever purchased a cheap pair of scissors or other tool that never worked properly?)

Systems can fail for expected reasons as well. Since all systems are exposed to such external forces as moisture and vibrations, these external forces begin to weaken the elements of a system. Many of these forces can be controlled. The paint on a metal swing set, which resists rust forming from moisture in the environment, illustrates how the designer and engineer properly considered external forces. In addition to the expected forces, systems fail when they are used improperly. For example, all products are rated for maximum loading. If the system is overloaded, it will fail at some point. Designers and engineers will use a planned safety factor in setting load limits.

6-4a

6-4b

6-4c

6-4d

Figs. 6-4a,b,c and d

Structures serve many useful purposes.
6-4a Courtesy of NASA
6-4b Courtesy of GMC Truck Division
6-4c Courtesy of Sony Corporation of America
6-4d Courtesy of Portland Cement Association

Fig. 6-5

Tacoma Narrows Bridge. Courtesy of Bettmann Archive

Safety Factor

A **safety factor** determines how much a product, or an element within a product, is overbuilt. Take a paper clip and bend it back and forth until it breaks. Note the number of bends before the paper clip broke. Now bend another paper clip. Did it break at the same number of bends? Why?

Knowledge of how materials and components or systems will work is important in determining a safety factor (S.F.). Safety factors are usually greater than 1 (S.F. > 1.0). A race car, for example, is designed

$$\text{S.F.} = \frac{\text{Ultimate stress (breaking point)}}{\text{Working stress (maximum expected load)}}$$

Formula for calculating a safety factor

to minimize the weight of all components and should theoretically fall apart as the car crosses the finish line—in first place! The designer attempts to make the car as light as possible so that it will go as fast as possible. The safety factor for a racing car is very close to 1.0.

We expect that most products will work properly and safely, even though we may be using them to full capacity. A 2-ton jack is not expected to fail if 4,001 pounds are lifted. A jack typically has a safety factor of around 1.5, meaning that it will lift up to 6,000 pounds safely—2,000 pounds above its rated capacity.

Forces on Structures

The relationship between science and technology is nicely illustrated by looking at the effect of forces acting on technological systems. Remember that the purpose of science is to discover and describe, whereas the purpose of technology is to invent or innovate. While an important symbiotic relationship does exist, for the purposes of this book, scientific principles will be applied without further explanation. For example, in discussing forces, the concept of weight will be used and expressed as either English (American) pound force (lb = slug × ft/s2) or the **International System (SI)** Newton force (*N* = kg × *m/s*2), commonly referred to as the metric system. Both *slug* and *kilogram* (kg) are measures of mass. It is understood that weight is relative to location, whereas mass, which is used in most theoretical scientific calculations, does not change with location in space.

Scientific notations will be used where appropriate (see Appendix A for additional information and formulas). When further clarification is necessary, the student should consult a physics or other science text book, or talk to a science teacher. In technology, scientific principles are important for documenting and predicting that specific designs or design elements will function as expected.

A practical example will help illustrate the relationship between science and technological activity (Figure 6-6). For scientific purpose, the block represents a situation to be used to apply Newton's second law, which states that the acceleration of an object is directly proportional to the resultant force acting on it and inversely proportional to its mass. In this example, the motion of the block is described by the formula $ax = F/m$, where a = acceleration, x = direction, F = force, and m = mass. For technological purpose, the block represents a load that will be applied to a structure. This load (force) may be in the form of a product that needs to be moved from a manufacturing plant to a truck for delivery. Unlike the scientific problem, in which there is only one correct answer, the technological problem creates many opportunities. The designer or engineer may look for ways to reduce the friction acting on the product. If a rope were employed in the solution, a material would be selected or a new material would be developed to withstand the tension force on the rope. Various mechanisms would be explored to simplify the work to be accomplished. The human worker, worker safety, product reliability, and the environmental impacts would all have to be considered. Finally, the design would be evaluated for production and marketing concerns.

Fig. 6-6

Scientific versus technological work

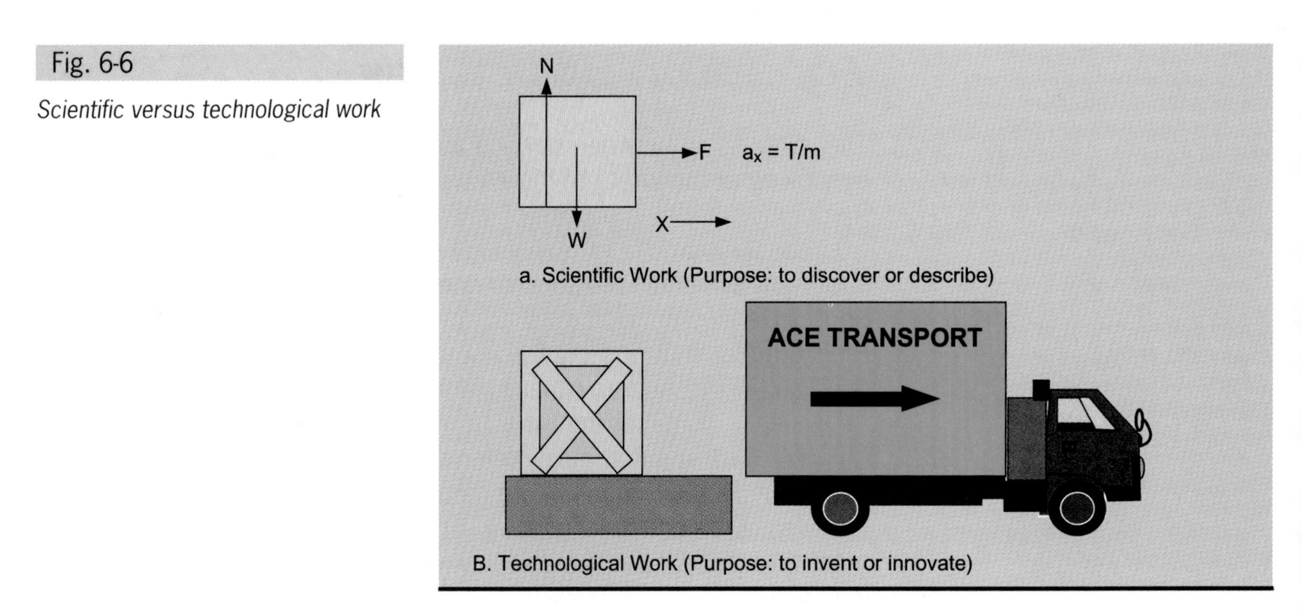

External Loads and Internal Forces

Two types of external loads must be considered in the design of a product. All external loads must be counteracted by internal forces, or the structure will fail. **Static loads** are those associated with a load at rest. An example of a static load is a person sitting in a chair and not moving. Static loads are easy to predict, because they are easily measured. In the chair example, by using the appropriate anthropometric table, we can determine the weight of the 99th percentile adult male.

Dynamic loads are characterized by forces in motion. An example of a dynamic load is a person sitting down in a chair. You may have younger brothers or sisters who throw themselves into a chair rather than sitting down in a chair. In this example, the dynamic load created is considerably higher than the static load. A chair or any product is best designed for expected dynamic forces.

Equilibrium

When an external load is applied to a device such as a chair, internal forces are necessary to counteract each load force. **Internal forces** are generated by the molecular structure of the material in question to counter external forces. **External forces** are the loads that are applied to the object in question. In the example of the chair, the wood or metal components have a molecular structure that push back against the load (the weight of the person sitting in the chair) with a force that is equal to the load. This condition is known as **equilibrium** (Figure 6-7). Notice that $F_1 = F_2$. If the load were greater than the force that could be generated by the molecular structure of the material in the chair, then the chair would fail or break. If the internal force of the chair were greater than the external force of the load (person), then the person would be shot into space! The concept of equilibrium is the basis of design and engineering in planning technological structures.

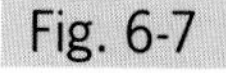
Fig. 6-7

Forces in equilibrium

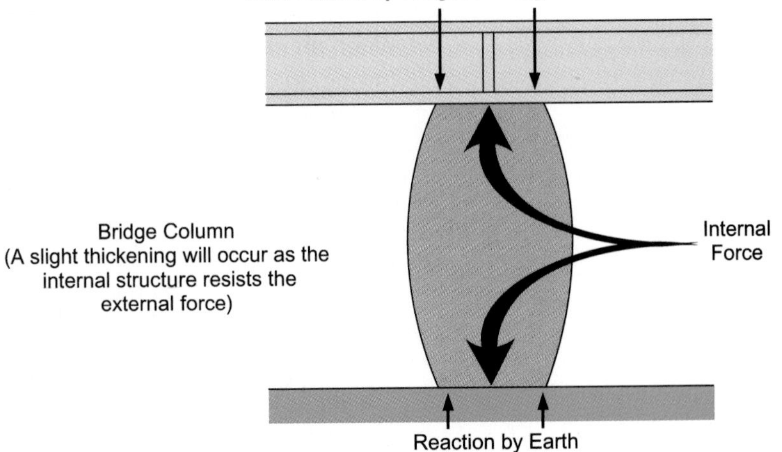

Fig. 6-8

Internal forces equal external load in a stable structure.

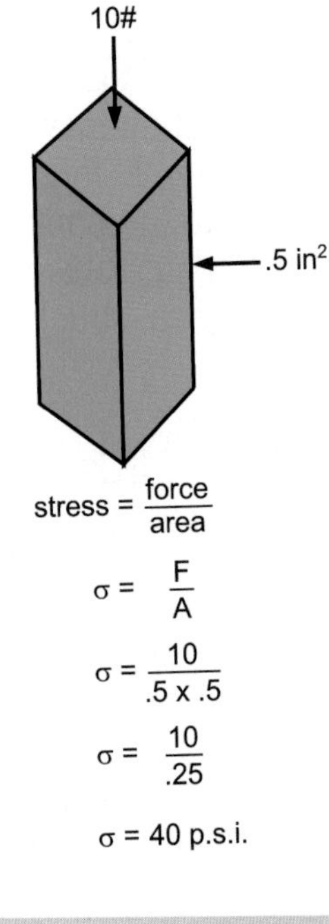

Fig. 6-9

Stress formula with sample problem

Stress and Strain

When a force or a load is applied to a structure, the internal forces help the material to resist being changed by the external load (Figure 6-8). Although changes in the material may not be evident, slight thinning or thickening of all materials occurs under a load. A concrete highway may show little change, whereas slight pressure on a water balloon will show a significant change.

Stress. The strength of a material is described by its **stress**. It is the force being exerted on the molecules of a material by compression or tension forces. Stress is calculated by dividing *F* (force) by *A* (area) (Figure 6-9). In SI measurement, stress is expressed in meganewtons per square meter (MN/m^2). In the conventional English system, stress is expressed as pounds per square inch (psi). Stress is represented by s or the Greek letter σ (sigma). Stress describes the strength of the material by noting when the material will fail or break.

Strain. **Strain** is used to describe the change in shape of a material caused by compression or tension forces (Figure 6-10). It is the distance that the molecules move apart. Whereas stress describes how hard it is to break a material, strain describes how far the material stretches under a load. Strain is usually expressed as a percentage of change in the original length. Strain is represented by the Greek letter ϵ (epsilon).

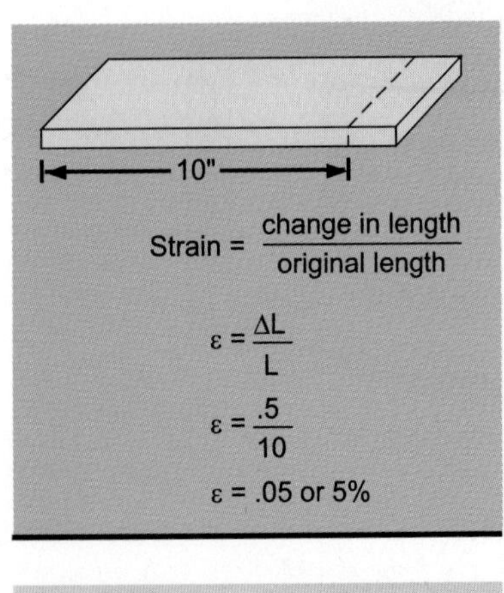

Fig. 6-10

Strain formula with sample problem

By combining the measures of stress and strain, a measure of the stiffness of a material can be made. This measure is known as **Young's modulus of elasticity**, or *E* (Thomas Young 1773–1829) (Figure 6-11). Three points are important on the stress and strain

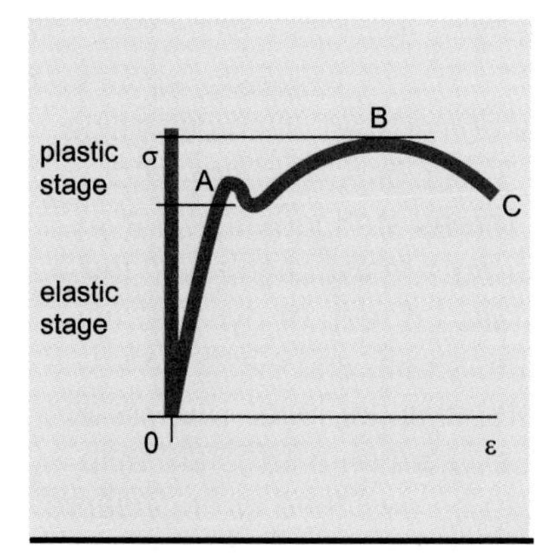

Fig. 6-11

Young's modulus of elasticity

graph. Each point represents a different stage in the molecular structure of the material. From point "O" to point "A," the material is said to be in an *elastic stage*. During the elastic stage a material under force will change shape; however, when the force is removed, the material will return to its original shape. A fiberglass pole used by a pole vaulter is a good example of a material that bends under force but returns to its original shape after a vault. The next stage occurs between points "A" and "B" and is known as the *plastic stage*. During this stage, the material under load changes shape permanently and will not return to its original shape after the load has been removed. If the fiberglass pole remained curved after the vault, it no longer would be useful. The final stage, occurring at point "C," identifies material failure or the *breaking point*. No material should be used outside its elastic stage.

Material (Approximate Values)	*Young's Modulus of Elasticity* (E)	
	MN/M²	**psi**
Rubber	7	1,000
Shell membrane of egg	8	1,100
Human cartilage	24	3,500
Human tendon	600	80,000
Wallboard	1,400	200,000
Unreinforced plastics	1,400	200,000
Plywood	7,000	1,000,000
Wood (along grain)	14,000	2,000,000
Concrete	17,000	2,400,000
Magnesium metal	42,000	6,000,000
Ordinary glasses	70,000	10,000,000
Aluminum alloys	73,000	10,400,000
Brasses and bronzes	120,000	17,000,000
Iron and steel	210,000	30,000,000
Aluminum oxide	420,000	60,000,000
Diamond	1,200,000	170,000,000

Fig. 6-12

Table of E values

From a design and engineering perspective, the *E* (modulus of elasticity) value is useful in the selection of appropriate material for a particular application (Figure 6-12): A material with a low modulus of elasticity, such as concrete, should not be used where it would be exposed to flexing motions, even though concrete is extremely strong (under compression). Mild steel or aluminum, on the other hand, combine both strength and flexibility.

Common Forces

There are five forces that are commonly considered when designing structures: 1) compression, 2) tension, 3) bending, 4) shear, and 5) torsion. As noted previously, external forces are loads that are applied to a structure, whereas internal forces resist changes in the structural shape. These forces can be static (at rest) or dynamic (in motion). The concepts of stress (strength), strain, and elasticity describe the physical changes in the material under load. The five different ways in which forces or loads are exerted on a structure help determine how the structural element is to be designed or engineered.

Compression. When the load is applied to the top of a structural element, the element is said to be in **compression** (Figure 6-13). In compression, the load is pressing down on one end of the element and the ground is pressing up on the other. If the element is to be in equilibrium, it must have sufficient strength (stress) to withstand the compression force. Compression loads try to collapse the element. The legs of a chair or table are under compression.

Tension. Frequently, a load is applied along the element in a pulling action (Figure 6-14). When this occurs, the element is said to be in **tension**. Tension forces try to pull the element apart. To maintain equilib-

Fig. 6-13

Compression

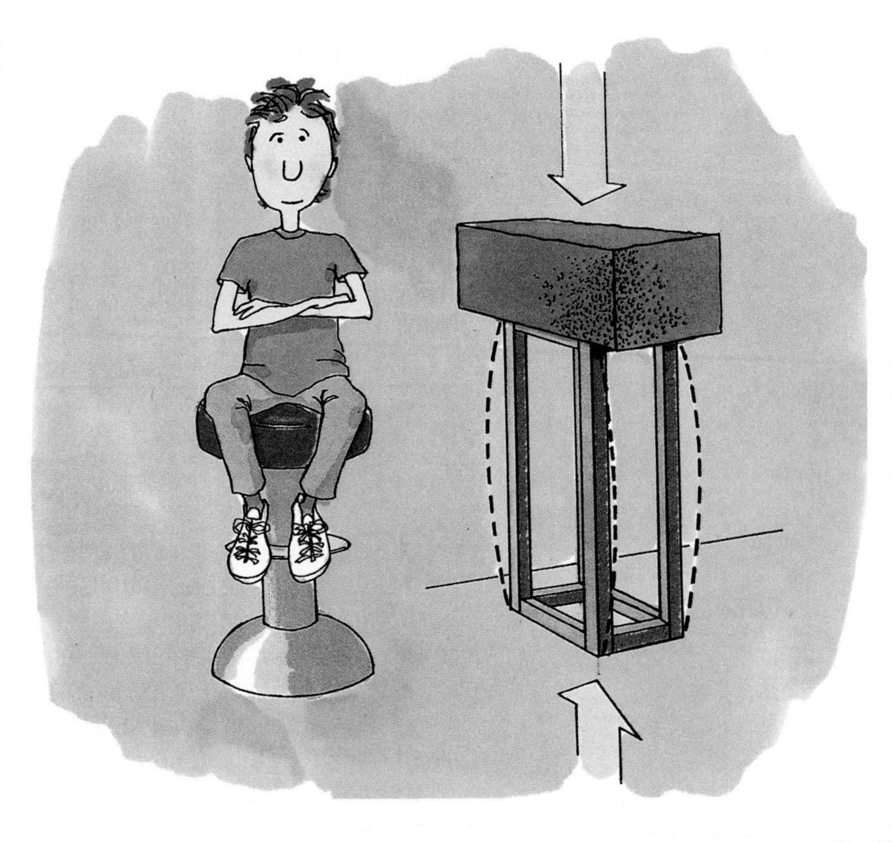

rium, the element must be able to resist the outward pulling forces. It is easiest to design for materials that are only under tension loading. The rope used in a tug-of-war is an example of an element under tension. In both compression and tension, the loads are acting along the length of the element. Some loads, however, act across rather than along the element.

Bending. When an element such as a bookshelf is loaded with books, the load is applied across the material (Figure 6-15). Under these conditions, a **bending** force is working on the element. Bending forces cause the material to be under both compression and tension (Figure 6-16). A neutral axis exists near the center of the element. Material on the inside of the bend is under compression; that is, the load is trying to force the molecules together. At the same time, the material on the outside of the bend is under tension. The bending action is trying to pull the molecules apart in this section of the element.

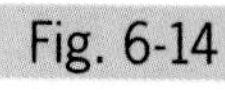
Fig. 6-14

Tension

Fig. 6-15

Bending

This action can be easily seen by taking a piece of thick foam rubber or similar material. A line representing the neutral axis is drawn on one edge. A series of lines, approximately ½ inch apart, are drawn perpendicular to the original line. If the foam rubber is bent, you will observe how the material is under compression above the neutral axis and under tension below the neutral axis.

Most materials used in shelving will bend to some degree under load. Plywood, which is often used in shelving, is made from thin layers of wood glued together. Why would this material more effectively resist bending (consider that each layer has a neutral axis)? If the supports are too far apart or too many heavy objects are placed on the shelf, the shelf material may be strained beyond its elastic stage and fail.

Shear. When forces are exerted on a material in the same plane but in opposite directions, the result is called **shear** (Figure 6-17). In the example, the forces are acting to cause the one section of the board to slide past adjacent board material, which is being forced to slide in the opposite direction. The two directions of force are being caused by the weight of the person and the reaction of the column support. Scissors use shear force to separate materials from thin paper to heavy metals. Most fasteners, such as bolts, nails, board, and welds, are under shear force.

Torsion. In applications such as drive shafts, the structural element is under **torsion** (Figure 6-18). Torsion describes forces that try to

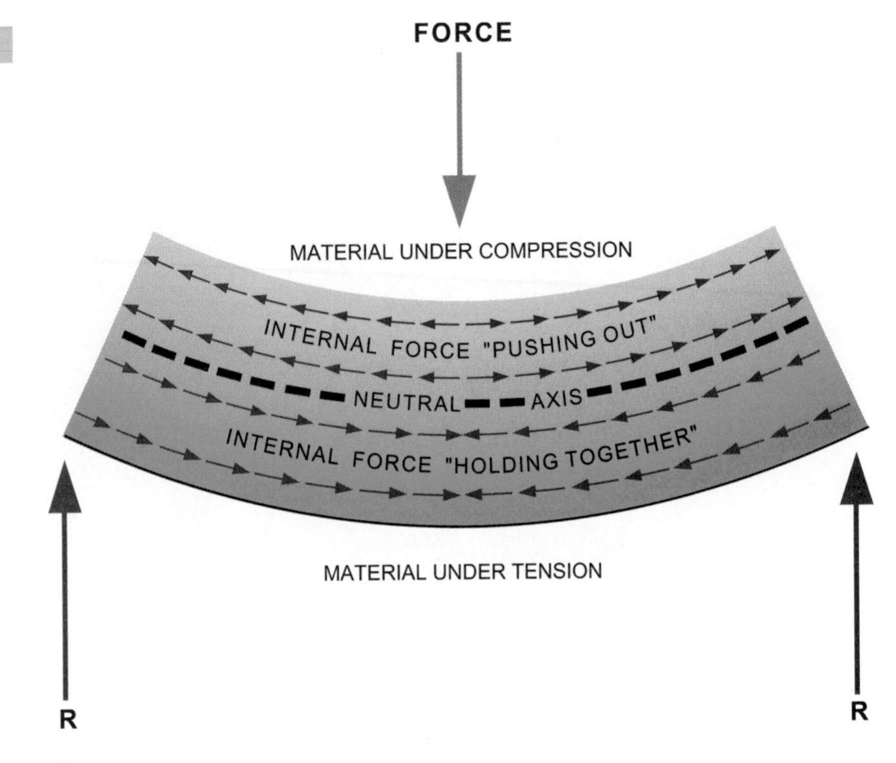

Fig. 6-16

A neutral axis exists near the center of a material being bent.

Fig. 6-17

Shear

Fig. 6-18

Torsion

twist the material apart. As with bending, there is a neutral axis, and materials within the element are under both compression and tension forces. Torsion (also known as *torque*) is most associated with mechanical systems, such as motors, engines, and the related moving parts. Torsion occurs in many other systems, including airplanes, bridges, and nearly all other technological systems.

The five types of forces—compression, tension, bending, shear, and torsion—typically occur in varying combinations with each other in structures. For example, a chair or a table leg may be primarily under compression but must also resist bending forces. When you are designing structural elements, it is necessary for you to determine the type of force and the amount of load at each point in the structure. Different types of structural components must be selected to maintain structural stability or equilibrium.

Structural Components

The job of all structural components is to support a load, maintain the relative position of elements, and resist failure. The load on the structure, including all internal and external forces, is attempting to make the structure fail. Designers and engineers must first determine the strength and type of force (compression, etc.) that will be present on structural components. Based on this knowledge, they may select or design an appropriate component.

Beams

Horizontal structural components that are designed to resist compression and bending forces are called **beams**. Early beams were large,

Courtesy of French Government Tourist Office

Eiffel Tower

Paris, France

The famous Eiffel Tower was built for the 1889 World's Fair. The designer, Alexandre Eiffel (1832–1923), wanted to see how tall a structure could be made from steel. The Eiffel Tower is 320 meters tall. Eiffel, a French civil engineer, also designed the Statue of Liberty on Liberty Island, New York.

solid materials, such as stone or timber (Figure 6-19). Through innovation, new beam designs have incorporated less material while at the same time making the beam stronger (Figure 6-20). This has improved

Fig. 6-19

Early beams

Fig. 6-20

Many natural materials were used to make early beam structures.

the beam weight-to-strength ratio. By using less material, natural resources are conserved and costs are reduced. New materials may be developed to overcome other problems, such as material weight, cost, strength, aesthetics, environmental safety, or availability of materials. In bridges, beams are used to make trusses and girders (Figure 6-21). In construction, beams are incorporated into all residential and commercial construction. Beams are used in nearly all structures.

Fig. 6-21

Trusses and girders

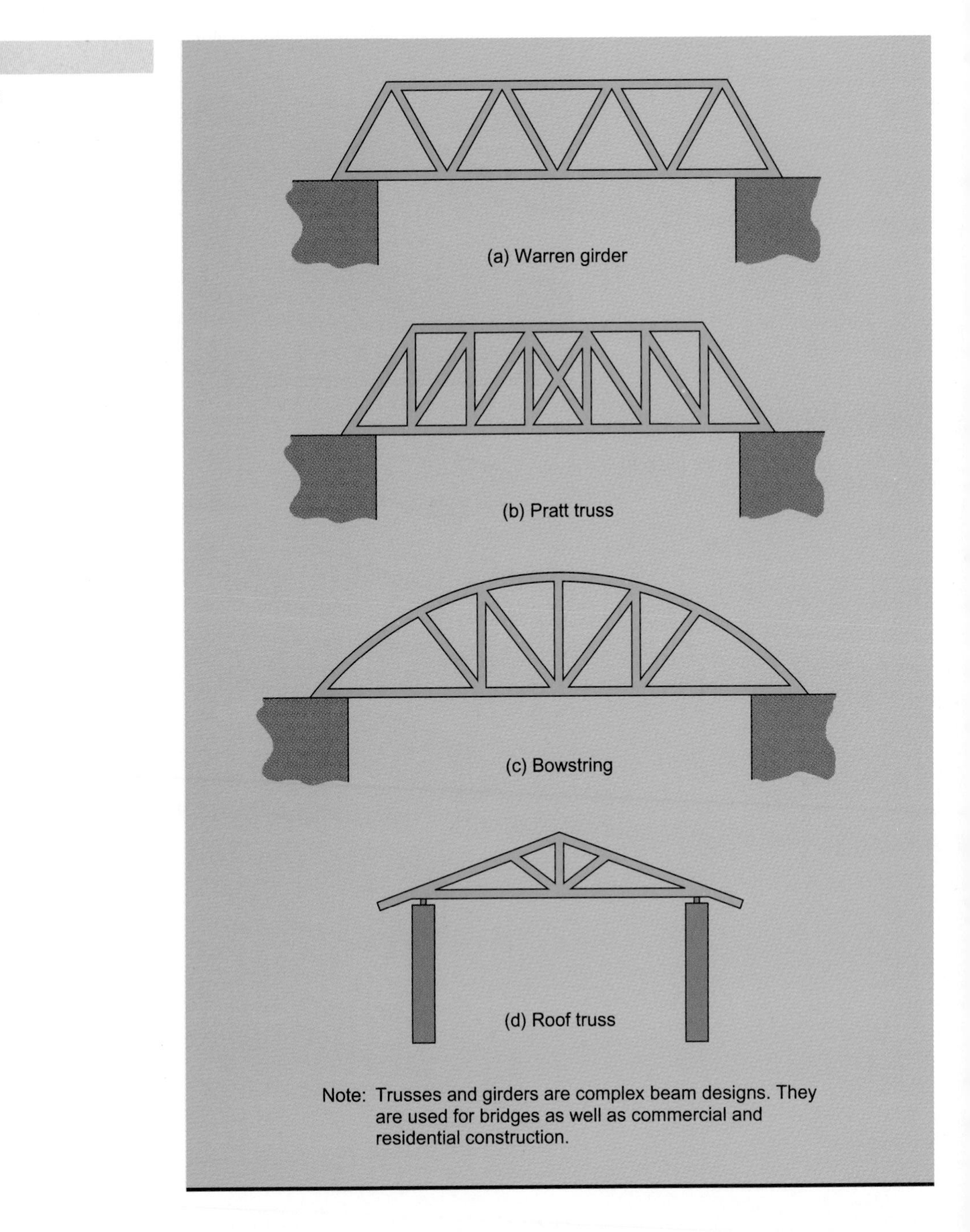

Struts and Ties

Structural components that resist compression are called **struts.** Struts are used to make bridge piers and columns in construction (Figure 6-22). **Ties** are structural components that must resist tension. Ties can be steel cables on a suspension or cable-stayed bridge, or they may be more rigid steel elements in truss bridges (Figure 6-23).

Fig. 6-22

A pier is an example of a strut in a bridge structure.

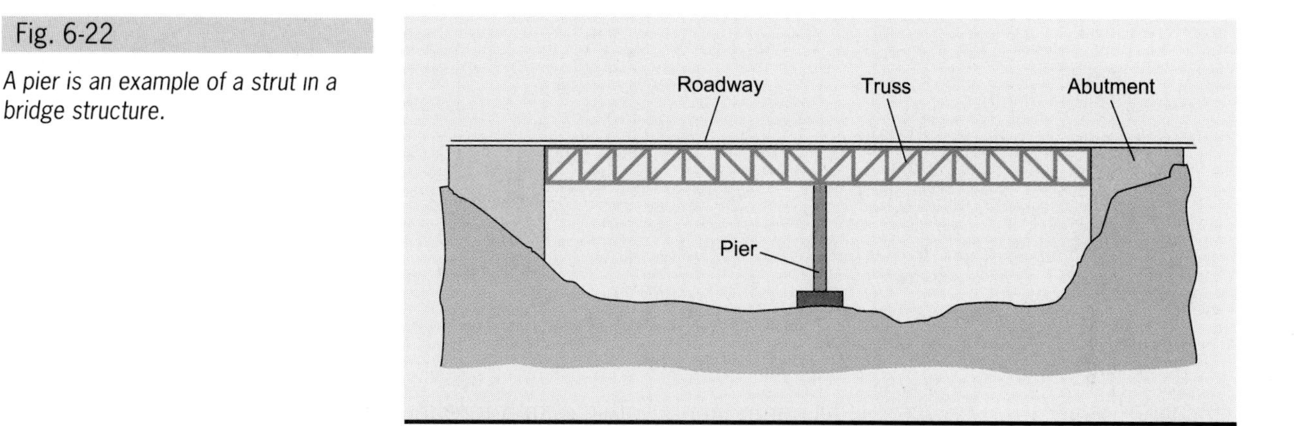

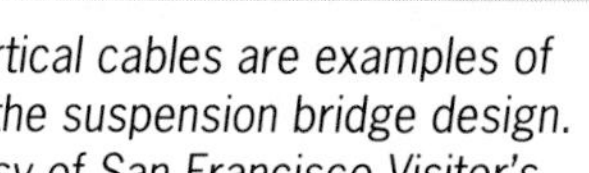

Figs. 6-23a and b

The vertical cables are examples of ties in the suspension bridge design. Courtesy of San Francisco Visitor's Bureau

6-23b below

6-23a above

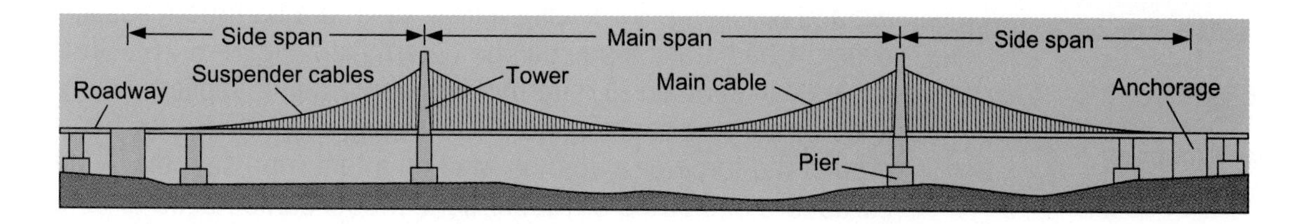

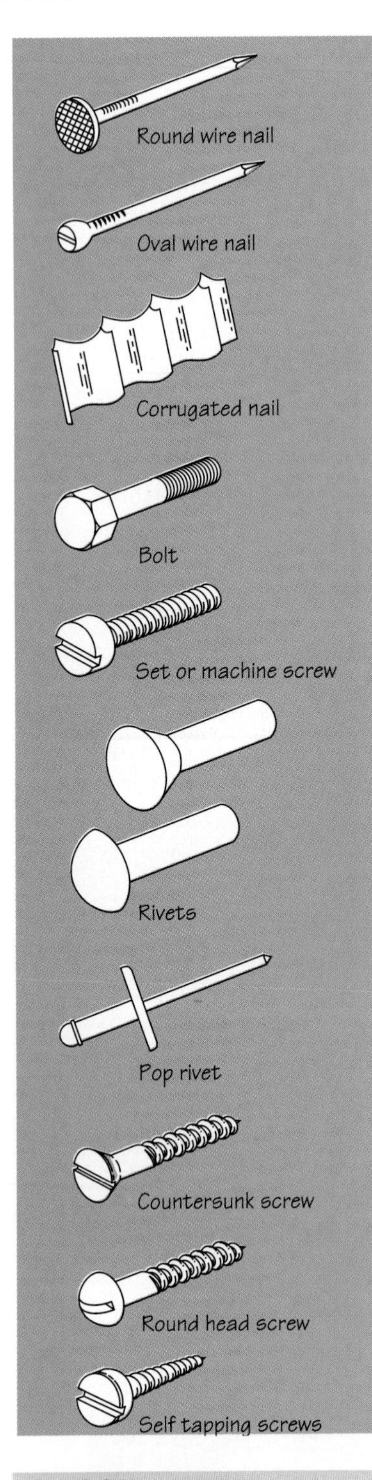

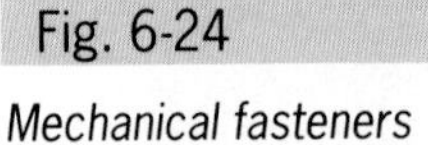
Fig. 6-24

Mechanical fasteners

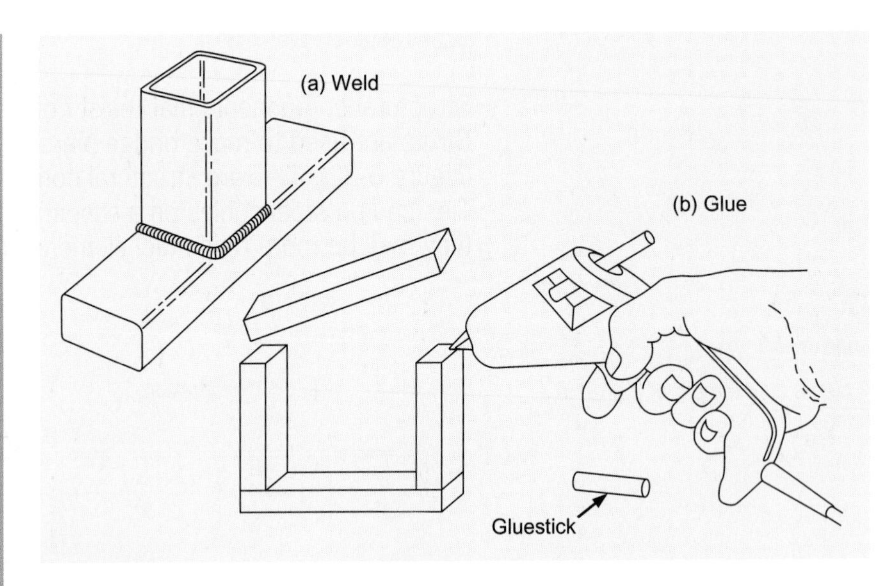

Fig. 6-25

Chemical fasteners

All components within a structure must be fastened together with joints. Mechanical fasteners consist of structural components such as rivets, bolts (including clamps), screws, and nails (Figure 6-24). Chemical fasteners include welds and glued joints (Figure 6-25). In most instances, fasteners are subject to shear forces.

Calculating Loads on Structures

Building a kitchen chair out of 2" × 4" timber does not make much sense and probably wouldn't sell. To be able to create an effective design, the designer must calculate the load and type of force acting on the structure. With this knowledge, the proper material and type of component (beam, etc.) can be selected. If unexpected forces are applied to the structure or if incorrect components or materials are used, it may fail.

Simple Testing of Loads

Basic load measurements can be made through simple *empirical testing* and *modeling*. By placing a chair next to a bathroom scale, typical dynamic and static loads can be determined (Figure 6-26). Spring scales can be used to determine the force being applied on a hanging lamp or a swing seat (Figure 6-27). Spring scales can also be used to predict the force of a 25-mph wind on a tent rope support. A simple cardboard and string or rubber band model can be used to determine which elements in a bridge or a truss are under compression and tension (Figure 6-28). In this test, components believed to be under com-

Fig. 6-26

Measuring dynamic and static loads with a bathroom scale

Fig. 6-27

Measuring a tension force with a spring scale

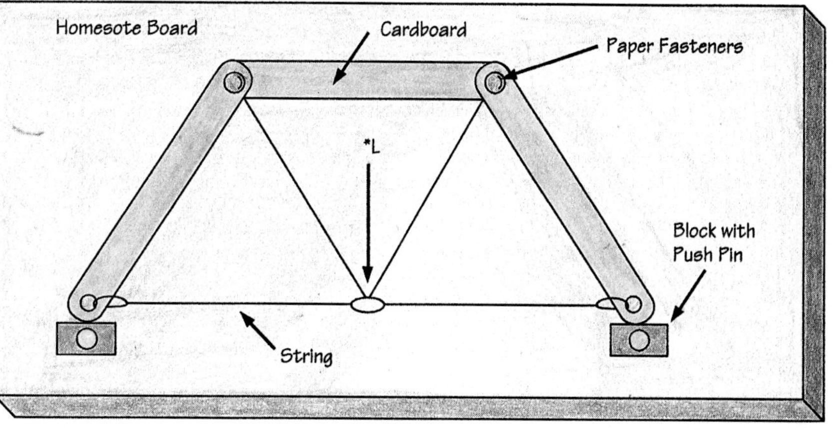

*When the Load is applied, does the string get tight or collapse?

Components:

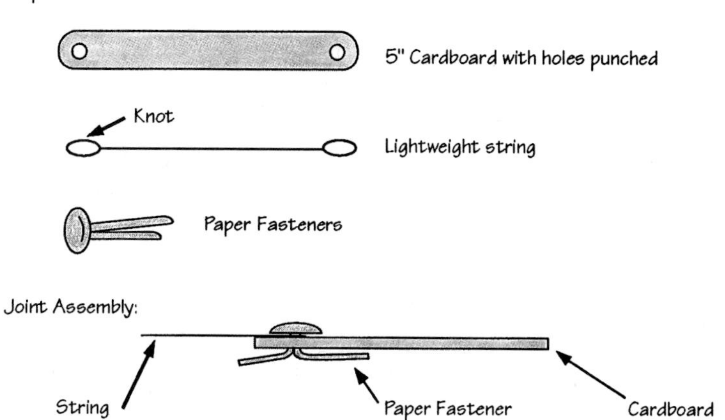

Fig. 6-28 right

Low-relief modeling of a truss design idea

pression are made of cardboard, whereas components believed to be under tension are made with string. If a string component collapses under load, the component is really under compression and must be replaced with cardboard. All elements that are not obviously under compression should be tested with string. It is not always appropriate to use physical modeling techniques. When appropriate, mathematical or graphical models should be used.

Using Mathematics to Calculate Loads

Predicting loads in structures usually involves mathematical or computer modeling. A simple moment calculation in Figure 6-29 demonstrates this point. By calculating a moment, the effect of a force applied to a door handle or a box wrench can be determined. Moment calculations are important measurements when a mechanism such as a lever handle for a tool is being designed. The force of the turning moment continues until equilibrium is achieved by a tightening of the nut. The force acting on a structural component must be counteracted by a reaction force if the structure is to remain in equilibrium (Figure 6-30).

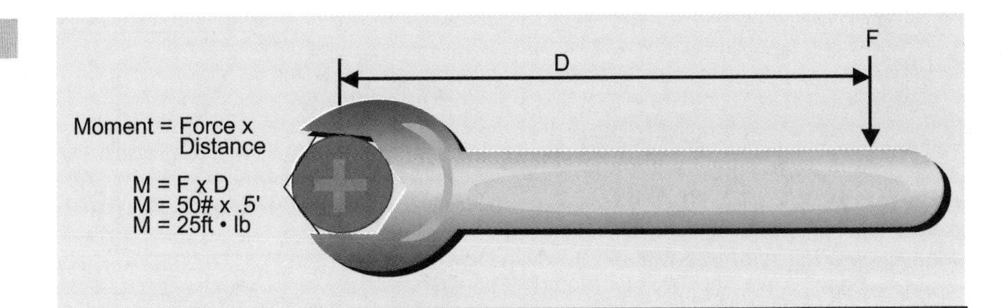

Fig. 6-29

Moment calculations

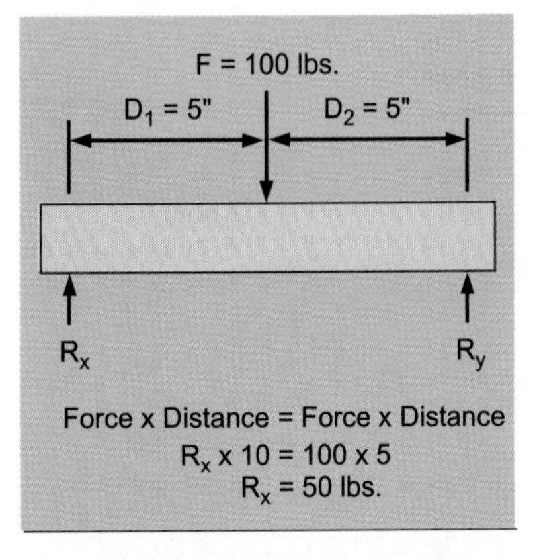

Fig. 6-30

Determining equilibrium by calculating moments and reactions on a beam design

The load in this example is called a **point load**. Many structures, such as a roof, have loads distributed over an entire surface. When a calculation is to include the weight of the material making up the structure, as well as loads that may be distributed over the entire surface, a uniformly distributed load (UDL) measure is used (Figure 6-31).

Additional questions about this structure can be answered using UDL measures. What is the shear force at R_y? What is the maximum expected load, assuming a 12-inch (30.5 cm) wet snow load? What is the load if a new shingle roof is put over the existing roof?

Vector Analysis. In order to determine the amount and type of load being applied to a wall bracket structure, vector analysis, incorporating algebraic and trigonometric functions, can be used. As can be observed in Figure 6-32, even the mathematical method involves some graphical analysis.

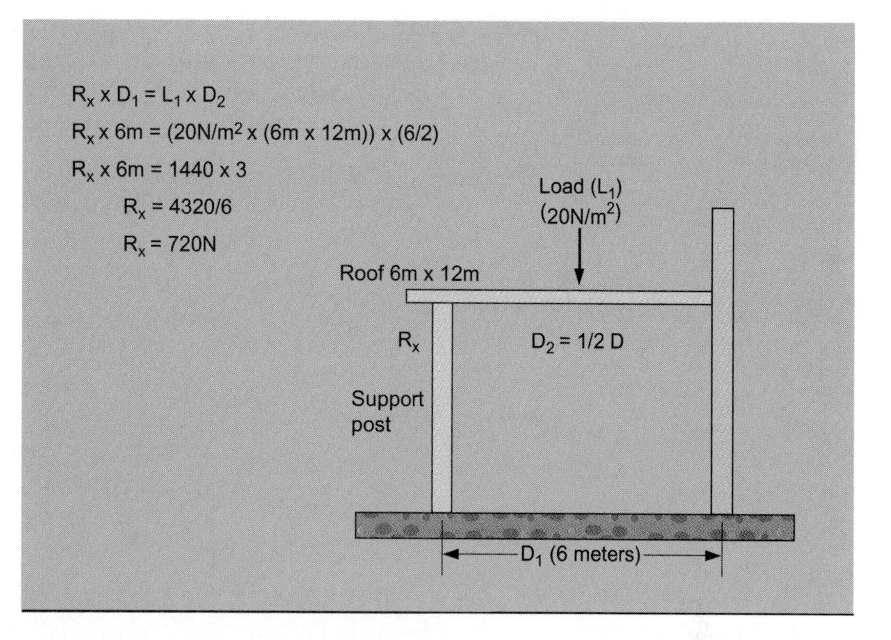

Fig. 6-31

Determining a reaction using uniformly distributed loads (UDL) calculations

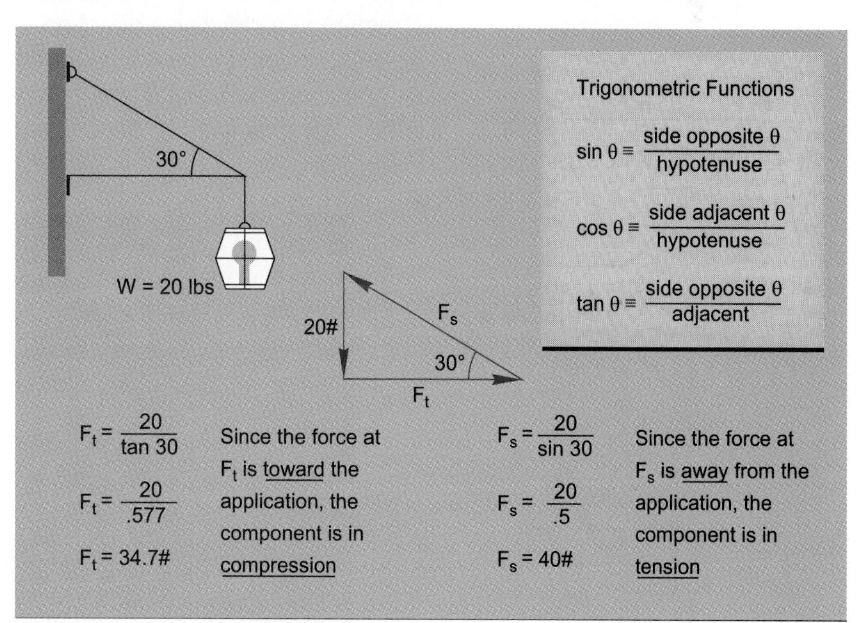

Fig. 6-32

Determining the tension and compression forces on a structure using trigonometric functions

Graphical Analysis

Vector polygons form the basis for answering questions using graphical analysis. **Vectors** are lines that describe both the *direction* and *magnitude* of a force being applied to the system. Polygons are figures having more than four angles. A bicycle frame will serve as the illustration for this graphical analysis (Figure 6-33a). Remember that, in order to solve a technological problem, it is necessary to know the magnitude of the force on the component within the system, as well as the type (compression) of force it is.

Fig. 6-33

6-33a Bicycle with rider
6-33b Bow's notation
6-33c Vector diagram (known vertical forces)
6-33d Vector diagram (unknown horizontal forces)
6-33e Free body diagram for joint #1
6-33f Vector diagram for joint #1
6-33g Combined diagrams showing force magnitude and direction for joint #1

Bow's Notation. Bow's notation is a method of identifying forces within a structure. It is a graphical method in that accurate drawings of the structure must be made to *scale* (Figure 6-33a).

Step 1: Calculate the reaction of R_x and R_y

$$R_x \times 44 = (9 \times 100) + (35 \times 33)$$
$$R_x \times 44 = 2055$$
$$R_x = 2055/44$$
$$R_x = 46.7$$

$$R_y \times 44 = (9 \times 33) + (35 \times 100)$$
$$R_y \times 44 = 3797$$
$$R_y = 3797/44$$
$$R_y = 86.3$$

Step 2: Label the scale drawing with Bow's notation (Figure 6-33b). Begin in the lower-left corner, and work around the outside of the structure in a clockwise direction. Letter each space between the external forces. Letter each internal space. Label all known forces, and number each joint. Note that each line can be identified by two letters; for example, *AE* describes the bicycle front fork.

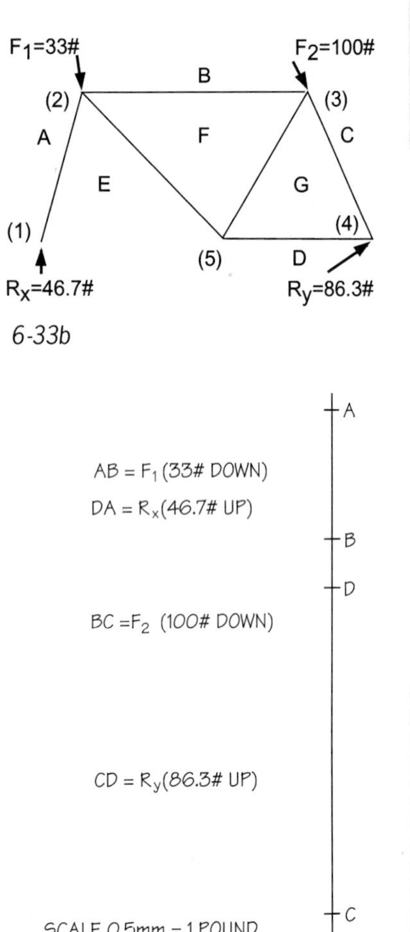

6-33b

6-33c Step 3

6-33a

Step 3: The actual calculations begin by drawing to a chosen scale (we have chosen 1 mm = 1 pound) a vector diagram of all the known forces (Figure 6-33c). Note that some forces are directed down (the weight of the rider), whereas other forces are directed up (the reaction of the ground through the tires). The up and down forces are in equilibrium ($F_1 + F_2 = R_x + R_y$).

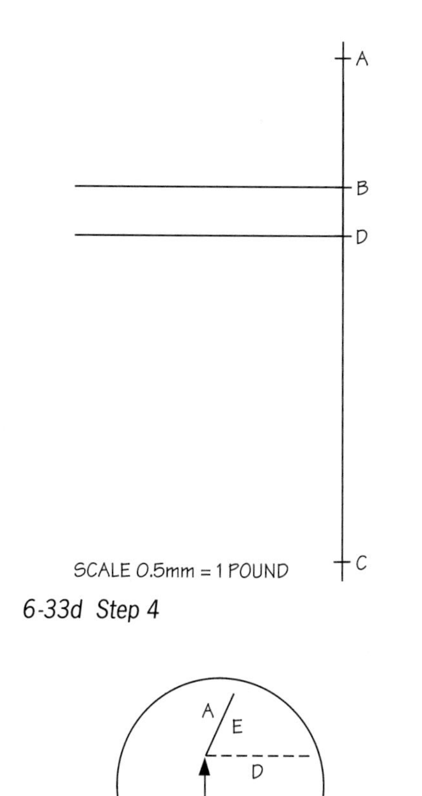

6-33d Step 4

6-33e Step 5

Step 4: Add to the vertical forces shown on the vector diagram all horizontal forces. In this example, lines *BF* and *DG* are horizontal at this point, so the actual length of the line cannot be determined (Figure 6-33d). The basic vector diagram will be used to measure the magnitude and type of force at each joint.

Step 5: Draw a free body diagram of the joint in question. Start in the lower-left corner with joint #1. The free body diagram will be used to determine the direction of the forces (compression or tension) (Figure 6-33e). Notice that we are only concerned with force R_x and line *AE*. If there were a connecting component between joints #1 and #5, then line *DA* would also be considered.

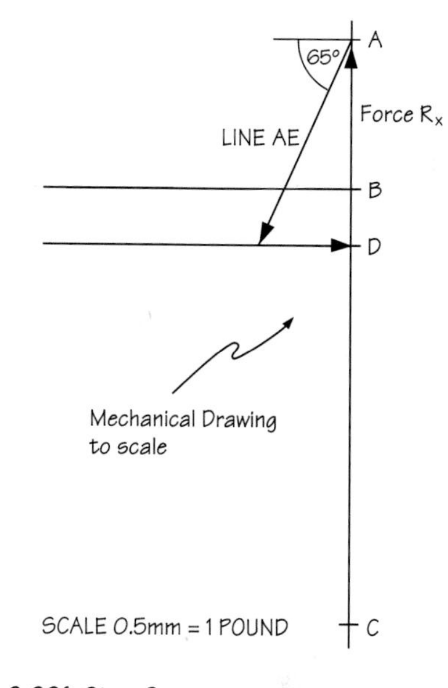

6-33f Step 6

Step 6: Draw on the vector diagram (Figure 6-33f) a line from point *A* running parallel to line *AE* found on the scale drawing (Figure 6-33b), in Figure 6-33f. Extend the line until it intersects the horizontal line at point *D*.

Interpreting the Figure. The length of line *AE* on the vector diagram is equal to the force on that component in the scale drawing. Since line *AE* is 26.5 mm long and the scale is 0.5 mm = 1 pound, then line *AE* represents a force of 53 pounds that will be applied to the front fork of the bicycle. Of equal importance is the type of force that the load (53 lbs) will be exerting on the structure. Note that the arrows have been drawn on the vector diagram in the direction of the force being applied to the system. Since the analysis is of joint #1, it is necessary to follow the force at R_x through the components in the free body diagram. Force R_x is being applied up the vertical line of the vector diagram. At point *A*, the force follows down line *AE* to point *E*, where it turns back towards point *D*. It is important at this point that the direction of the force identified in the vector diagram be transferred to the free body diagram (Figure 6-33g). If the arrow (direction of the force) is pointing toward the application point (the joint in this example), then the component is under compression. If, on the other hand, the arrow is pointing away from the application point, then the component is under tension.

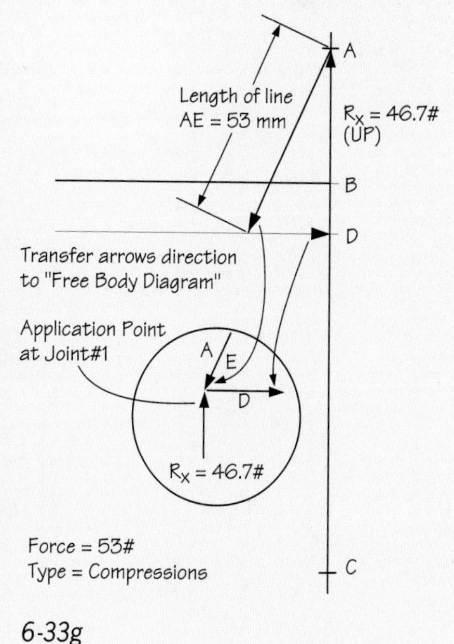

6-33g

Step 7: Repeat Steps 5 and 6 for each of the remaining joints: joint #2, lines *EF* and *BF* (Figure 6-33h); joint #3, lines *FG* and *CG* (Figure 6-33i); joint #4, line *DG* (Figure 6-33j).

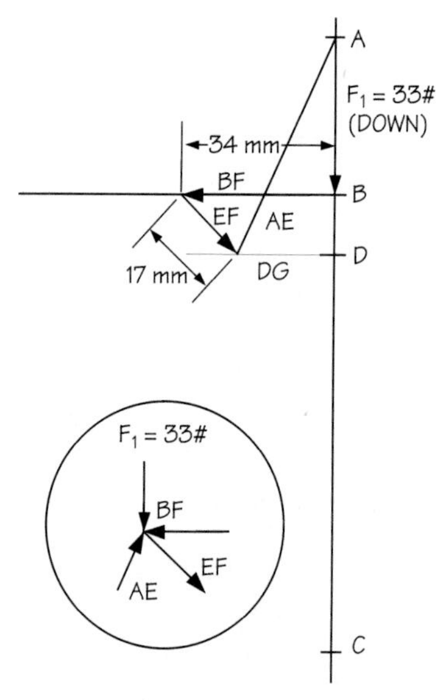

6-33h Forces at Joint 2

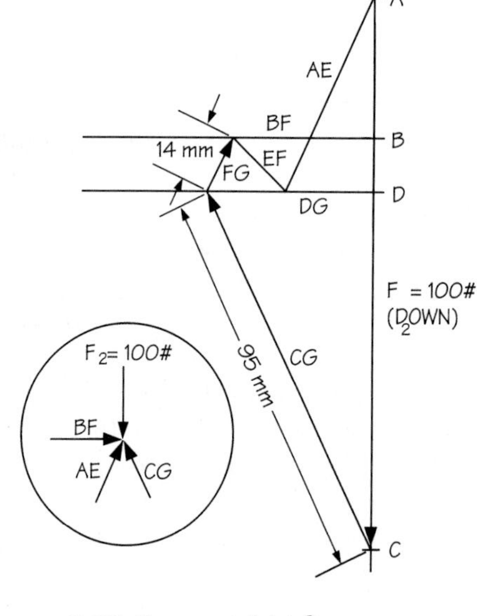

6-33i Forces at Joint 3

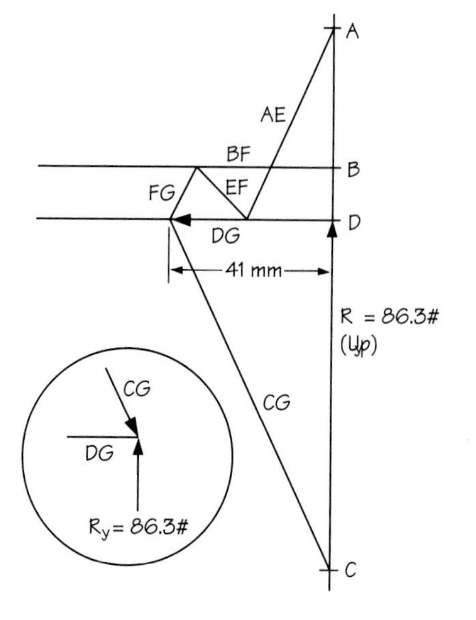

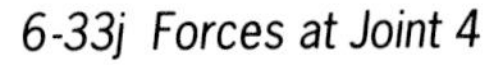

6-33j Forces at Joint 4

Fig. 6-33 continued

6-33h Combined diagrams showing force magnitude and direction for joint #2
6-33i Combined diagrams showing force magnitude and direction for joint #3
6-33j Combined diagrams showing force magnitude and direction for joint #4
6-33k Summary chart of component and forces for the bicycle frame

Step 8: Make a chart of all components showing magnitude and type of force (Figure 6-33k).

Component	Force	Type
AE	53#	Compression
BF	34#	Compression
EF	17#	Tension
FG	14#	Compression
CG	95#	Compression
DG	41#	Tension

6-33k

If we consider the bicycle structure and observe that no component ties joint #1 with joint #5, what conclusion can be made about the forces at joint #2? Many other new questions may be found as a result of asking the question, "What magnitude and type of forces are being applied to the bicycle structure being considered?"

Summary

Structures are exposed to both internal and external forces. Understanding these forces and how they affect the structural components is necessary when you are designing any structure. By using either a mathematical or graphical analysis, you can determine the magnitude and type of force being applied on each component of a structure. With this necessary information, proper materials, types of components, safety factors, and other necessary elements can be selected for the new design. Chair legs, sign-mounting brackets, a bicycle frame, or a structural component that you are designing can be selected that will function properly, will be aesthetically pleasing, and, very important, will not fail!

Observation/Analysis/Synthesis

1. Make a simple one-view sketch of ten different structures in your environment that serve different purposes or functions.
2. Select a type of bridge from this chapter or choose one of your own, and do a simple paper analysis. Use cardboard for all components that you believe will be in compression and string for all components that you believe will be in tension. Tape all joints. Apply a load, and evaluate your model.
3. Design a beam using 1 ounce of wooden stirring sticks. The beam should span a 12-inch open space. The beam must be designed to resist bending and torsion forces while being supported only at the two ends.
4. Design and make a model of a high-tension power line tower that will support two 1-pound loads as far apart as possible and as high from a tabletop as possible using no more than twenty sheets of 8½" × 11" bond paper and 36 inches of cellophane tape. Designs should be finished and displayed before you begin construction of the tower.

Fig. 6-34

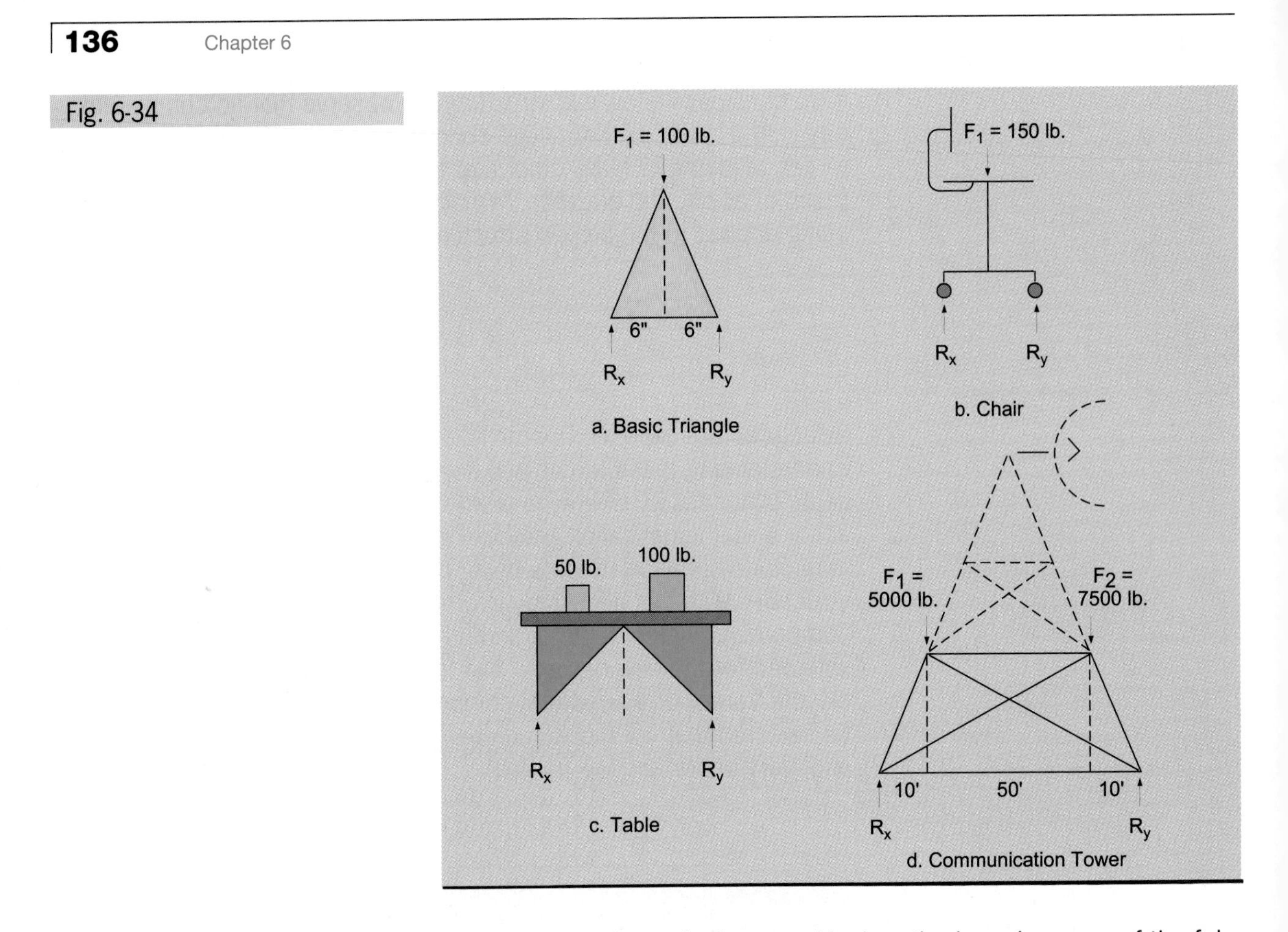

5. Using a mathematical or graphical method, analyze one of the following structures with a static load. Present documentation showing the results of your analysis.
6. Use material presented in this chapter to develop and analyze any structural component associated with your specific design problem.

CHAPTER

7 Designing Mechanical Systems

Introduction

Mechanisms and mechanical systems can be found in many products and devices. Mechanisms help extend human capability by creating some desired output motion or force. This chapter will introduce mechanical concepts, including basic calculations used to describe how a system will function.

Throughout history, people have used mechanisms to help solve technological problems. Leonardo da Vinci (1452–1519), an Italian Renaissance painter of the famous *Mona Lisa* painting and many other important art works, was also a designer and inventor of many mechanisms. Over 4,200 sketches still exist today of mechanisms he designed. Most of da Vinci's designs could not be made, because the materials and production practices necessary to execute the designs were not developed until many years after his death. Some of his ideas included an experimental flying machine (400 years later the Wright brothers made the first successful flight in 1903), a movable bridge, parachute, and a construction crane. Da Vinci did not formulate any scientific laws or principles; rather, he was an excellent observer of the natural world.

Mechanisms have played an important role in shaping society. In 1769, James Watt (1736–1819) patented the principle of the separate condenser for a steam engine. A Scottish engineer, Watt is best known for his improvement of the steam engine (Figure 7-1). The steam engine is considered one of the primary forces behind the Industrial Revolution in Europe during the eighteenth century and in the United States during the early nineteenth century. Other innovations by James Watt include crank movements (so that the steam engine could turn wheels), a throttle valve, a governor, and many other innovations to make the steam engine more useful.

Fig. 7-1

James Watt's steam engine (1769). Courtesy of the Trustees of the National Museum of Science and Industry, London

Mechanisms and Machines

Science describes six basic machines. These machines include the *lever*, the *wheel and axle*, the *pulley*, the *inclined plane*, the *wedge*, and the *screw* (Figure 7-2). Historically, these machines were primarily associated with moving heavy loads, such as the stones for the Egyptian pyramids. In the late nineteenth century, a German engineer by the name of Franz Reuleaux developed the fundamental concepts of modern mechanisms known as kinematics. As a branch of solid mechanics, *kinematics* deals with relative motion. Kinematics and kinetics (the action of forces on bodies) together form the basic elements of engineering dynamics. Reuleaux identified six mechanical elements upon which every mechanism and machine were based. These elements include: (1) the *lever and crank*; (2) the *wheel and gears*; (3) the *cam*; (4) the *screw*; (5) *things that transmitted tension or compression*, such as belts, chains, and hydraulic fluid lines; and (6) *things that transmit intermittent motion*, such as the ratchet.

Machines and Tools. *Machines* and *tools* are mechanical devices that work by transmitting or converting energy. We usually think of a machine as being more complex than a tool and more independent of the human user. A lawn mower, hair dryer, computer, washing machine, electric hand drill, food processor, printing press, and the Watt steam

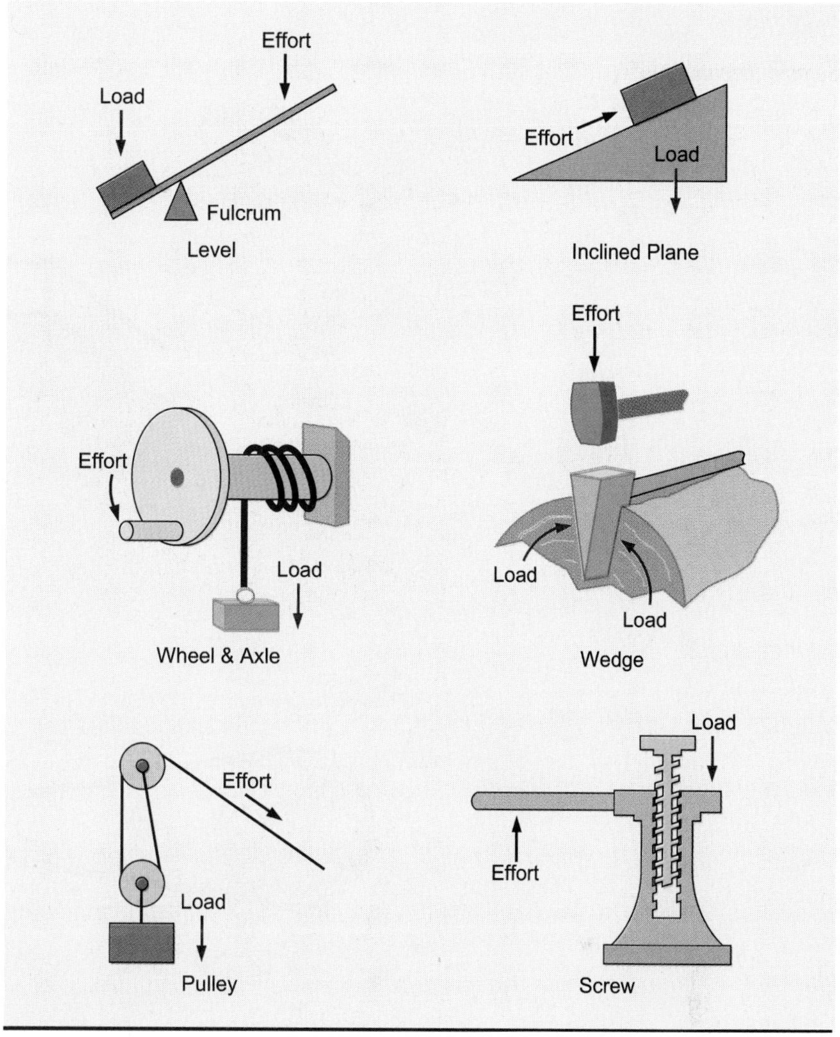

Fig. 7-2

The six basic machines of science

engine are all considered machines. Some machines, such as the steam engine, diesel or gasoline engine, or hydroelectric turbine, are considered prime movers because they are intended to convert energy directly into mechanical motion. All machines are made up of mechanisms. Would a clock be considered a machine or a mechanism?

Mechanisms. A **mechanism** takes an input motion or force and creates a desired output motion or force. A mechanism may consist of a single lever, crank, wheel, gear, cam, or screw, or, more typically, a combination of these elements. The analog clock is considered a mechanism, because its purpose is to create a desired output motion (movement of its hands). Most mechanisms serve unique purposes, such as clamping, lifting, locating, opening, coupling, making fine adjustments, or folding (Figure 7-3). Many toys incorporate mechanisms to create interesting motion and often serve as an excellent problem-solving situation for students (Figure 7-4).

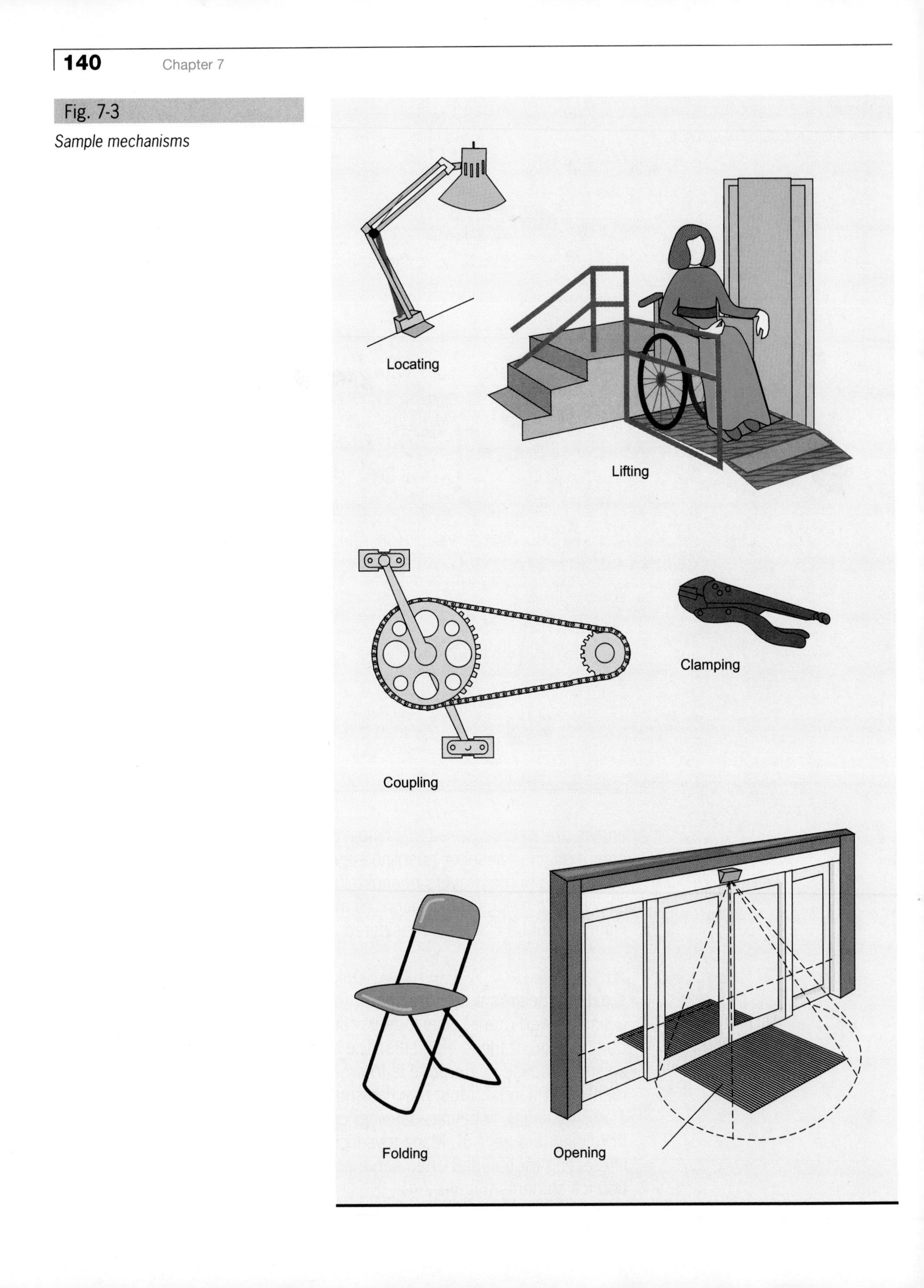

Fig. 7-3

Sample mechanisms

Kinematics

The study of motion, without regard to the force or mass of the things moving, is called **kinematics**. It is a branch of engineering mechanics. Kinematics can give us information about the movement within mechanical systems and provides us with an organized way to study, record, and design these systems.

Fig. 7-4

Mechanical toy

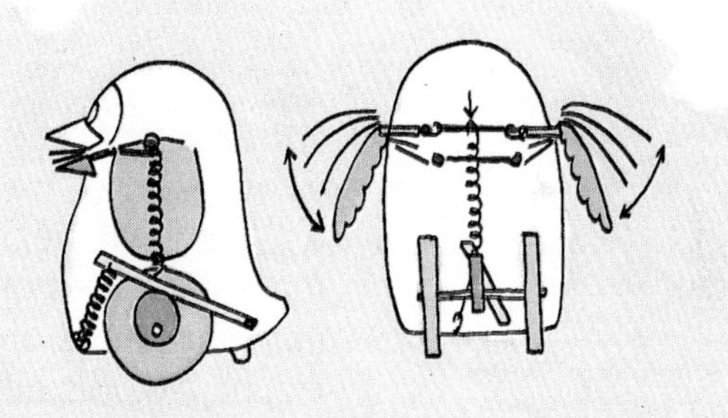

Fig. 7-5

Sample problem and design brief for a can crusher. (Full documentation of one student's approach appears in Appendix C.)

Identification of Problems

Americans accumulate over 240 million tons of solid waste each year. The waste is polluting the environment. My community requires recycling of a number of items, including aluminum cans. Every time someone has a soda, that person can choose either to clutter the kitchen counter or to take the can to the recycling bin located outside. Opening the door to take the can outside lets a lot of cold air into the house and wastes energy, but the empty cans on the counter look messy.

Design Brief

Design and build a hand-operated device that will reduce the volume of an aluminum can by at least 75 percent. The device must be attractive (salable), fit on or hang above a counter or on the back of a cabinet door, and must be usable by a person 12 years of age or older. The device must be safe to operate (does not require safety glasses, gloves, etc.) and must provide a relatively safe product (no extremely sharp or jagged edges). The design must represent an efficient and proper use of materials and production practices.

Kinematic Diagrams

As your ideas begin to take shape, it is important that you be able to make estimates of how the mechanism will function. Because of the complex nature of a mechanical drawing, such as the one shown in Figure 7-6, it is difficult to visualize and analyze the critical movements of the mechanism. To solve this problem, designers and engineers use *kinematic* or *skeleton diagrams* (Figure 7-7). These simple diagrams serve a similar purpose to electronic schematic diagrams (see Chapter 8) in that they show only the essential elements of the mechanism. Kinematic diagrams allow preliminary visualization. When a more detailed analysis of the system is needed, a kinematic diagram is made to scale. From this scale diagram, an analysis of pivot locations, fixed angles, link length(s), motion(s), displacement, force, and other factors can be calculated.

Kinematic diagrams are referenced, beginning with all fixed or ground links labeled with a *1*. The remaining links are numbered consecutively. All joints are referenced with letters *A*, *B*, and so on. "Input" motion and "output" motion are also noted. Figure 7-8 lists commonly used kinematic diagram elements.

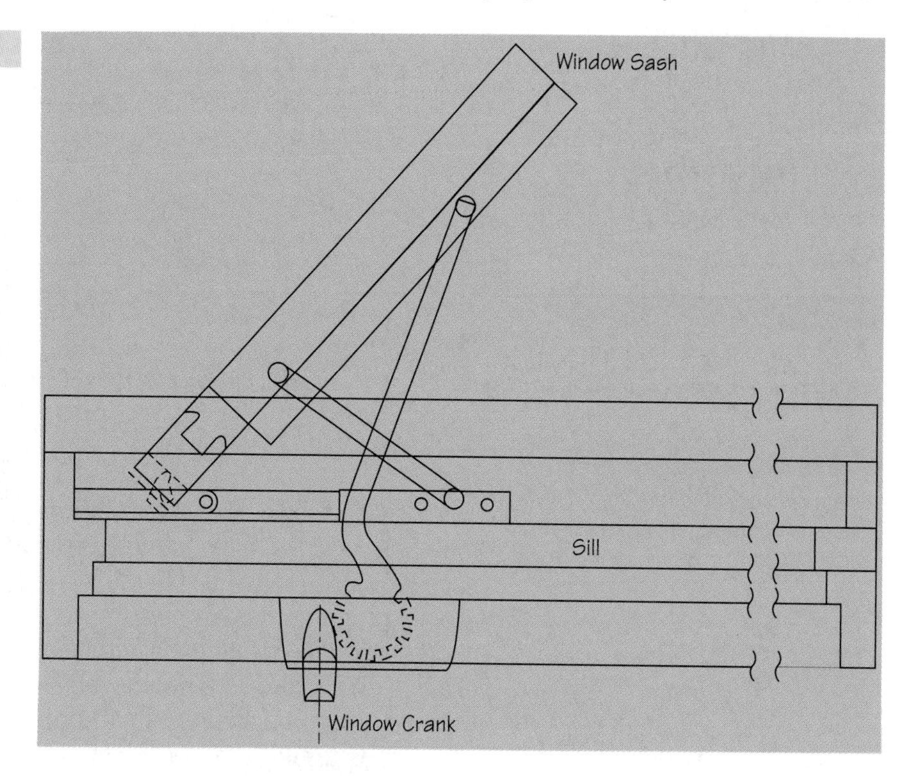

Fig. 7-6

The operation of the mechanism is not effectively shown in the mechanical drawing of a casement window.

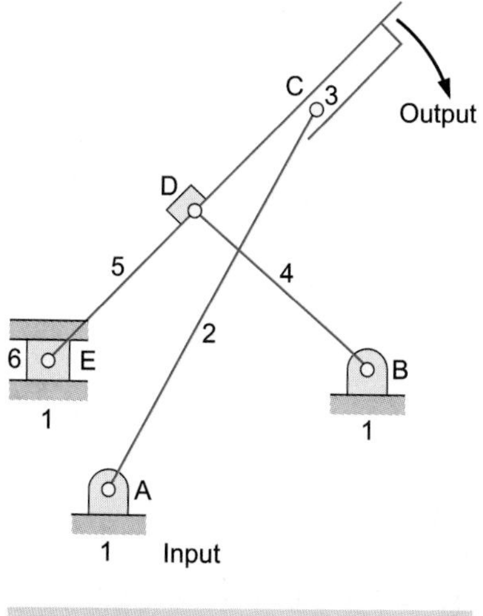

Fig. 7-7

Kinematic or skeleton diagram of a casement window opening mechanism

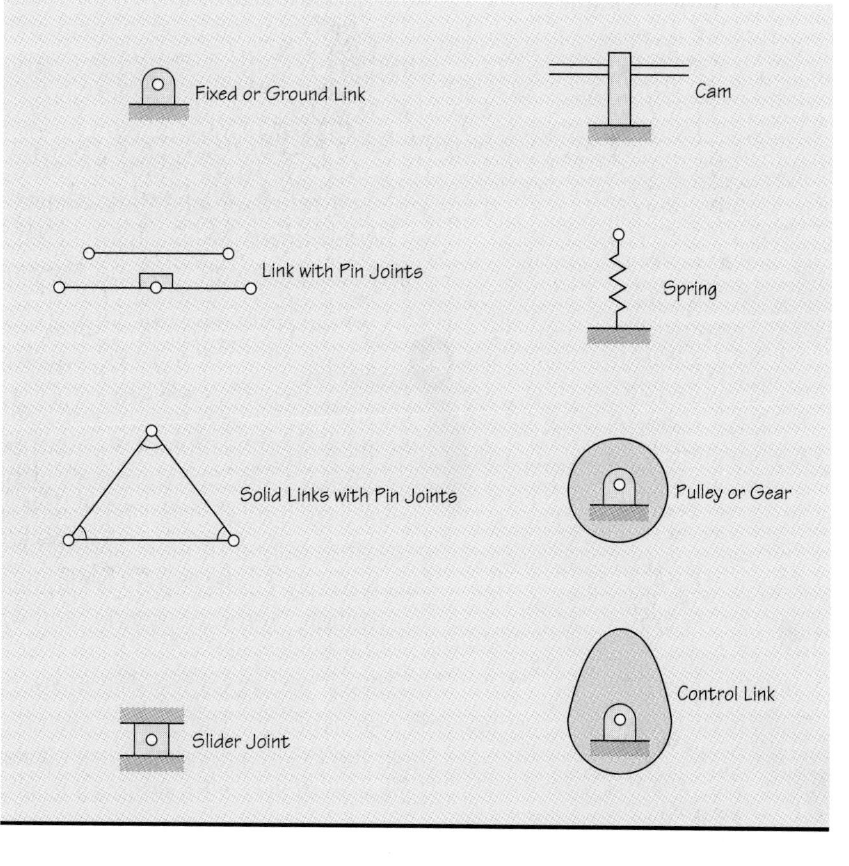

Fig. 7-8

Kinematic diagram symbols

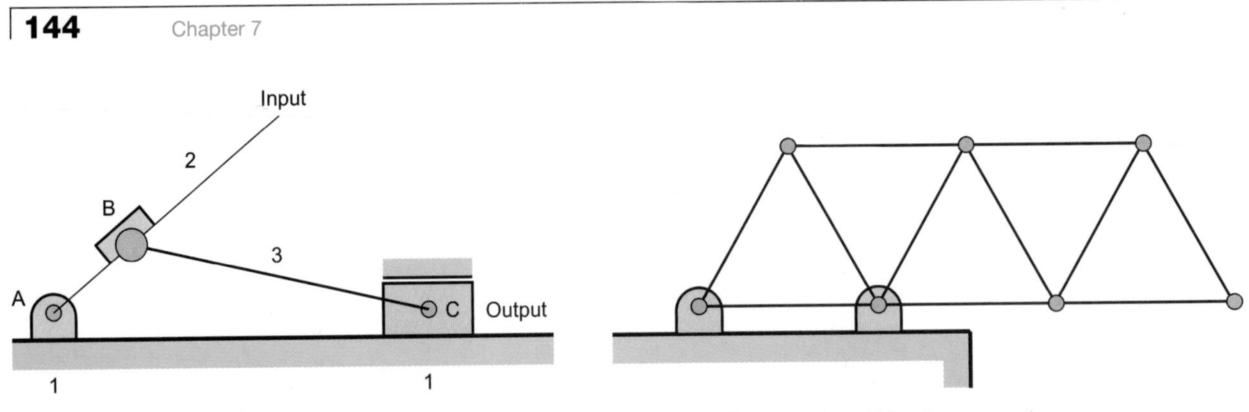

a. Mechanism (Capable of predictable movement)

b. Structure (Rigid - not capable of movement)

Fig. 7-9

Kinematic chain showing the difference between a mechanism and a structure

Kinematic Members

The individual rigid elements that are used in a mechanism are called *members* or *links*. Links may be solid elements or nonrigid members, such as cable, chain, or pneumatic and hydraulic fluid lines. All links are held together by elements called *joints*. A combination of interconnected links and joints is called a *kinematic chain*. If one end of the chain is connected to a nonmovable object, called a *fixed* or *ground link*, the kinematic chain is said to be a mechanism (Figure 7-9a). As previously stated, a mechanism is a device that takes an input motion or force and creates a desired output motion or force. If the combination of elements, after being fixed, is not capable of movement, it is no longer referred to as a mechanism, but is more correctly called a structure (Figure 7-9b). (See Chapter 6.)

All mechanisms must have a predictable output motion or force based on the known input motion or force. Linkages and slides are classified as lower forms of kinematic chains. Rollers and cams are classified as higher forms of kinematic chains. More about these forms of mechanical elements will be presented later in this chapter.

Motion

Because mechanisms deal with motion, it is important to understand the different kinds of motion possible. A mechanism was previously defined as a device that takes an input motion (and force) and creates a desired output motion (and force). Four common types of motion are typically controlled by a mechanism: 1) linear motion, 2) reciprocal motion, 3) rotary motion, and 4) oscillating motion (Figure 7-10).

Linear Motion. **Linear motion** is a straight-line motion that occurs in one direction. The toaster switch and the drawer that slides out to hold the CD in a compact disk player are examples of linear motion. Linear motion can be returned to a starting position.

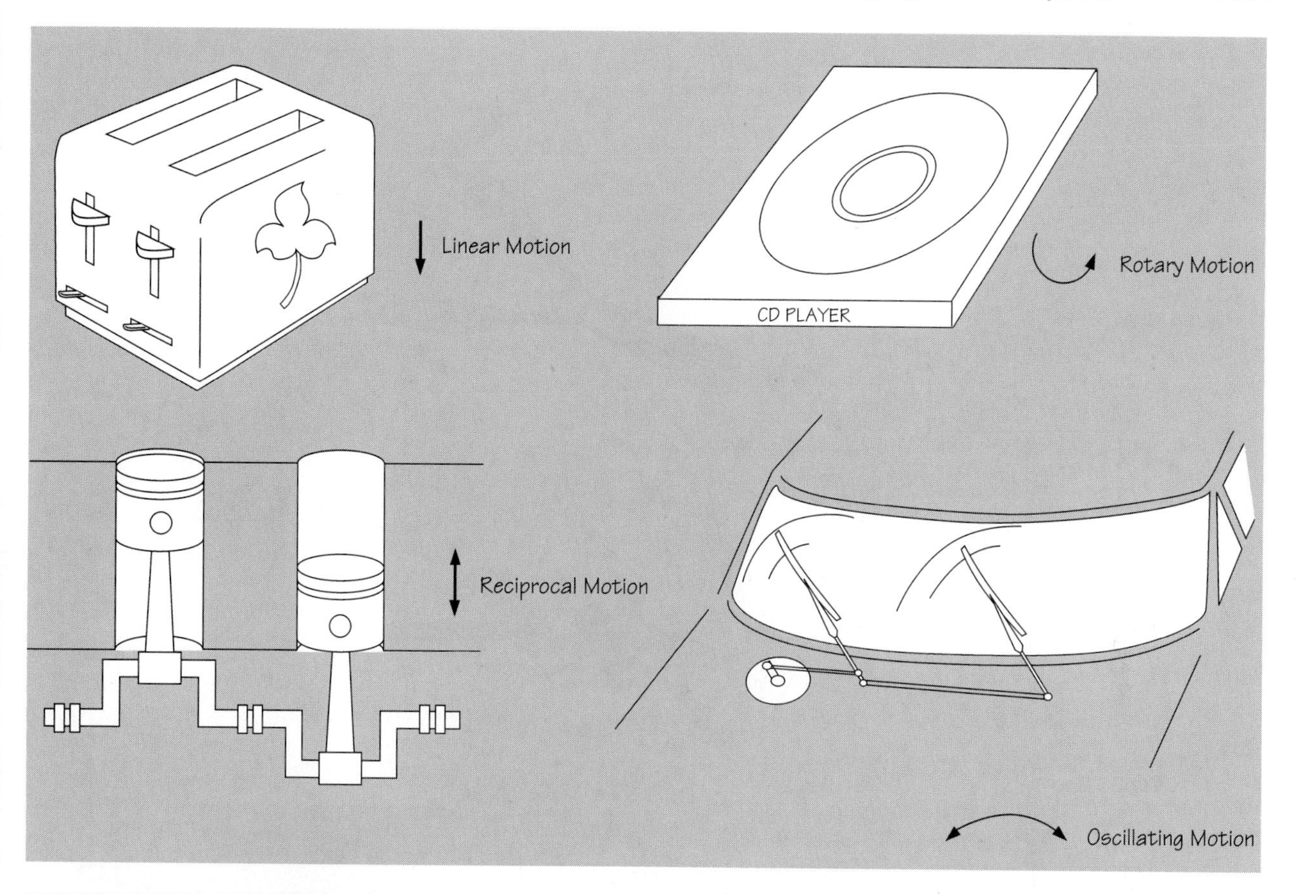

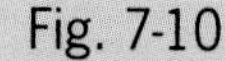
Fig. 7-10

The four common types of motion

Reciprocal Motion. **Reciprocal motion** is back and forth linear motion. The piston of internal combustion engines and the needle of a sewing machine both have a reciprocal motion.

Rotary Motion. **Rotary motion** occurs around an axle or center point, such as the rotating movement of the steering wheel of a car or the handlebars of a bicycle. Rotary motion is very efficient and is the most common form of mechanical motion generated by a prime mover. Rotary motion is often used in machinery, such as the huge printing presses used to print daily newspapers.

Oscillating Motion. **Ocillating motion** involves a back and forth movement in an arc. A clock pendulum and the agitator in a top-loading washing machine move in an oscillating motion.

Mechanisms use one form of input motion and create a modified or different output motion. The output motion can be continuous or nearly continuous. Many mechanisms create an output motion that changes relative speed throughout the motion cycle, which can cause other problems. Mechanisms are also capable of intermittent motion. Ratchets are the oldest form of mechanisms to create intermittent motion. Leonardo da Vinci used ratchet mechanisms on many of his designs.

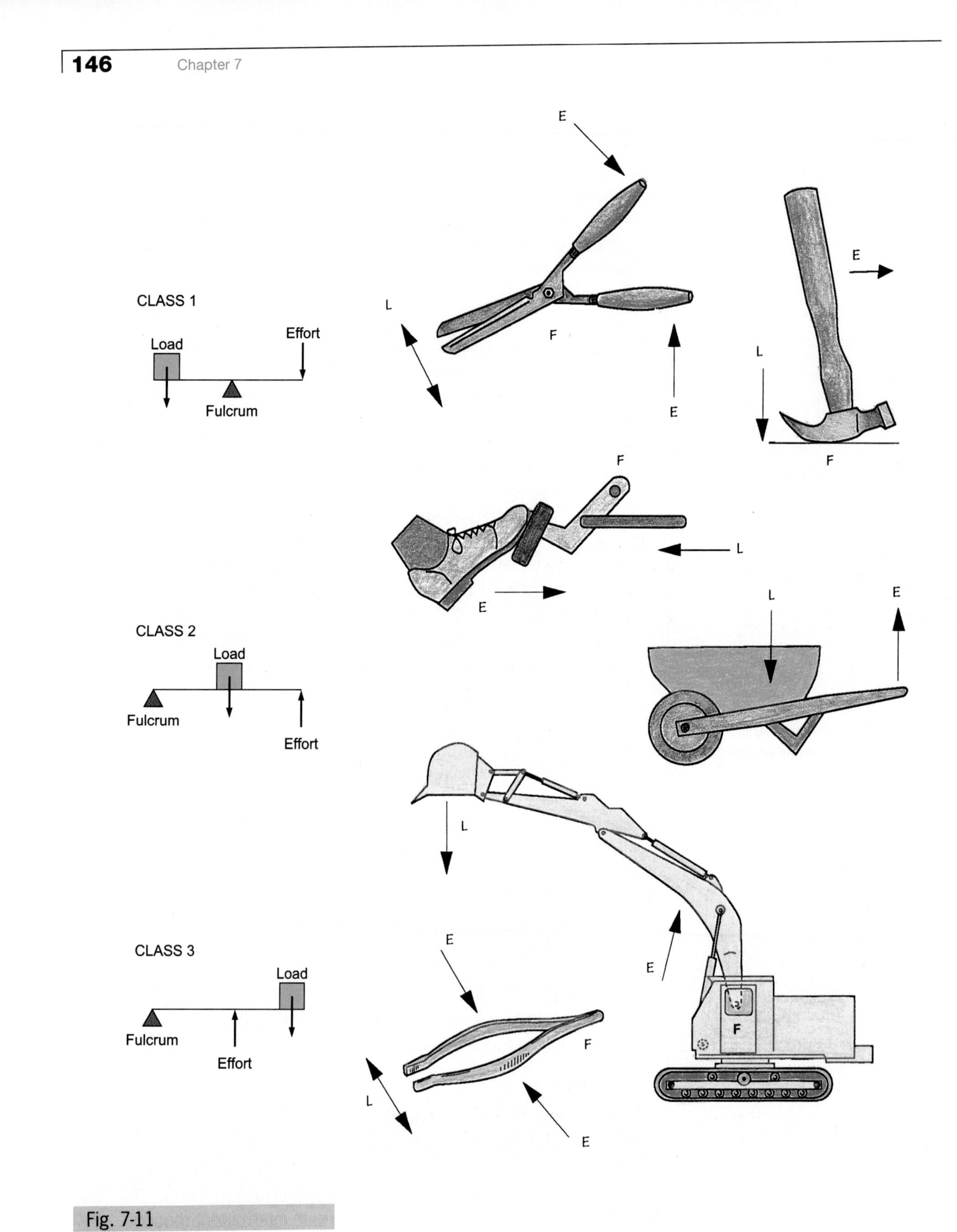

Fig. 7-11

Examples of Class 1, 2, and 3 levers

Levers and Linkages

The lever is one of the basic machines and is a simple mechanism that has been used to extend human capability throughout history. Lever mechanisms are grouped by class. The lever class is determined by the placement of the *fulcrum*, or pivot point, relative to the *effort* and *load* (Figure 7-11).

A *Class 1 lever* has the fulcrum between the effort and load. A playground seesaw, pliers, and scissors are all examples of Class 1 levers. *Class 2 levers* have the load placed between the fulcrum and the effort. A wheelbarrow, nutcracker, and an appliance hand truck are all examples of Class 2 levers. A *Class 3 lever* has the effort exerted between the fulcrum and the load. The garden shovel and bathroom tweezers are Class 3 levers.

Changing the relationships between the load, effort, and the fulcrum allows the lever to be used to gain mechanical advantage or to gain motion (Figure 7-12). In Chapter 6, you were introduced to the concepts of equilibrium and moments. Levers are in equilibrium when the effort (force) times the distance from the fulcrum is equal to the load (force) times the distance from the fulcrum. A moment, you will recall, is equal to $F \times D$. Remember that a mechanism is used to create an output motion and force. When the lever is used to increase output force, then the amount of input motion will need to be increased. When the lever is used to increase output motion, then the amount of input force will need to be increased. These trade-offs can be calculated mathematically.

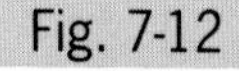

Fig. 7-12

Gaining motion by moving the fulcrum closer to the effort

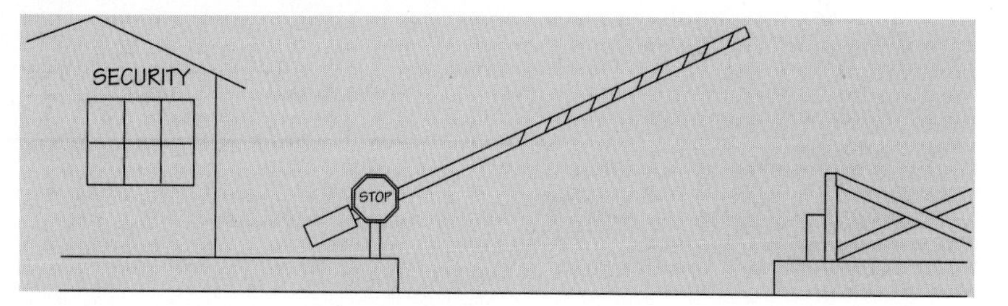

Mechanical Advantage

The measure for **mechanical advantage (MA)** is the ratio between load and effort (Figure 7-13). A mechanical advantage greater than 1 (MA > 1) means that a gain in output force has been achieved. A mechanical advantage less than 1 (MA < 1) means that the input force will be greater than the output force. Because Class 3 levers always have a MA less than 1, they would not be used to gain output force. Class 3 levers are typically used where a gain in motion is desired or to conserve space.

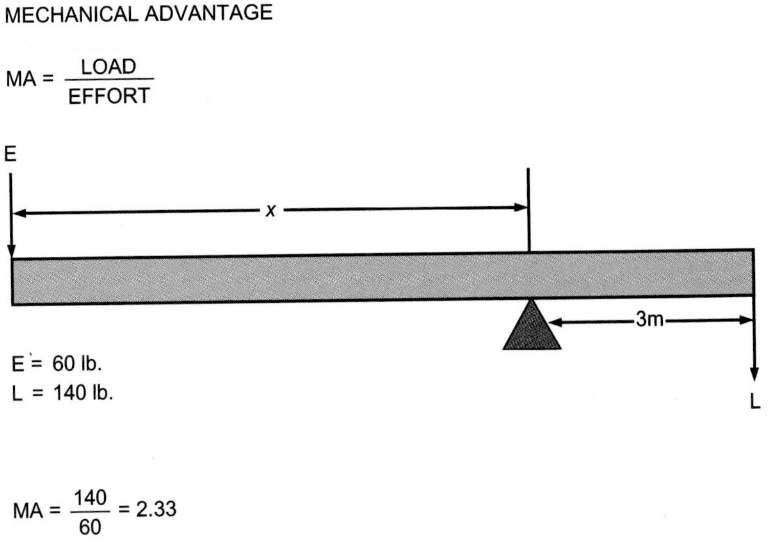

Fig. 7-13

Determining mechanical advantage of a lever

MECHANICAL ADVANTAGE

$$MA = \frac{LOAD}{EFFORT}$$

E

x

3m

L

E = 60 lb.
L = 140 lb.

$$MA = \frac{140}{60} = 2.33$$

Fig. 7-14

Determining the velocity ratio of a lever

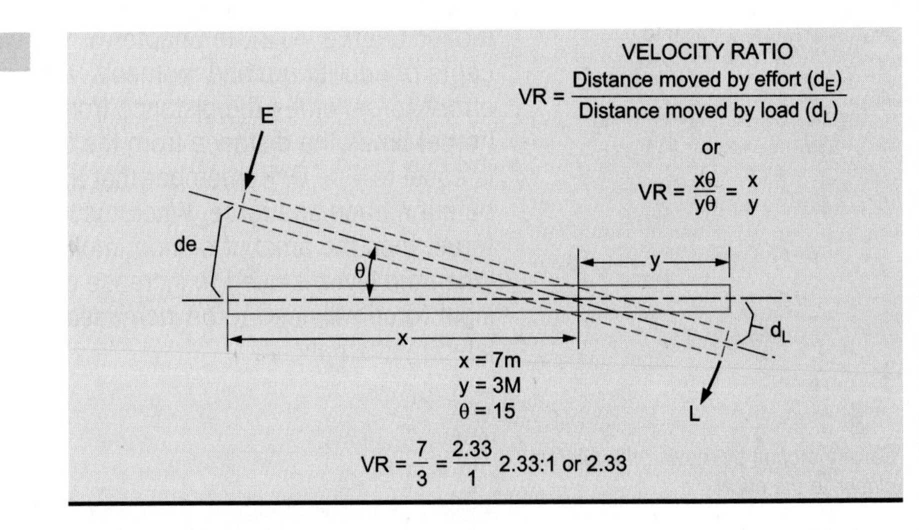

Velocity Ratio

For each gain in mechanical advantage, an efficiency in motion is lost. In science as in technology many trade-offs exist. **Velocity ratio (VR)** is the relationship between the distance moved by the effort and load (Figure 7-14). In order to gain mechanical advantage, the effort must be moved over a greater distance. An example of this is when a hoist is used to lift a heavy load. By using a number of pulleys, a hoist can have a very high mechanical advantage. The trade-off, however, is that you may have to pull the hoist chain 25 feet to lift the load 1 foot. A VR greater than 1 ($VR > 1$) means that the effort is moving a greater distance than the load.

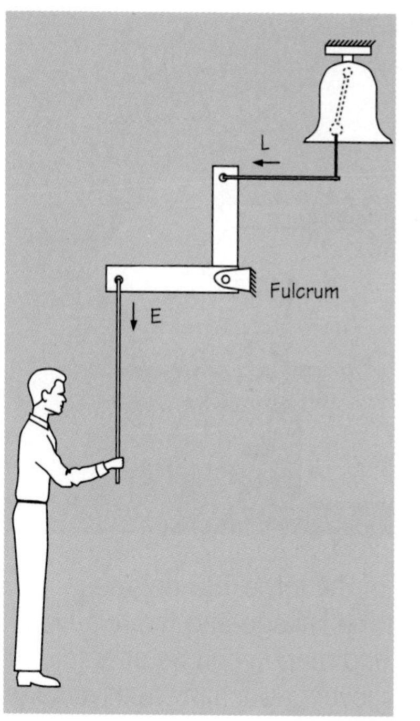

Fig. 7-15

An example of a bell crank linkage

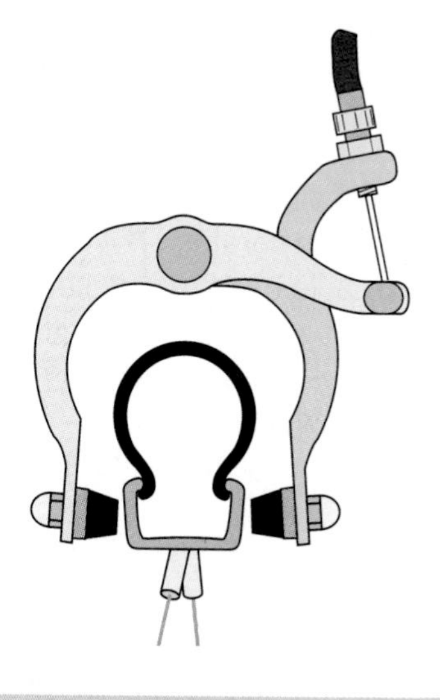

Fig. 7-16

Bicycle braking system using a bell crank linkage

Linkages

Linkages are very important to mechanisms, because they transmit the motion or force to the desired output location. Linkages can change the direction of the force, change the length of motion of the force, or split the motion and force over multiple paths. Linkages can make things move in just about any way desired.

Bell Crank and Reversing Linkages. The bell crank is a basic linkage that got its name because it was used to ring bells (Figure 7-15). The **bell crank** linkage changes the direction of force 90 degrees. Double bell cranks are also used. A bell crank linkage is used as part of the braking system on many bicycles (Figure 7-16).

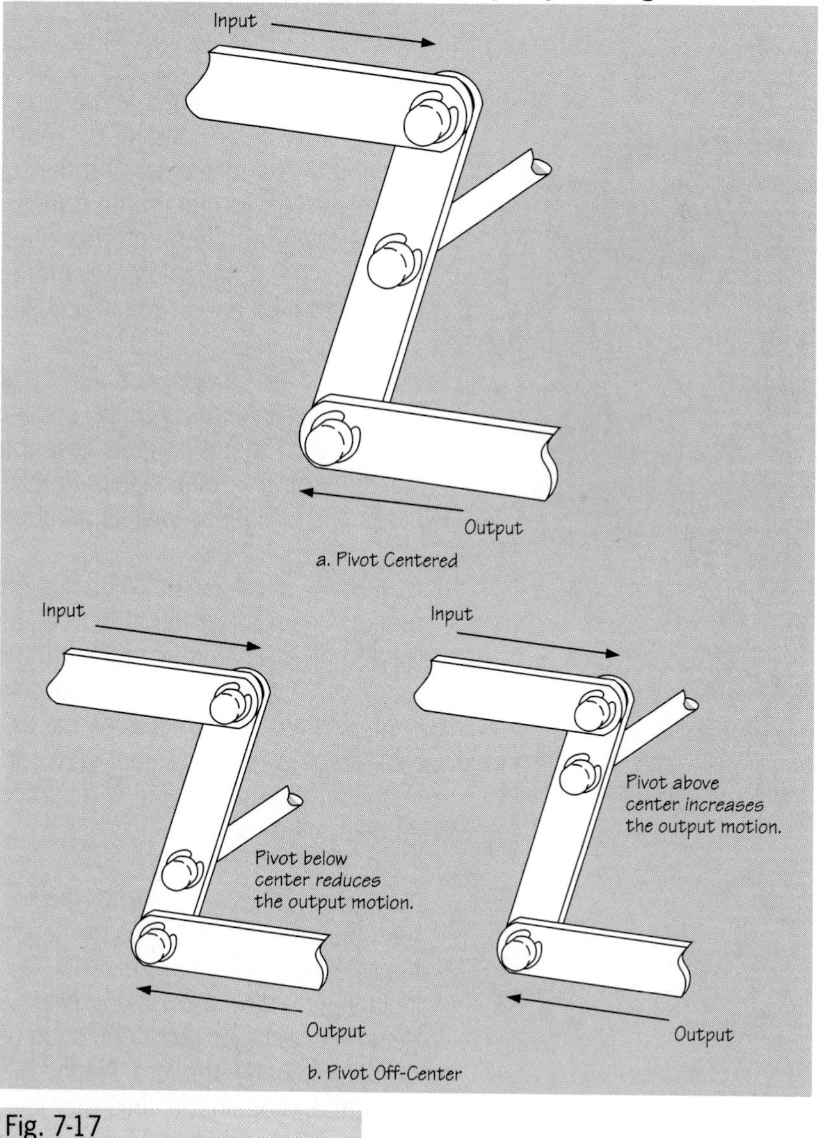

Fig. 7-17

Motion-reversing linkage

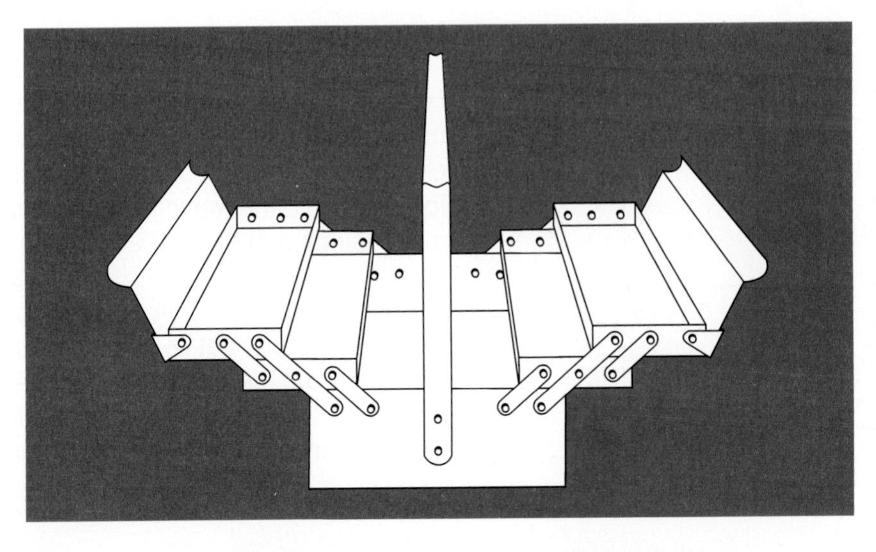

Fig. 7-18

A tool box design using parallel linkages

Reversing linkages change the direction of the force 180 degrees (Figure 7-17). By changing the length of the linkage and the relative position of the fulcrum point, the force and motion can be altered. Linkages work on the same principle as levers. Mechanical advantage and velocity ratio can be calculated for each linkage used in a system. Compare the mechanical advantages of the braking mechanisms of a ten-speed bike and a mountain bike. How are they different?

Parallel Linkages. A variety of fascinating mechanisms incorporate parallel linkages. *Parallel linkages* are based on the *geometric parallelogram*. These linkages allow component parts to move while maintaining a parallel relationship. Parallel linkages are used in scissor-type gates and tables, as well as pantographs (Figure 7-18).

Fig. 7-19

Treadle-operated sewing machine from the mid-eighteenth century. Courtesy of Smithsonian Institute

Treadle Linkage. When it is desirable to change an oscillating motion to a rotary motion, a *treadle linkage* is incorporated. The treadle linkage got its name from the foot-operated treadle used to turn sewing machines, wood lathes, and other similar devices prior to the twentieth century (Figure 7-19). Today, the treadle linkage is used in such common areas as the operation of a car's windshield wipers. In this instance, the mechanism is reversed, causing a rotary motion to output an oscillating motion.

Toggle Linkages. A *toggle linkage* is considered a clamping linkage, because it generates tremendous force as semivertical angle Ø nears 90 degrees (Figure 7-20). Many mechanisms use toggle linkages, including clamps and locking pliers. Toggle linkages are also used in general consumer products, such as card tables, folding chairs, and collapsible baby strollers. Notice that, if the toggle linkage is taken past straight and against a solid member, the linkage is locked in place (Figure 7-21). Additional force will need to be exerted in order to take the linkage [straight] to open the toggle linkage.

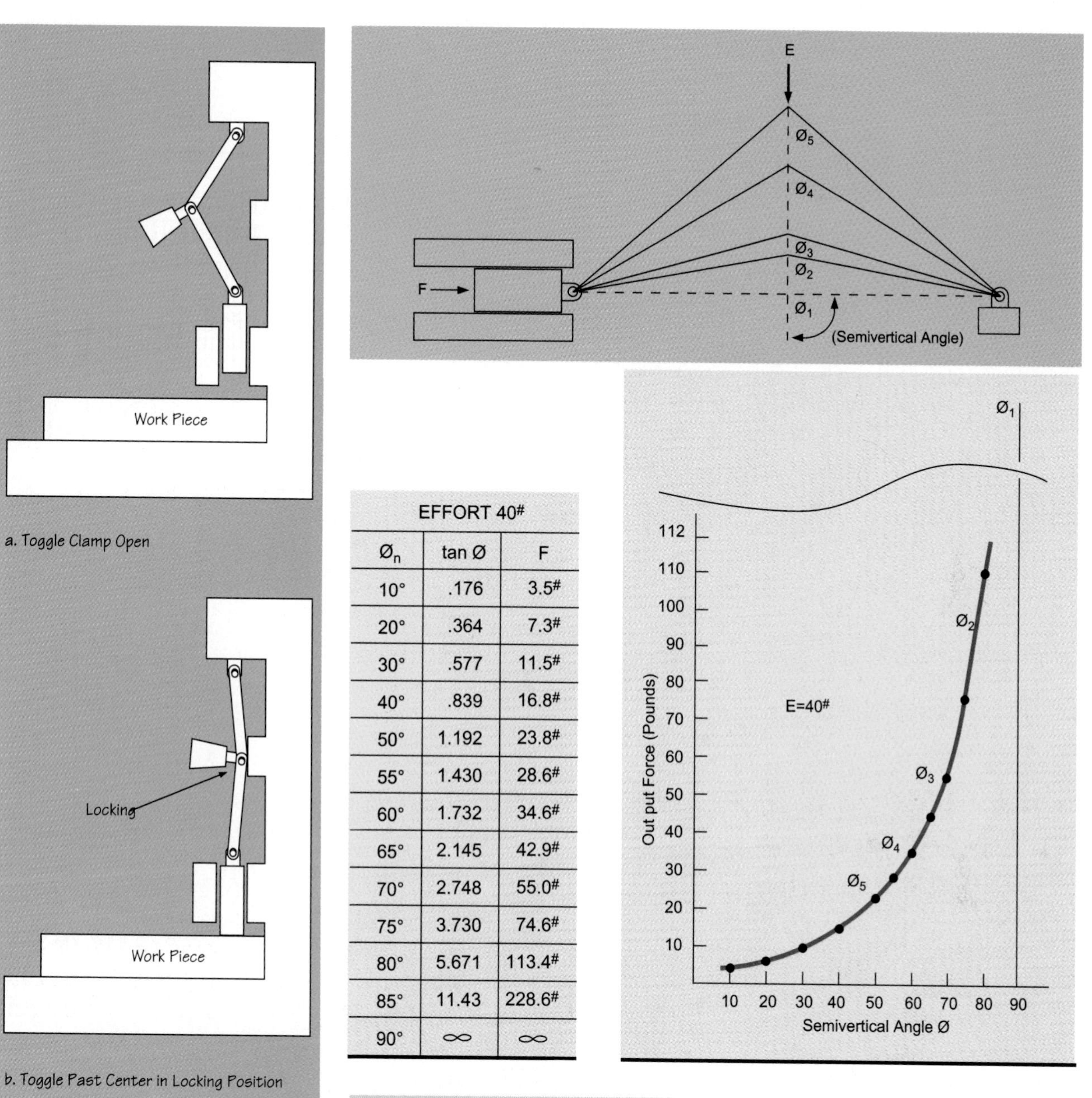

EFFORT 40#		
$Ø_n$	tan Ø	F
10°	.176	3.5#
20°	.364	7.3#
30°	.577	11.5#
40°	.839	16.8#
50°	1.192	23.8#
55°	1.430	28.6#
60°	1.732	34.6#
65°	2.145	42.9#
70°	2.748	55.0#
75°	3.730	74.6#
80°	5.671	113.4#
85°	11.43	228.6#
90°	∞	∞

Fig. 7-21

Locking of a toggle mechanism

Fig. 7-20

Chart showing the changing force generated by a toggle mechanism at different angles

Rotary Mechanisms

Gears, pulleys, cams, and other related mechanisms are considered higher-order kinematic pairs. These *rotary mechanisms* transfer or change an input rotational motion and force to an output motion and force. The output can be either rotary or reciprocating motion, depending on the rotary mechanism employed.

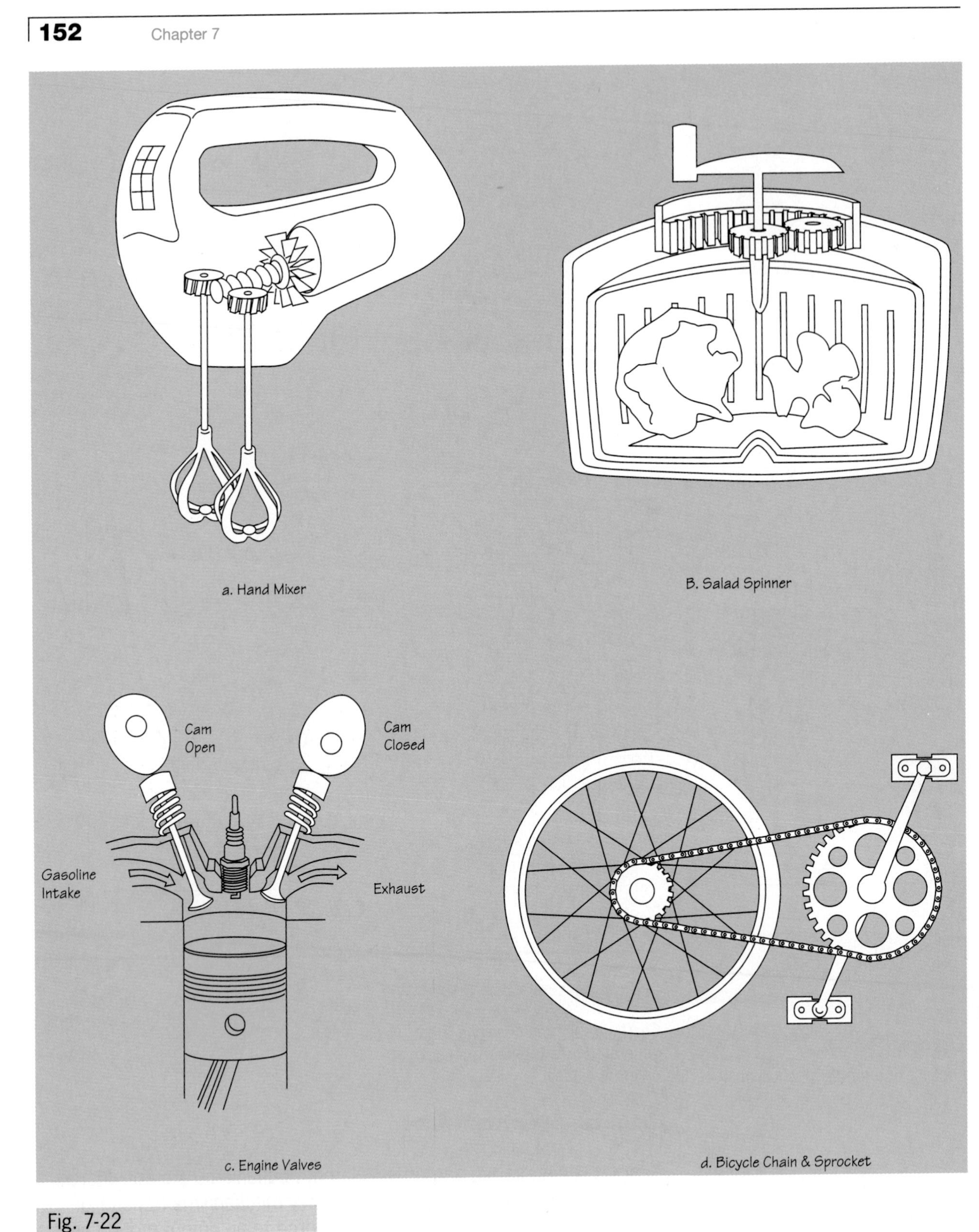

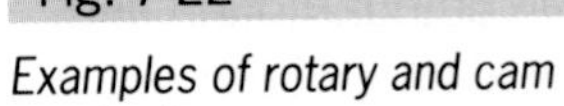
Fig. 7-22

Examples of rotary and cam mechanisms

Fig. 7-23

Driver and driven gears

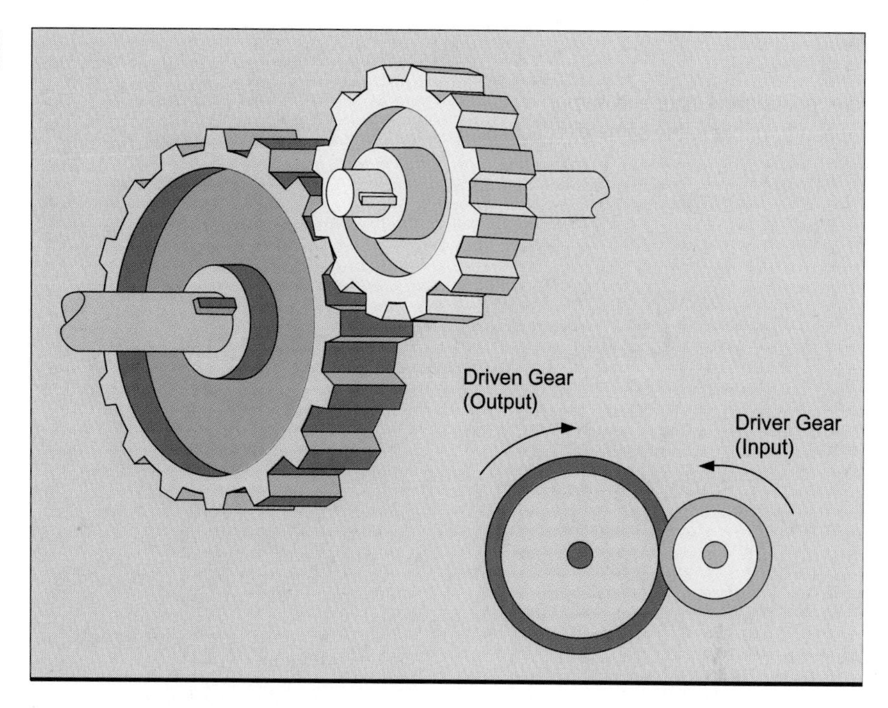

Gears. *Gears* are toothed wheels fixed to a axle. Remember that the wheel and axle is one of the basic machines. One gear fixed to an input axle is called the *driver gear*. The other gear fixed to an output axle is called the *driven gear* (Figure 7-23). When a number of gears are connected together, the system is known as a *gear train*. The relationship between input motion and force and output motion and force is the same as for levers, and can be calculated using the following formulas:

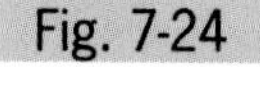
Fig. 7-24

Determining gear ratios

Gear Ratio

$$GR = \frac{\text{Number of driven teeth (output)}}{\text{Number of driver teeth (input)}}$$

$$GR = \frac{16}{12} = 1.333:1$$

(Gear ratio for gear train in Figure 7-23)

When two gears mesh, the motion of the driver gear turns the driven gear in the opposite direction. When both the input (driver gear) and output (driven gear) are required to turn in the same direction, an *idler gear* is needed (Figure 7-25).

Sometimes large speed or force changes are needed. In this situation, where a high- or a low-velocity ratio is required, *compound gear*

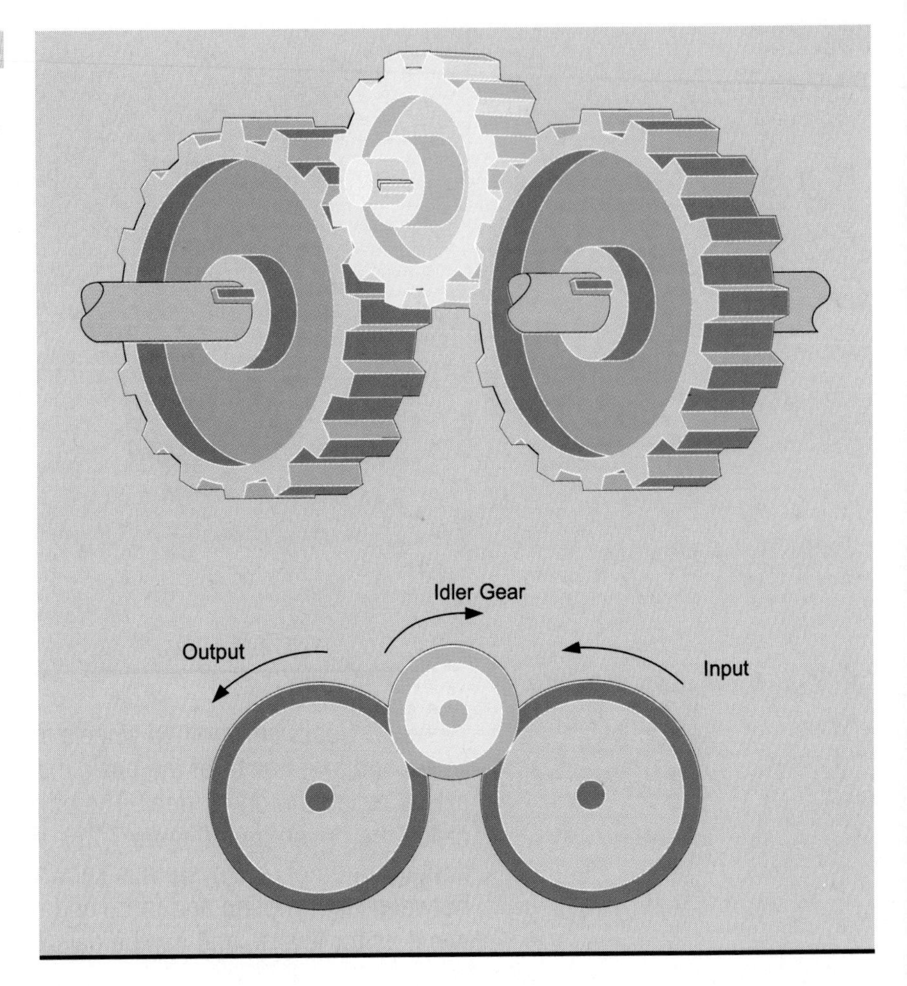

Fig. 7-25

Idler gears allow input and output motion to occur in the same direction.

trains are used. Compound gear trains have two gears of different sizes on one shaft, one of which acts as an input or driven gear and the other, as a driver gear to the next gear in the train. This system makes compact gearing mechanisms possible. To determine the input to output gear ratio in compound gear trains, the GR for each pair of gears must be determined (Figure 7-26).

Gear Types. The most basic type of gear is called the *spur gear*. This gear is relatively easy to manufacture and is one of the earliest forms of gears (Figure 7-27).

The *helical gear* is a more technologically advanced gear design (Figure 7-28). This gear is cut at an angle, and its shape is part of a helix. These gears are designed to run quieter at high speeds and to transfer more torque than other gear designs. This is because helical gears have more teeth area in contact. In transmission systems, where there is a gear train, helical gears generally shift more easily than spur gears.

Fig. 7-26

Determining gear ratios for compound gear trains

COMPOUND GEAR RATIO

$$AB = \frac{24}{40} \times BC\,\frac{16}{40} \times CD\,\frac{8}{40} = \frac{6}{125}$$

6:125 or 1:20.83

It is common practice to express gear ratios as 1:n for speed increasing gear trains and n:1 for speed reduction gear trains. In this example, the output (driven) gear turns 20.83 revolutions (speed increasing) for each revolution of the input (driver) gear.

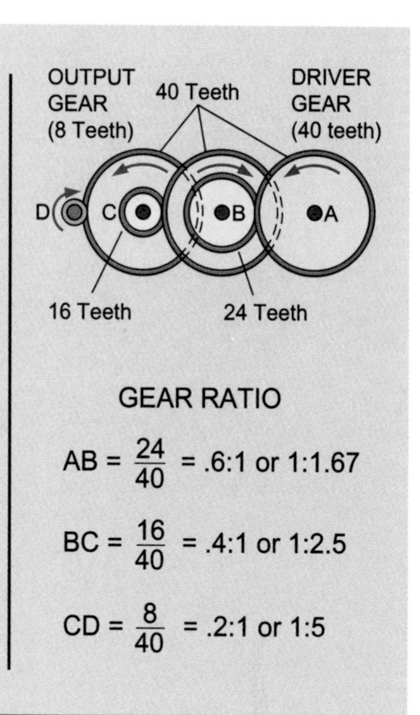

Fig. 7-27

Spur gear from Stanserhorn Cable Car, Switzerland

Fig. 7-28

Helical gears

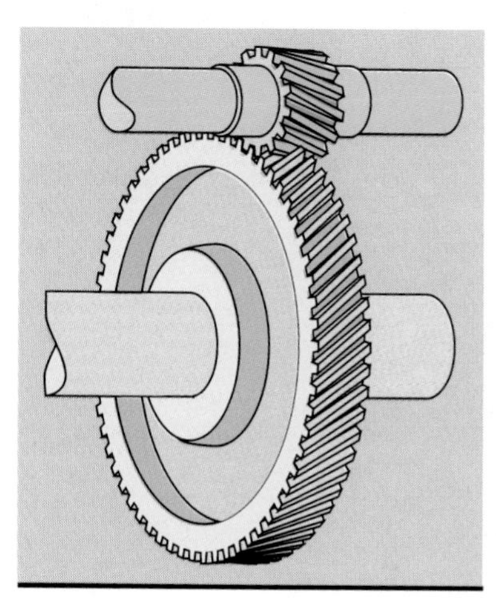

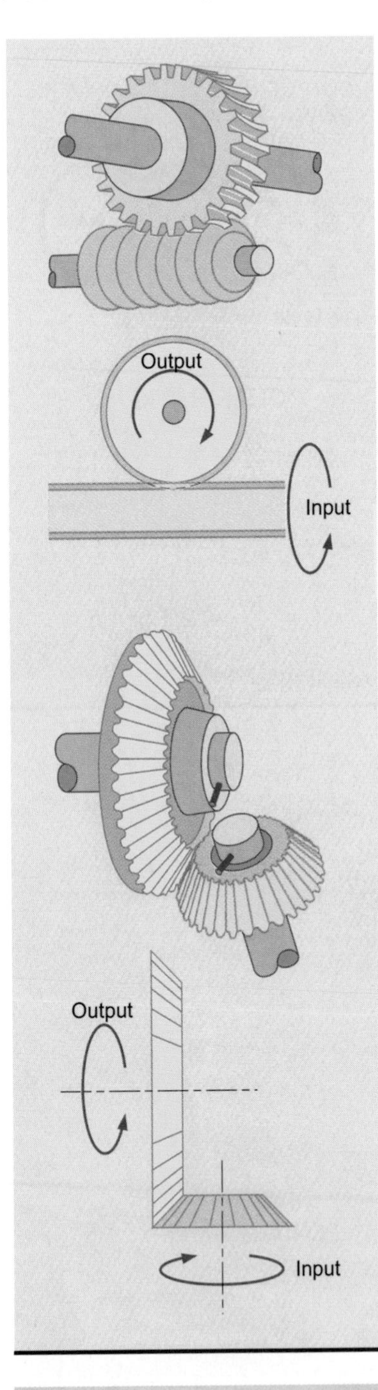

Fig. 7-29

Worm and bevel gears change the rotation axis.

Worm and Bevel Gears. Both the worm and bevel gear are used to change motion direction, typically at 90 degrees. The *worm gear* looks something like a screw but is really a single-gear tooth wrapped around a driver axle. This gear system is capable of making large speed reductions. Figure 7-29 shows a worm gear system with a 30:1 velocity ratio.

In the *bevel gear,* VR changes are made by changing the size of the driver and driven gears. The bevel gear system is used for smaller velocity ratio changes than the worm gear system. A form of bevel gear system is used in the rear axle of rear wheel drive cars and trucks to change both the speed of rotation and the direction of the rotary motion of the drive shaft. These systems generally have a ratio of 3.00:1 to 4.10:1.

Rack and Pinion. The *rack and pinion* gear system changes a rotary input motion to a linear output motion (Figure 7-30). In this gear system, one gear is produced as a flat strip. Rack and pinion gears can be found in many automotive steering systems and some machines in the technology lab.

Gears are made of a variety of materials. While most gears are manufactured in steel, other metals such as brass are used. Plastic gears are also common. Plastic gears are not as strong as their metal counterparts but are often less expensive, run more quietly, and do not require as much lubrication.

Other Rotary Systems. Pulley and sprocket systems use a variety of belts and chains to convert a rotary input motion and force to a desired output motion and force (Figure 7-31 a,b,c).

These systems work under the same measure of mechanical advantage and velocity ratio as the gear. *Pulley and belt* systems are quiet to operate, and they are capable of changing both the speed and direction of the output motion. In most pulley and belt systems, adjust-

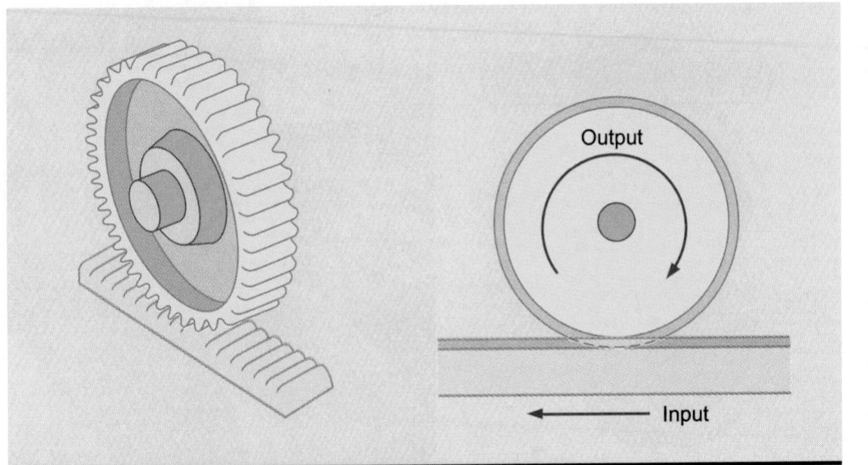

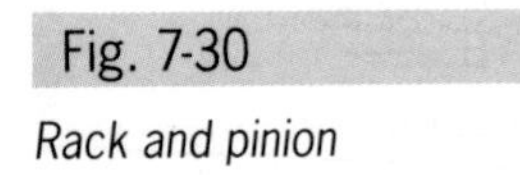

Fig. 7-30

Rack and pinion

Figs. 7-31a, b and c

Examples of pulley and sprocket systems.
7-31a Courtesy of Trek Bicycle Corp.
7-31b Photo by Tom Carney
7-31c Courtesy of Ford Motor Co.

7-31a

7-31b

7-31c

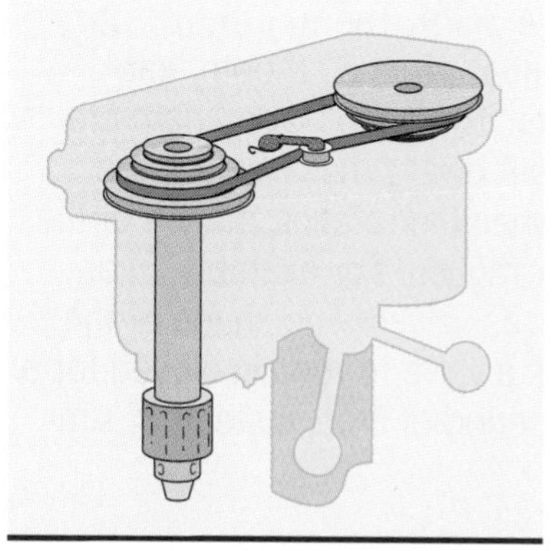

Fig. 7-32

The idler pulley puts tension on a belt to prevent slippage.

ments must be possible to hold the belt in proper tension. Sometimes a spring-loaded or adjustable idler wheel is used (Figure 7-32).

Sprocket and chain systems are not as quiet as pulley and belt systems but they are capable of transmitting greater force (Figure 7-33). Chains must be kept well lubricated, or rapid wear occurs. In addition to changing the output speed of rotary motion, sprocket and chains can be used to change rotary to linear motion. A *conveyor system* is a good example of rotary to linear motion.

Cams and Crank Slider Mechanisms. Cam and *crank slider* mechanisms convert rotary input motion to a desired reciprocating or oscillation output motion. An eccentric is the simplest cam shape

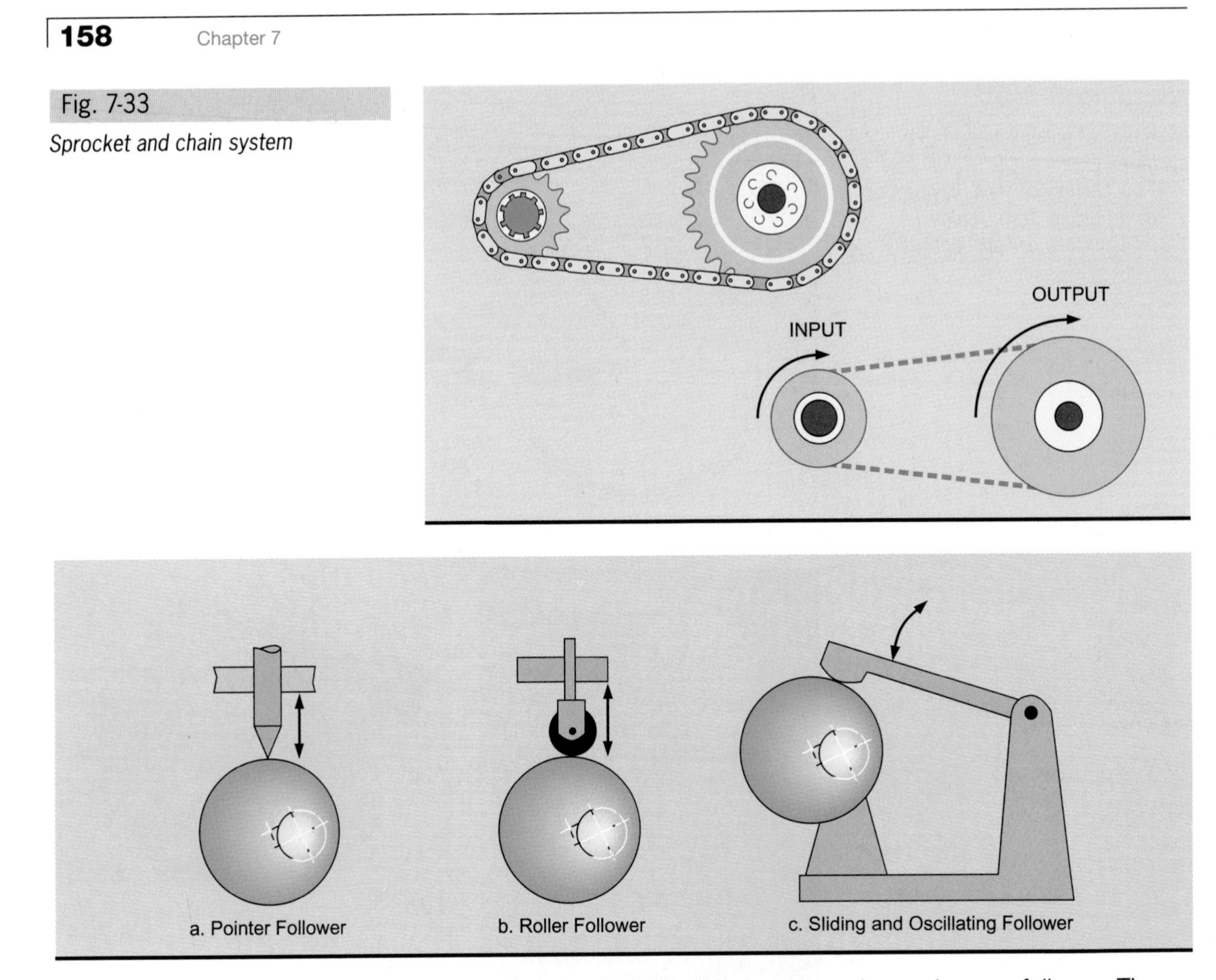

Fig. 7-33

Sprocket and chain system

Fig. 7-34

Eccentric cams with different types of followers

and provides a smooth reciprocating motion to the cam follower. The *follower* can be blunt or a shaft with a wheel (Figure 7-34).

Cam Motion. The input and output motions of a cam are represented by a *displacement diagram* (Figure 7-35). The **eccentric cam**, a circular disk with the axle placed off-center, gives a smooth, harmonic, linear output motion to the follower. The *pear-shaped cam* creates an intermittent linear motion with a set dwell time. **Dwell** is the time, expressed in degrees of rotation, when the rotating motion of the cam does not cause a change in follower motion. For example, in an automobile engine, a cam opens and closes valves. The shape of the cam, however, causes a dwell that allows a valve to remain closed for a period of time and to remain open for a period of time, instead of simply snapping open and shut.

Cams can be used to obtain all kinds of interesting motion, such as in a child's toy. A *heart-shaped cam* is another commonly used cam. Can you determine its motion and think of an application for this cam design?

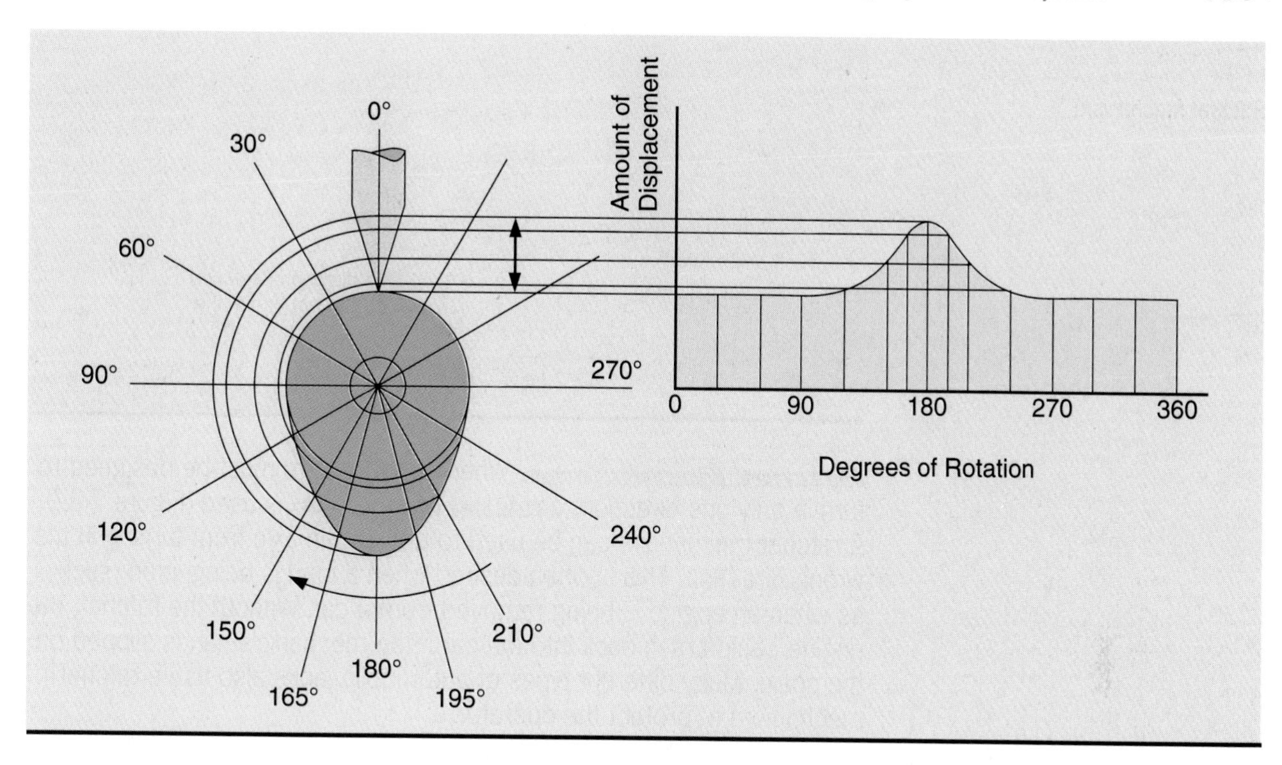

Fig. 7-35

This illustration shows the motion (displacement) of the follower created by the cam shape.

Fig. 7-36

Crank and slider mechanism used to create motion for a simple toy

Crank and Slider Mechanisms. The *crank and slider mechanism* is used to change between rotary and reciprocating motions (Figure 7-36). In a VCR, the crank and slider mechanism converts the rotary input motion from a small electric motor to the linear action that accepts or ejects the tape cassette. A prime mover, such as a diesel or gasoline engine, uses a crank and slider mechanism to convert a reciprocating motion caused by the engine pistons moving back and forth in a cylinder into a rotary output motion. This action is similar to the treadle linkage mechanism discussed earlier in the chapter.

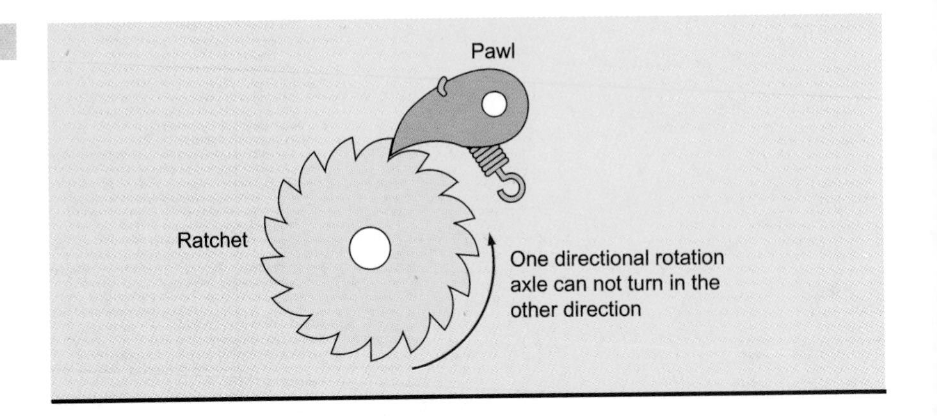

Fig. 7-37

Ratchet mechanism

Ratchet Mechanisms. When a mechanism must be designed to turn in only one direction, a **ratchet mechanism** is used (Figure 7-37). A ratchet mechanism can be used to prevent an axle from turning in the wrong direction. This is often desired when a load is being lifted, such as when an engine is being removed from a car. Without the ratchet, the engine could crash back into the car if the mechanic's hands slipped on the chain. Many different types of automobile jacks also use a ratchet mechanism to protect the operator.

A ratchet mechanism can also be used to convert linear motion to a rotary motion. In this design, the *pawl* is used to push against the teeth of the ratchet. The amount of rotary motion is determined by the amount the pawl is moved.

Clutches and Brakes

Clutches. When the motion created by a mechanical system is to be momentarily disrupted or stopped, clutches and brakes are incorporated (Figure 7-38). A **clutch** is a form of coupling that can be easily connected and disconnected. A *friction clutch* can be engaged or disengaged while either the input and/or output shaft is moving. A *centrifugal clutch* used in a lawn mower or a moped is a type of friction clutch that engages when the input shaft is rotating at a predetermined speed. A *positive clutch* interlocks the input and output shaft and can only be used if both shafts are stopped or are turning at the same speed.

Brakes. *Brakes* use friction to reduce the speed of a mechanism (Figure 7-39). Both *drum* and *disc brakes* are commonly used. Mountain bikes and most modern automobiles use a form of disc brake. Disc brakes generate greater braking force and dissipate heat more efficiently than drum brakes. Some automobiles use drum brakes on the back wheels because they are easier to connect mechanically to an emergency brake. Less braking force is needed on the back brakes of a vehicle as the load shifts forward during braking.

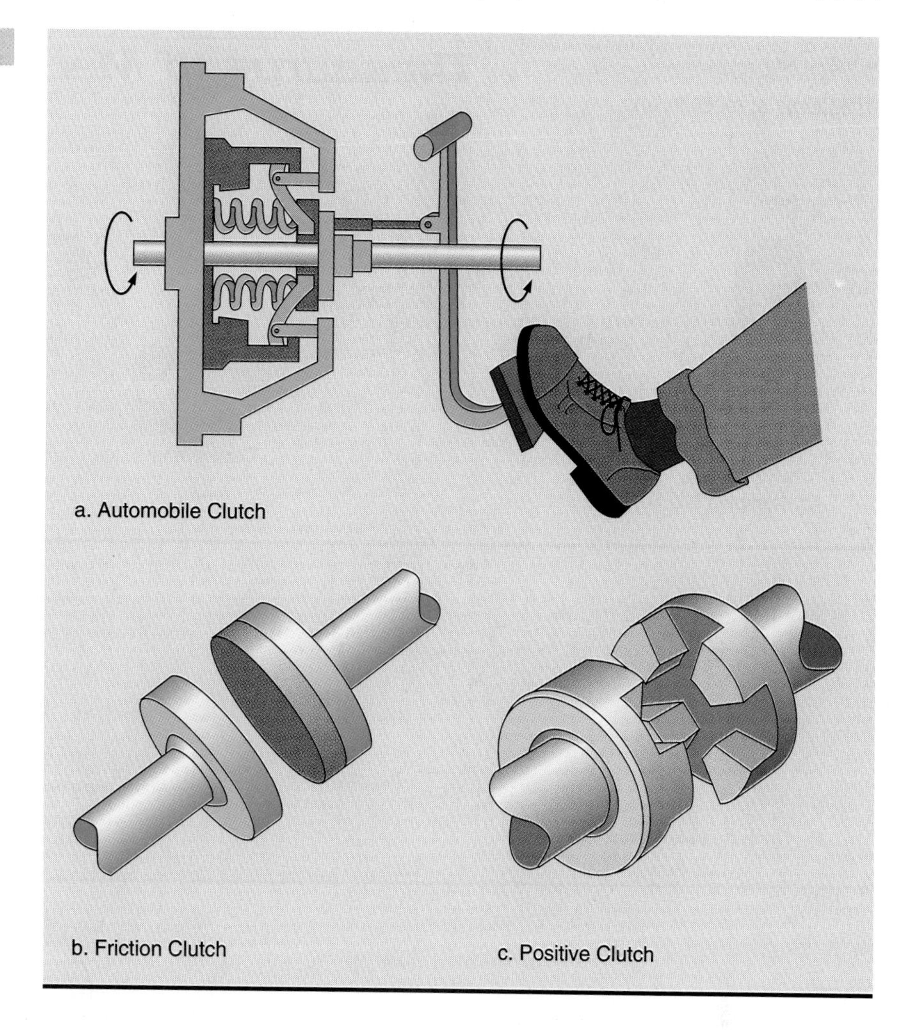

Fig. 7-38

Clutch mechanisms are used to connect rotating shafts.

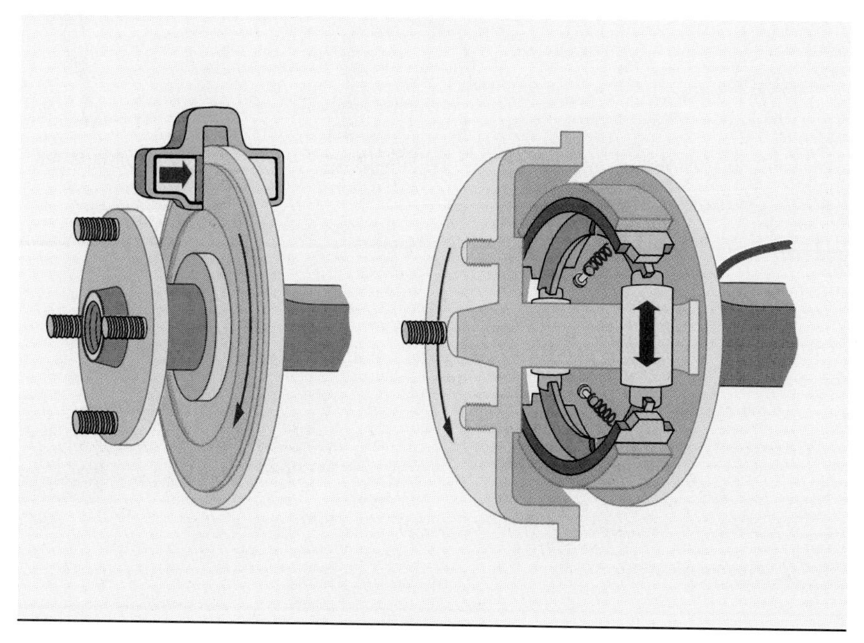

Fig. 7-39

Brake mechanisms

Fig. 7-40

Thesaurus of mechanisms

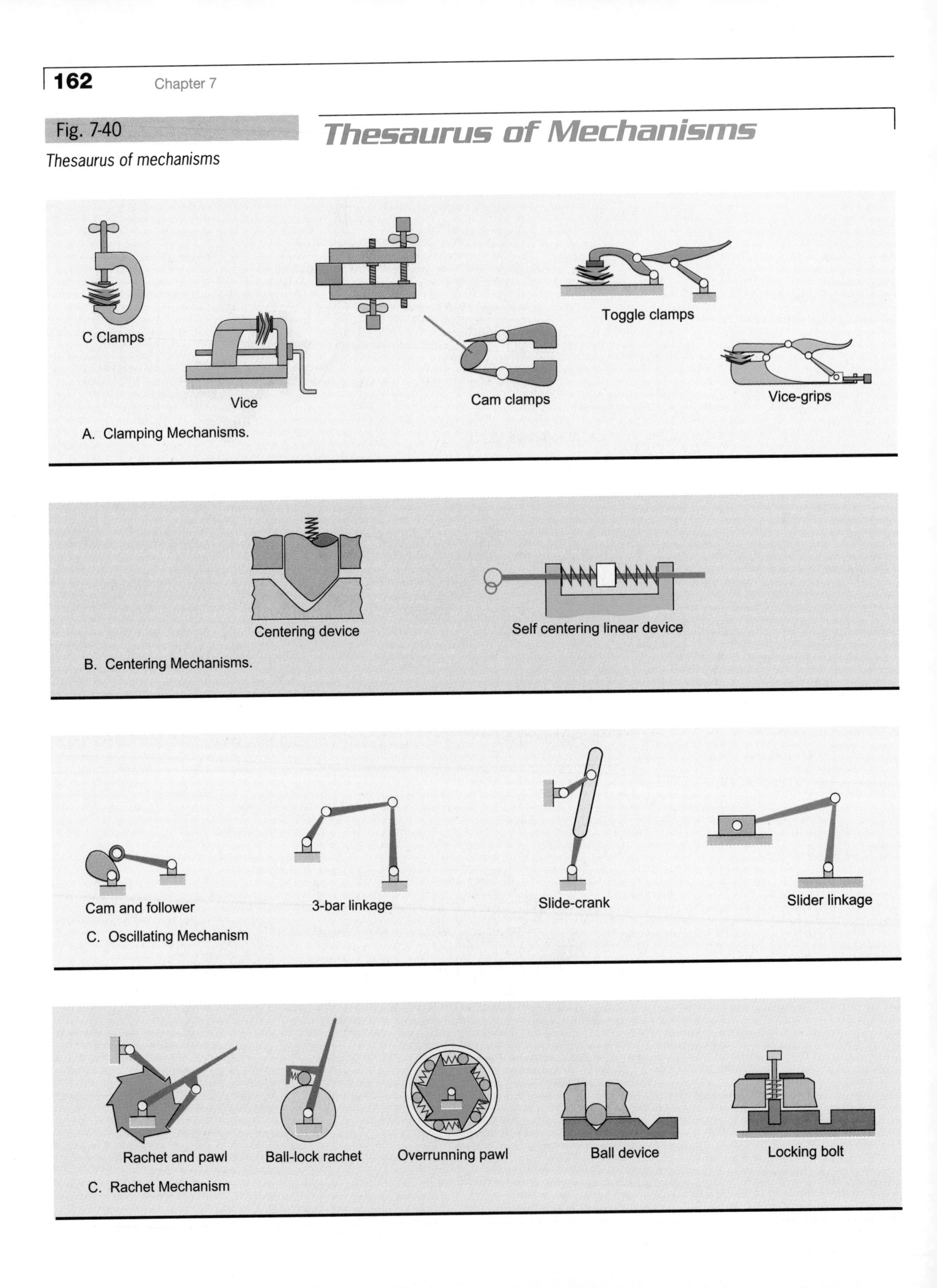

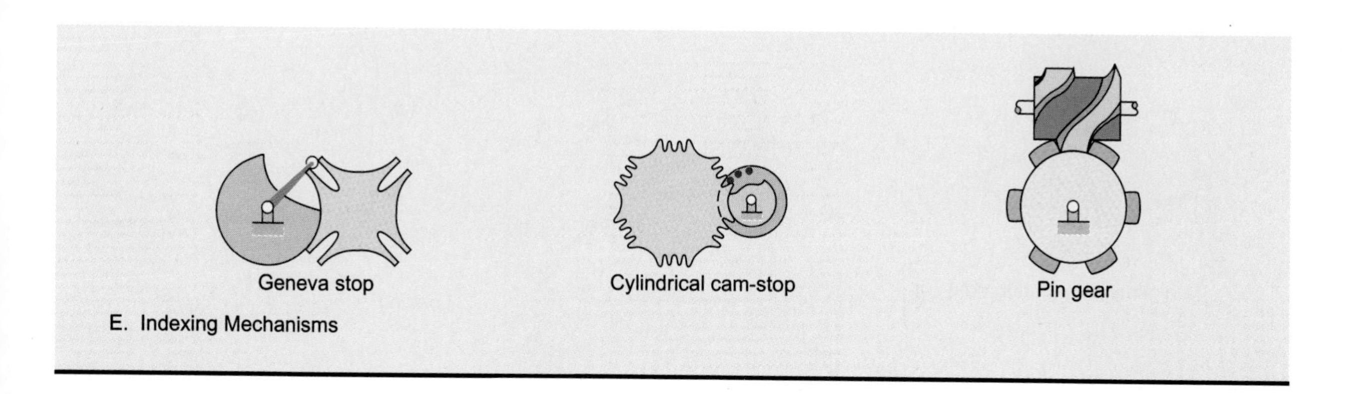

E. Indexing Mechanisms

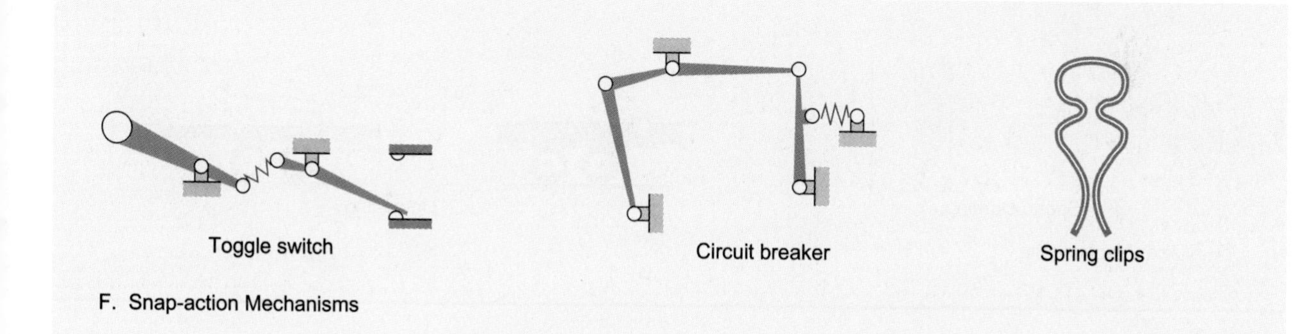

F. Snap-action Mechanisms

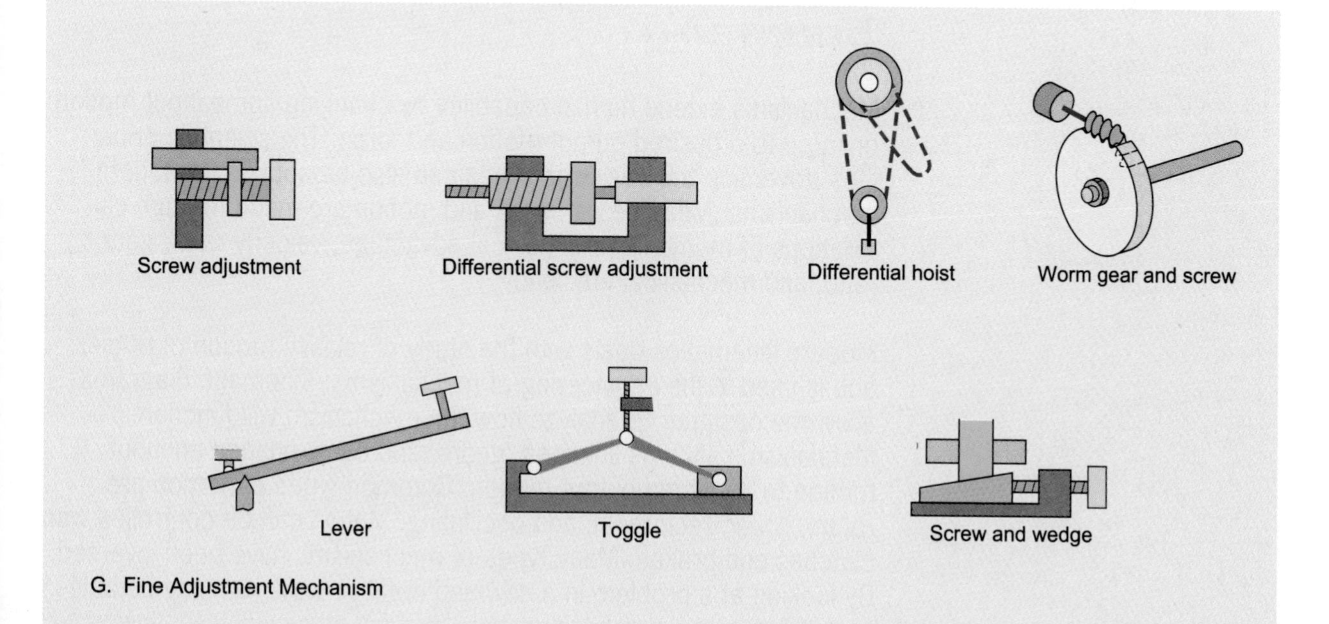

G. Fine Adjustment Mechanism

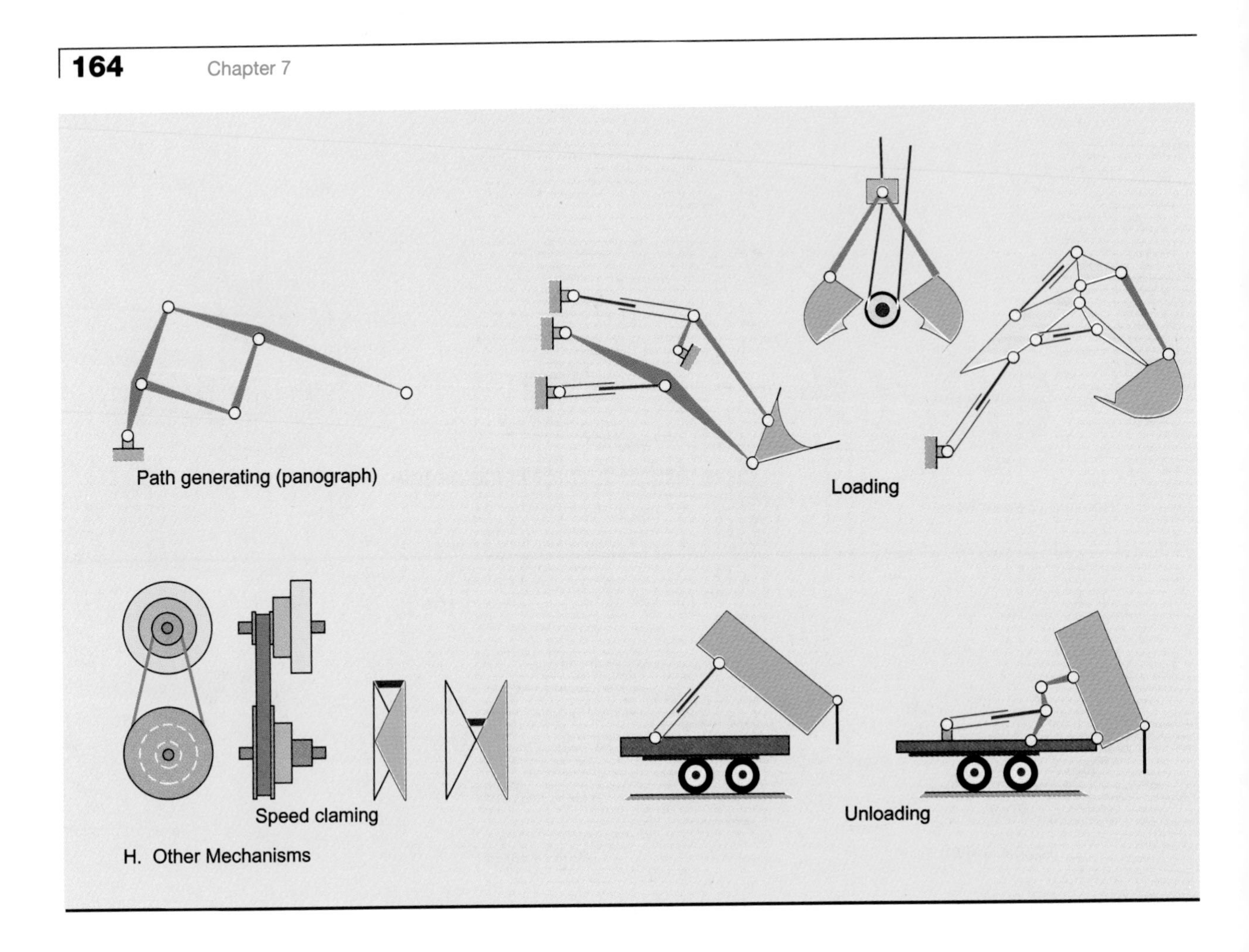

Summary

Mechanisms extend human capability by changing some input motion or force to a desired output motion and force. The scientific principles governing the basic machines can also be applied to modern mechanisms. Measures of force and motion are made through calculations of moments, mechanical advantage, velocity ratio, gear ratio, and mechanical efficiency.

Modern kinematics deals with the study of relative motion of bodies and is used in the engineering of mechanisms. Kinematic diagrams allow the designer to analyze how the mechanism will function. Mechanisms such as linkages, gears, and cams change an input motion to a desired output motion. Common types of motion are rotary, linear, reciprocal, and oscillating. Motion can be controlled with clutches and brakes. Many types of mechanisms have been invented. By looking at a problem in a new and creative way, you may be able to develop a new mechanism.

Observation/Analysis/Synthesis

1. Make a kinematic diagram of a lower and higher kinematic form of mechanism. Use some existing product, such as a home device, tool, or machine. Identify the input motion and force, as well as the output motion and force.
2. Select one of the mechanisms identified in Problem 1, and calculate its mechanical advantage and velocity ratio.
3. Using a modeling system, such as Lego™, Fischertechnik™, or Meccano™, design, construct, and test a mechanical system. Some suggestions are: wind thread on a bobbin; simulate the motion of a sewing machine needle; model the motion of a steam engine.
4. Design a cam that keeps a valve closed twice as long as it is open. The linear motion should be 1 centimeter.
5. Use low-relief modeling (Chapter 10) to show the escape mechanism in a mechanical clock.
6. Use the material in this chapter to design and document a mechanism that is part of an individual or group design problem.

CHAPTER

8 Designing Electronic Systems

Introduction

Most people realize the important role that electronics plays in our modern Technological Age. Fortunately, you don't need to understand all the electrical and electronic theories before you can use electronic systems to help you solve problems.

Electronic systems are a series of components that work together to control, monitor, or measure. As you learned in Chapter 1, a system consists of inputs, processes, and outputs (Figure 8-1). The input section of a simple microelectronic system may consist of one or more **sensors**. These sensors convert some physical phenomenon, such as heat, light, humidity, radiation, or magnetism, into an electrical **signal**. A signal is a term to describe electricity that contains some kind of information. The information may be as simple as a different voltage level of electricity or as complex as encoded information, such as that found in a telephone wire or computer cable.

Modifying and conditioning the input signal are the microelectronic devices that process or control an output electrical signal. Devices that *switch*, *time*, *compare*, and *amplify*, or any combination of these,

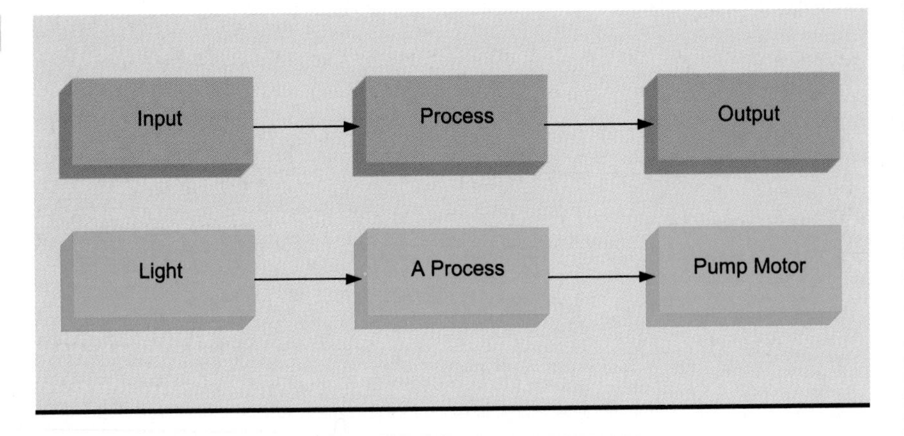

Fig. 8-1

An electronic system consists of inputs, processes, and outputs.

provide the control between the input and the output. We will limit our discussion of electronic processing to simple timing and switching in this text.

Output devices take the processed electronic signal and convert it back into some usable form. Output devices may be **transducers** or **actuators** that convert electrical signals into physical phenomenon. Examples of transducers are loudspeakers, buzzers, and lights. Actuators are devices that cause movement, such as motors and solenoids. Outputs may also be **displays,** which are devices that provide information for the human user. Displays may be as simple as a lamp or as complex as a computer display screen.

On a simple level, an electronic system may switch on a solar water heater pump when the sun is shining (Figure 8-2). In this system, a sensor detects the presence of light, because it contains a material that changes electrical properties when exposed to a light source. Because of the small amount of energy (in the form of photons from the sun) falling on the sensor, the electrical signal that results is very weak. This weak signal could not operate a pump motor, because the electrical requirements of a water pump are quite large. Therefore, the control part of this system must switch a rather large electrical current from another source to operate the pump successfully.

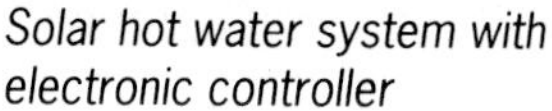
Fig. 8-2

Solar hot water system with electronic controller

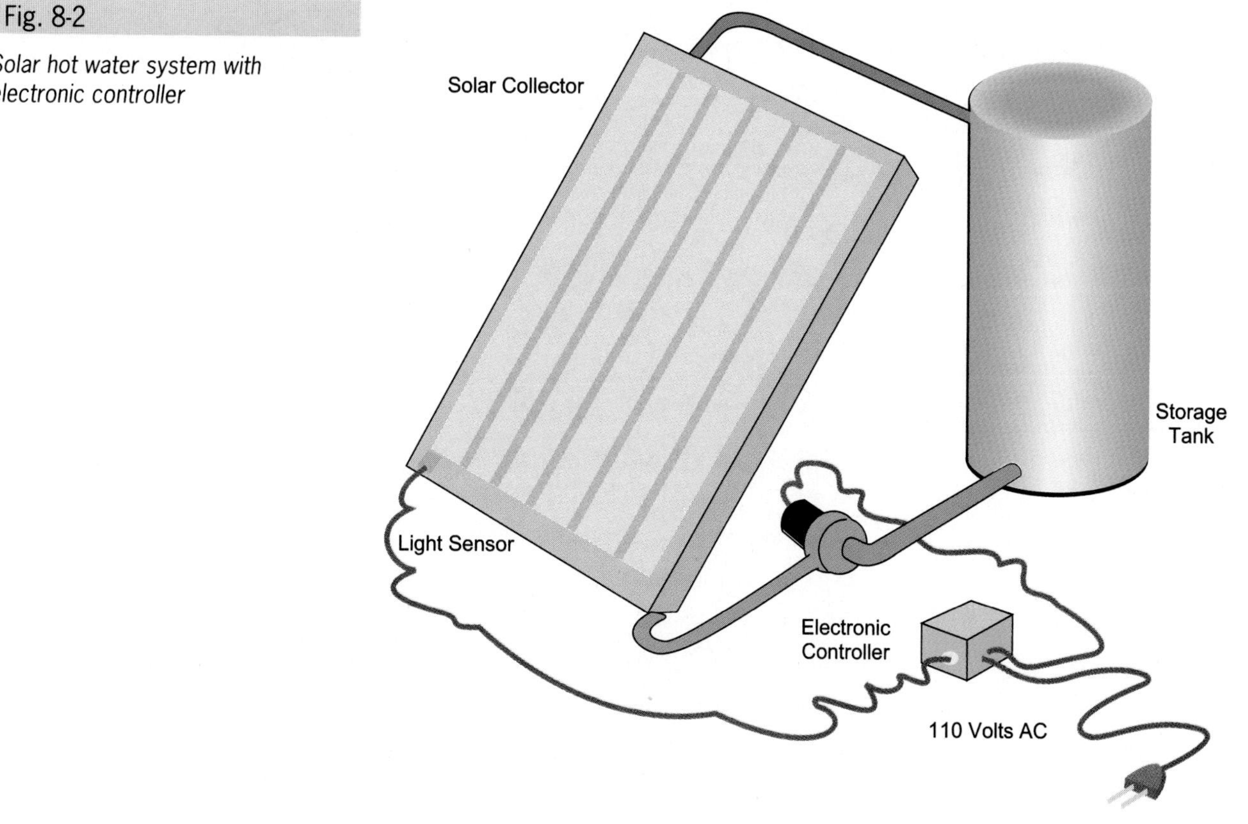

Important Terms

A few terms need to be defined to understand electronic systems. The first term is *electricity*, a word that we all use but that few of us understand. Electricity is the *movement of electrons* through a wire or other material that can easily carry electrons. Electrons are made to move by the powerful attracting and repelling forces produced by batteries, generators, and other devices that make electricity.

Direct current (DC) is electricity that flows in only one direction. Batteries, photovoltaic cells (solar cells), and some types of generators produce direct current. *Alternating current (AC)* is electricity that

Fig. 8-3

Ions are atoms that have gained or lost one or more electron.

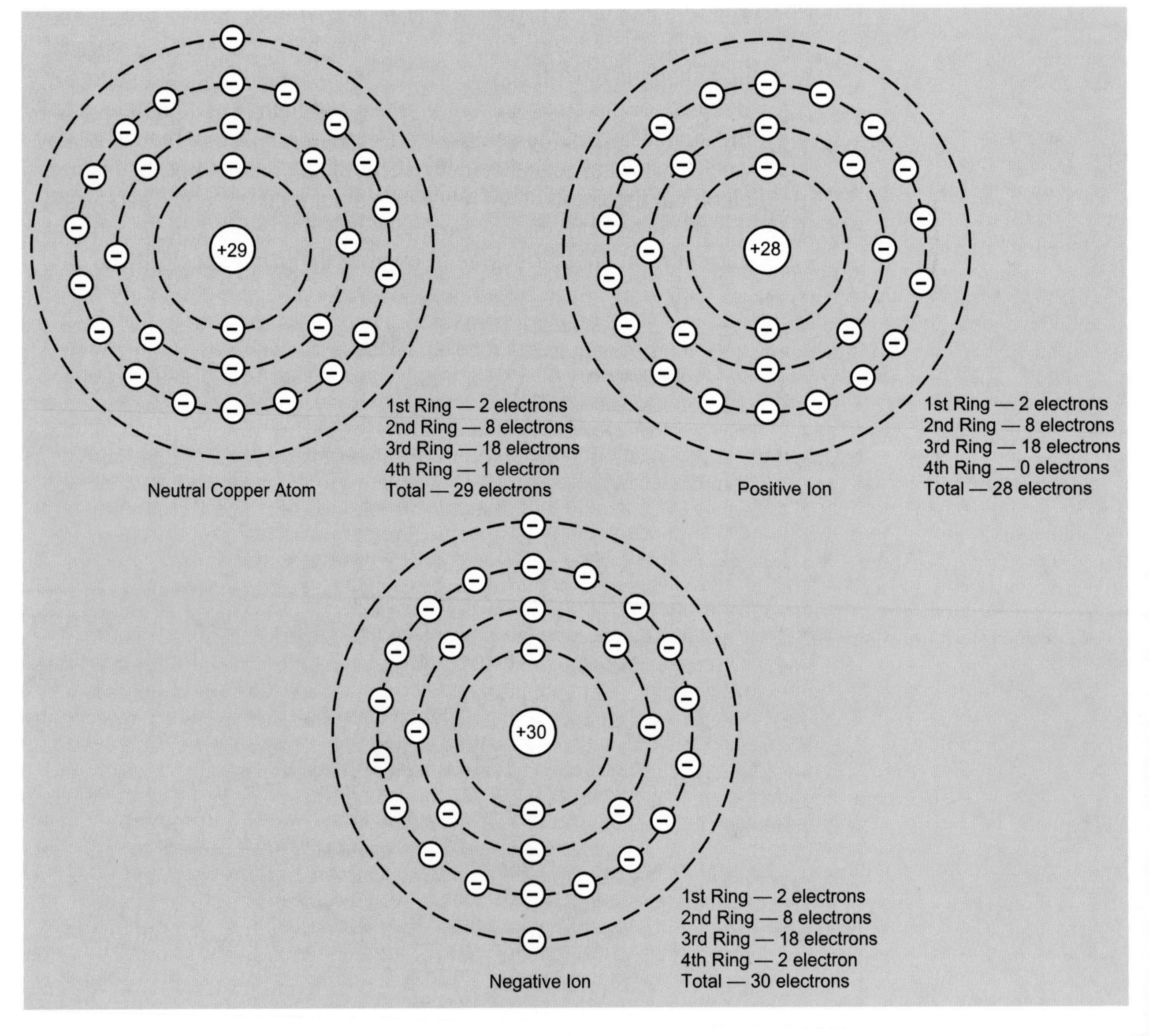

is constantly switching the direction of flow. The wall receptacles found at home provide *alternating current,* and most appliances, lights, and power tools operate on AC. While the basic principles of electricity apply to both DC and AC, this chapter will only deal with direct current.

To make electrons move requires that a **potential difference** exist between two points, such as the opposite ends of a wire in a simple circuit. The chemical action in a battery can provide this potential difference by forcing positive ions to the "+" terminal and negative ions to the "−" terminal. An **ion** is an atom or molecule that has lost or gained one or more electrons and is no longer neutral or in equilibrium (Figure 8-3).

As in other examples of nature, the atom will return to neutral if given the chance. When a path is provided between the two battery terminals, the extra electrons in the negative ions will travel to the positive ions, where a deficiency of electrons exist. This flow of electrons will continue until the battery can no longer convert chemical energy to electrical energy or until the path is disconnected. The flow of electrons has the capacity to do work.

Conductors are materials that easily carry electricity, and **insulators** are materials that do not carry electrons, or carry them very poorly. Copper wire is a very good conductor, as is silver and gold. Aluminum is a fairly good conductor, but it oxidizes over time and this can cause problems. Good insulators are glass, most plastics, and most ceramic materials.

Describing the Flow of Electricity

Three related terms that describe electricity are voltage, current, and resistance. **Voltage** is the *pressure* that drives electrons through a conductor and is measured in *volts (V).* It results from the potential difference between a positive and a negative terminal. **Current** is the *quantity* of electrons that flow through a conductor. Current is measured in *amperes (A).* It is common in microelectronics to be dealing with very small current values; therefore, you will often see current expressed in *milliamperes (mA).* A milliampere is equivalent to 1/1000th or 0.001 ampere. Current is affected by the potential difference between two terminal points, but it is also affected by the resistance to flow offered by different materials.

Resistance, the third value, is the *opposition* to the flow of electrons through a conductor. The atomic structure of a material affects its resistance, as does the length of the conductor and its temperature.

Resistance is measured in **ohms (R)** but you will often see the Greek letter omega (Ω) used to express ohms. The prefix *kilo-* is used to express resistance in the thousands of ohms and *mega-* for millions of ohms. It is not uncommon to find components in the kilohm or megaohm range.

The units of volts, amperes, and ohms are mathematically related in **Ohm's law**.

Ohm's Law

1 **volt** of pressure through 1 **ohm** of resistance results in 1 ampere of **current**. There are three ways of stating the relationship:

Voltage = Current × resistance or $V = I \times R$

Current = Voltage/resistance or $I = \frac{V}{R}$

Resistance = Voltage/current or $R = \frac{V}{I}$

V = Voltage in volts I = Current in amperes R = Resistance in ohms

Ohm's law is used to calculate an unknown value of either voltage, current, or resistance when the other two values are known. In the example presented, "Ohm's Law Problem," a typical problem involving Ohm's law is solved.

Ohm's Law Problem

Find the current in an automobile light bulb that has a resistance of 16 ohms and operates on 12 volts.

Step 1 Choose the correct formula:

$$I = \frac{V}{R} \quad \text{or} \quad \text{Current} = \frac{\text{voltage}}{\text{resistance}}$$

Step 2 Substitute known values in the formula:

$$I = \frac{12 \text{ volts}}{16 \text{ ohms}}$$

Step 3 Solve for unknown value:

$$I = \frac{12}{16} = 0.75 \text{ amperes}$$

Simple Ohm's law calculations

Fig. 8-4

A wristwatch with a pressure microsensor. The digital face provides altitude data. Courtesy of Casio

Fig. 8-5

Photovoltaic cells convert light energy to electrical energy. This panel of cells produces 12 volts and is capable of 30 watts of power.

System Inputs: Sensors

Electronic sensors detect the presence of some physical property or phenomenon. Some sensors may also detect the amount or quantity of the physical phenomenon. The presence of light, heat, magnetism, humidity, motion, strain, acceleration, position, and other properties are commonly detected by electronic sensors. Modern microelectronics has revolutionized the sensitivity and size of these input sensor devices.

Recent advances in sensor technology, called *microsensors*, have made it possible to incorporate a digital altimeter in a normal-sized wristwatch. No toy, this device is accurate to ±20 feet to over 13,000 feet in altitude (Figure 8-4). Other similar sensors monitor modern automotive engines and provide important operating information for the on-board computers. These computers control the fuel–air mixture and timing of the engine to minimize harmful emissions and to maximize fuel economy.

Properties and Sensors

Electronic systems need input from the real world to provide any useful function. Sensors provide this doorway between the physical world and the electronic world by converting certain physical properties into electrical signals. There are two basic types of sensors for our purposes here. The first type of sensor produces a voltage in response to the presence of some physical property. The second type of sensor changes internal resistance in response to some change of physical property.

Light. Light travels through tiny particles known as *photons*. Photons are smaller than electrons, and their existence is only theoretical. To detect the presence of light, a photovoltaic cell can be used as a sensor (Figure 8 -5). **Photovoltaic cells** are devices that produce a small amount of voltage when exposed to light. These are quite sensitive and are available in a wide variety of shapes, sizes, and outputs.

Another device sensitive to light energy is the **photoresistor**, or **light-dependent resistor (LDR)**. This device changes resistance characteristics according to the amount of light falling on it (Figure 8-6). If an LDR is connected in line with a small voltage source, the output current of the line will vary with the light hitting the sensor. LDRs are widely available and inexpensive.

Heat. Another form of electromagnetic radiation is heat. A **thermocouple** produces a small voltage when exposed to heat. It usually requires a rather high level of heat, ususally over 500 degrees F, to provide a signal and therefore is not very sensitive to small fluctuations in temperature.

The **thermistor** is another device used for detection of temperature changes (Figure 8-7). It works by changing resistance as it is heated. Thermistors are most often used to detect the presence of temperature change because they are widely available in a number of heat sensitivity ranges.

Sound. Sound travels through the atmosphere as a series of pressure changes. The *crystal microphone* produces voltage when these pressure changes cause a diaphragm to move and distort a quartz crystal, called "piezoelectric effect." It produces a relatively high voltage output and is therefore quite sensitive to low sound levels. The *dynamic microphone* produces a voltage when sound, striking a diaphragm, moves a coil through a magnetic field (Figure 8-8).

Position. A change from one location to another or a change in position can be detected by a sensor. A *switch* detects position, but most "on/off" switches can only detect one of two possible positions. However, there are a variety of switches useful in electronics work (Figure 8-9).

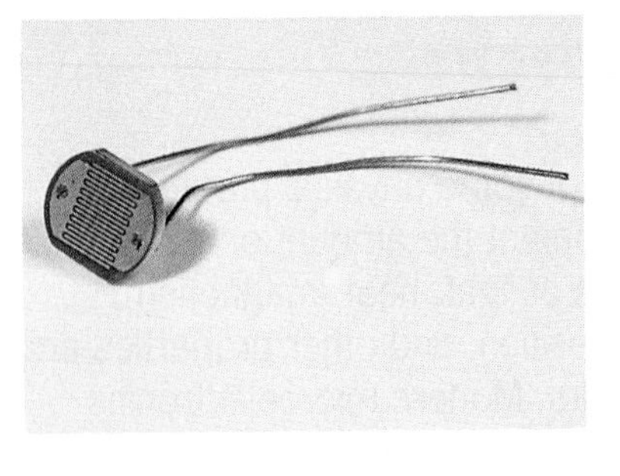

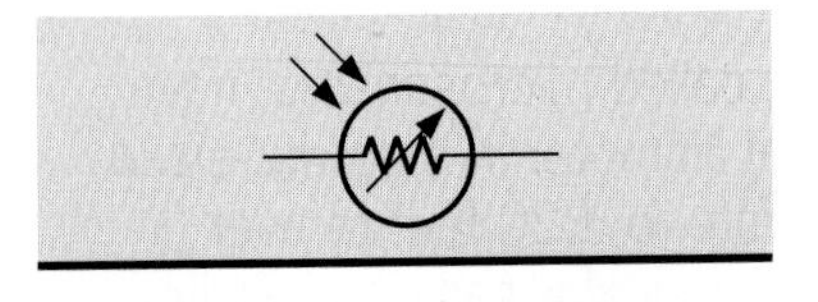

Fig. 8-6

A photoresistor, also known as a light-dependent resistor (LDR), changes its resistance with light level.

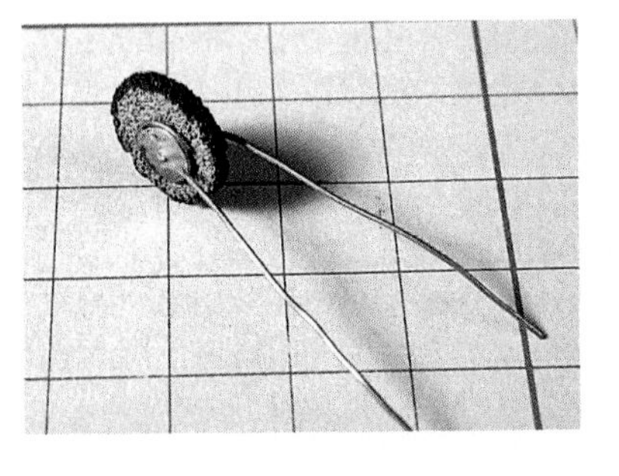

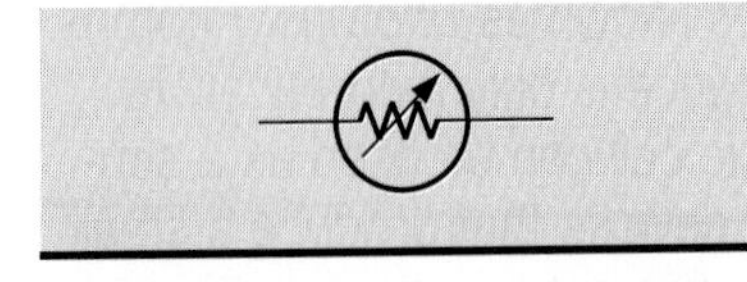

Fig. 8-7

A thermistor changes resistance with temperature.

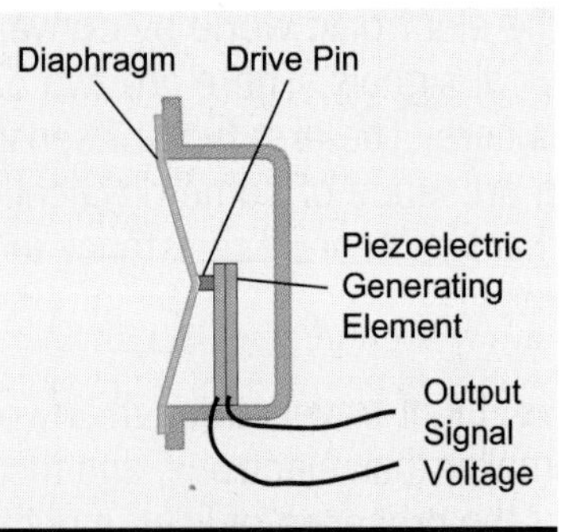

Basic structure of a piezoelectric microphone.

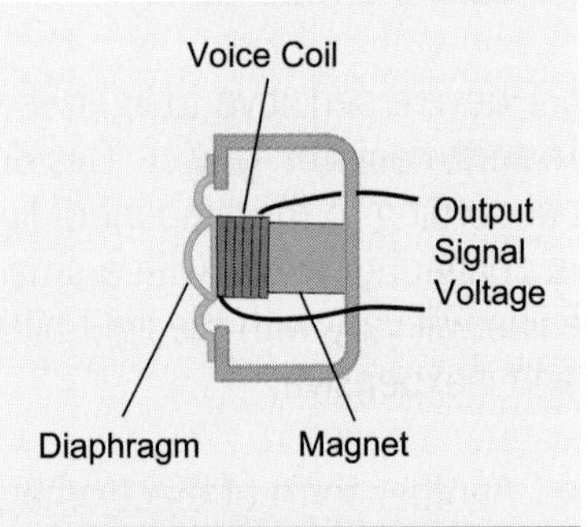

Basic structure of the dynamic (moving-coil) microphone.

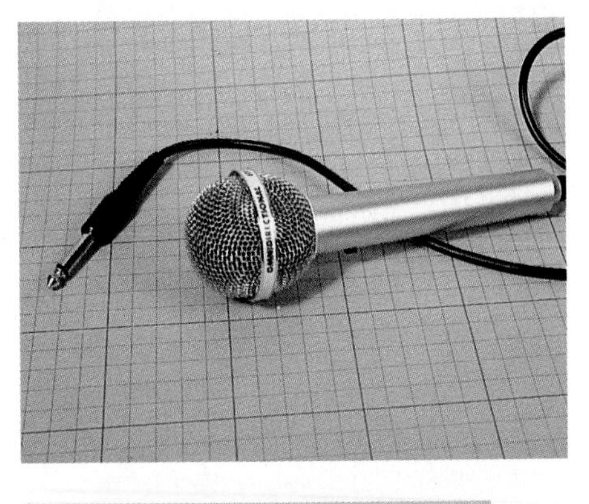

Fig. 8-8

Microphones change sound to an electrical signal.

Fig. 8-9

There are many types of switches for a wide range of applications.

On the other hand, a volume control on a radio can detect a variety of possible positions, either straight-line position, as a slide control, or rotary position. These variable resistors are called **potentiometers (pots)** (Figure 8-10). A potentiometer has a piece of carbon or other resistance material with an electrical connection on each end. In addition, there is a terminal on a movable "wiper" that rubs on the carbon material. The wiper's position moves with the potentiometer's sliding or rotating arm, and this changes the amount of carbon material, and thus the resistance, between the wiper and each end of the carbon piece. Potentiometers come as either a linear potentiometer, which means that the resistance is directly proportional to the position, or as an audio taper, a type of potentiometer that changes resistance logarithmically with position.

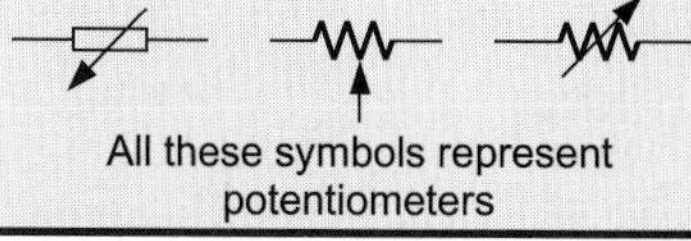

Fig. 8-10

Potentiometers are variable resistors.

System Outputs

Output devices may be grouped into *displays, actuators,* and *transducers*. Displays provide information, actuators make movements, and transducers convert an electrical signal into a physical property.

A light bulb is a simple display device. It can be used to show that a device, such as a computer, is on when it is lit. A motor is an actuator, because it rotates (movement) when electricity is applied. A speaker is

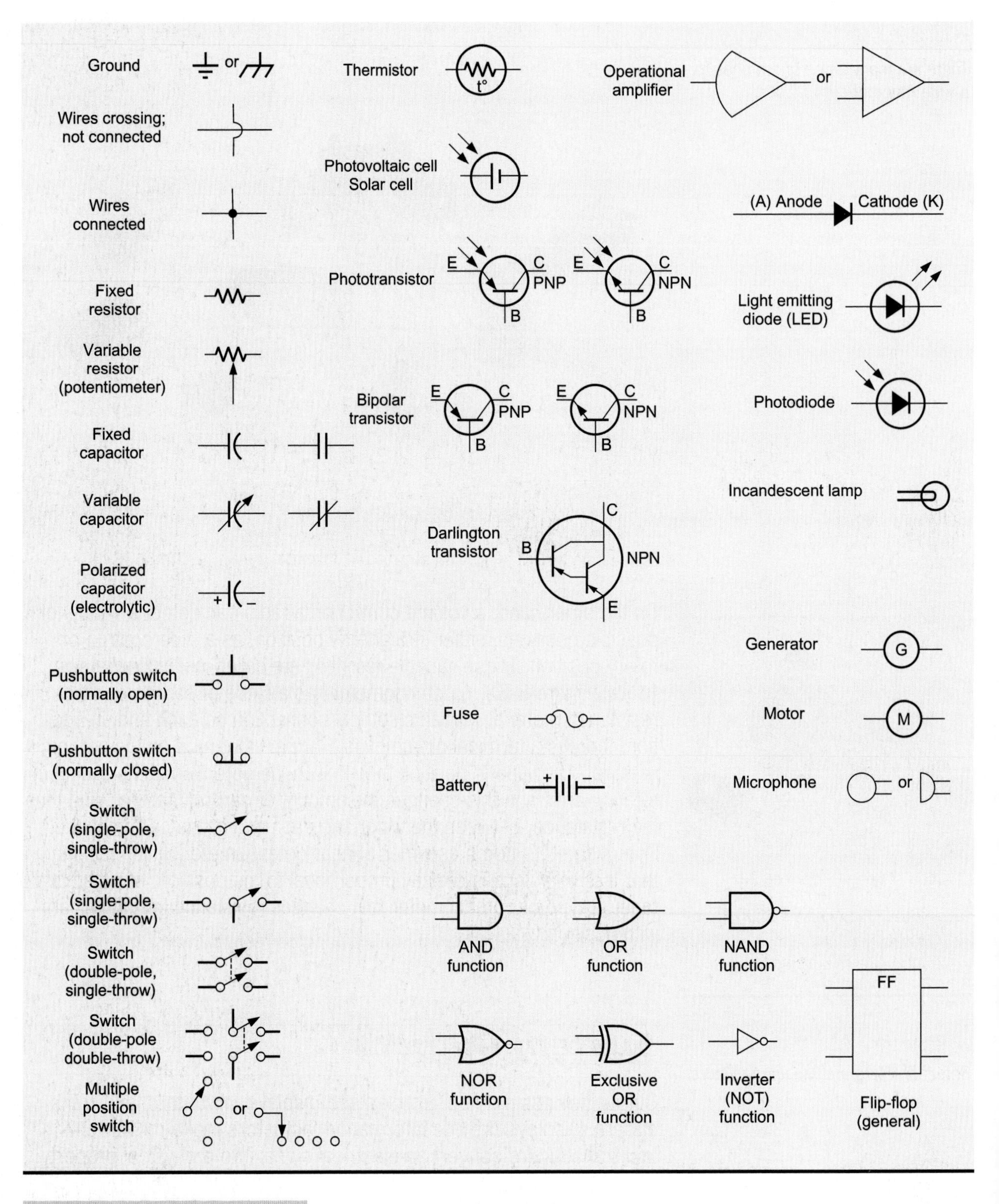

Fig. 8-11

Schematic symbols for common electronic components.

a transducer, because it converts a voltage signal into sound, a physical property. Here are some examples of these devices:

Displays	**Actuators**	**Transducers**
Meters	DC, AC motors	Loudspeaker
Lamps	Stepper motor	Headphone
LED	Electromagnet	Horn
Seven-segment	Relay	Buzzer
Solenoid	Solenoid	Heater

Displays

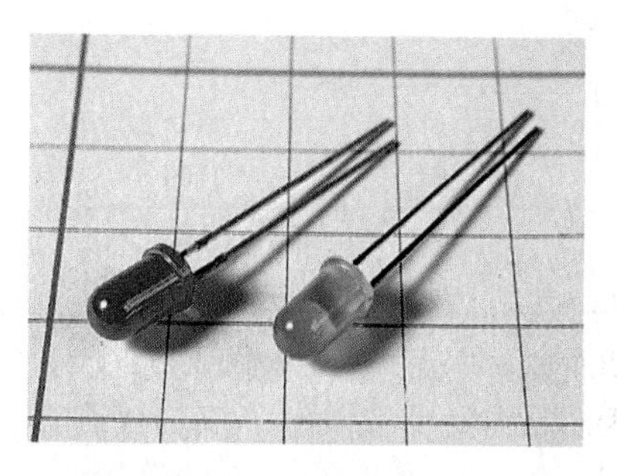

Fig. 8-12

Light-emitting diodes

Light-Emitting Diodes. **Light-emitting diodes (LED)** are devices that give off light using very little electrical current. They come in red, green, amber, and infrared (which is invisible to the human eye), and in a variety of shapes and sizes (Figure 8-12). They must be connected to a direct current (DC) power supply in such a way that the negative terminal is attached to the negative lead or **cathode** of the LED. Most LEDs have a flat spot next to the cathode wire (Figure 8-13). The other lead, the positive wire, is called the LED **anode**, and it is connected to the positive terminal.

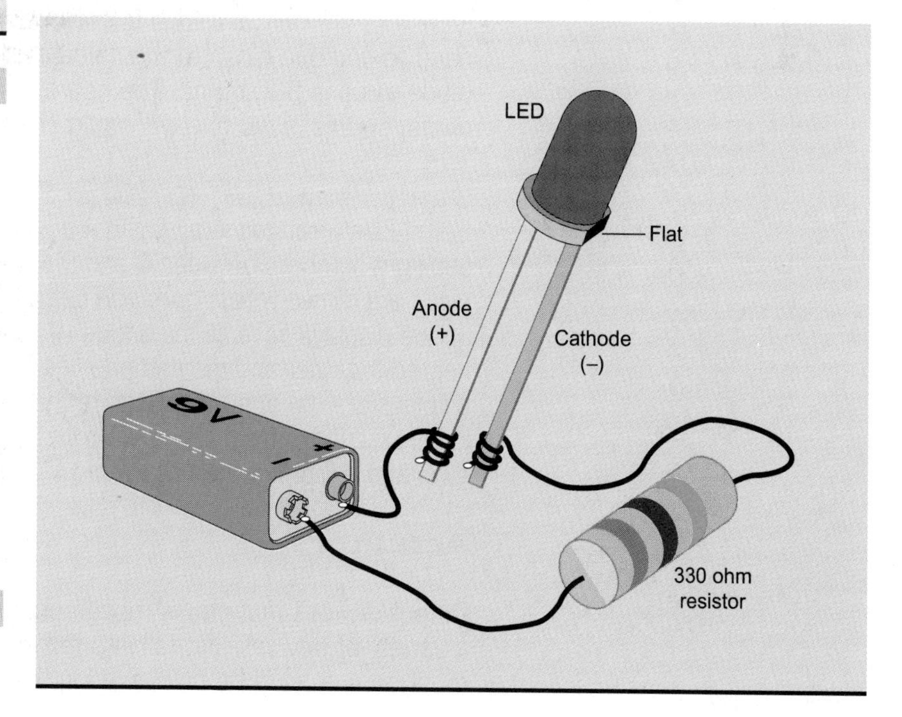

Fig. 8-13

A resistor must be used in an LED circuit to limit the current.

Fig. 8-14

Calculating the value of a resistor for an LED

Design Problem

A limitation of the light emitting diode (LED) is that it cannot tolerate current exceeding 20 milliamperes (0.020). What is the value of the resistor that will restrict the current to the LED to 20mA if you are using a 6 volt battery?

Use Ohm's Law to find resistance:

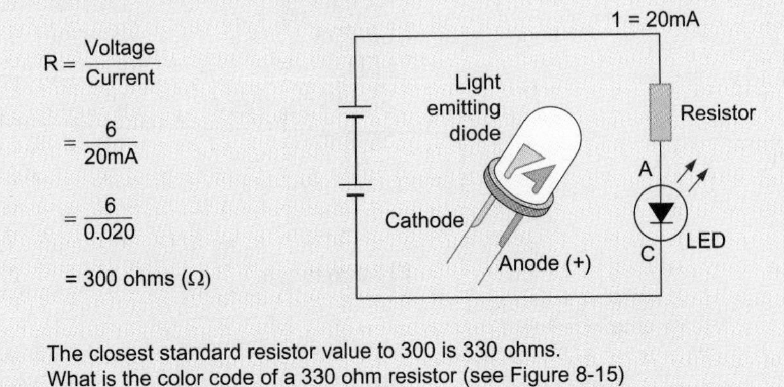

A light-emitting diode requires a current-limiting resistor, because it can only tolerate about 0.020 amperes (A) (20 milliamperes) of current. When used with a DC voltage supply of between 5 and 9 volts, a 330-ohm resistor is usually sufficient to limit the current through the LED. Figure 8-14 shows the Ohm's law calculation to determine the current flow through the LED and resistor.

Resistors are only available in certain standard values, so you will need to find one close to the calculated value. The calculated resistance value is 300 ohms. The nearest standard resistor value is 330 ohms, so this is the one we would choose.

Seven-Segment Displays. *Seven-segment displays* are made from LEDs arranged in a way that powering combinations of the LEDs can light numbers between 0 and 9 (Figure 8-16). Seven-segment displays are of two types: *common cathode* and *common anode*. Common anode displays have all the anode (+) connections of the LEDs connected together and the cathode (−) leads separate. Common cathode displays are the opposite. Each of the seven segments is labeled with a lowercase letter from *a* to *g*. To light a 3 you will need to light the *a*, *b*, *g*, *c*, and *d* (but not the *f* or *e*).

Actuators

Output devices that cause movement are called *actuators*. There are a variety of devices that fall into this category, but only a few of the most common will be discussed in this section.

Fig. 8-15

Resistor color code

COLOR	FIRST BAND	SECOND BAND	THIRD BAND (MILTIPLIER)	FOURTH BAND
BLACK	0	0	1	–
BROWN	1	1	10	–
RED	2	2	100	–
ORANGE	3	3	1000	–
YELLOW	4	4	10,000	–
GREEN	5	5	100,000	–
BLUE	6	6	10.000,000	–
VIOLTE	7	7	100,00,000	–
GRAY	8	8	1,000,000,000	–
WHITE	9	9	.1	5
GOLD	–	–	.01	10
SILVER	–	–		20
NONE	–	–		

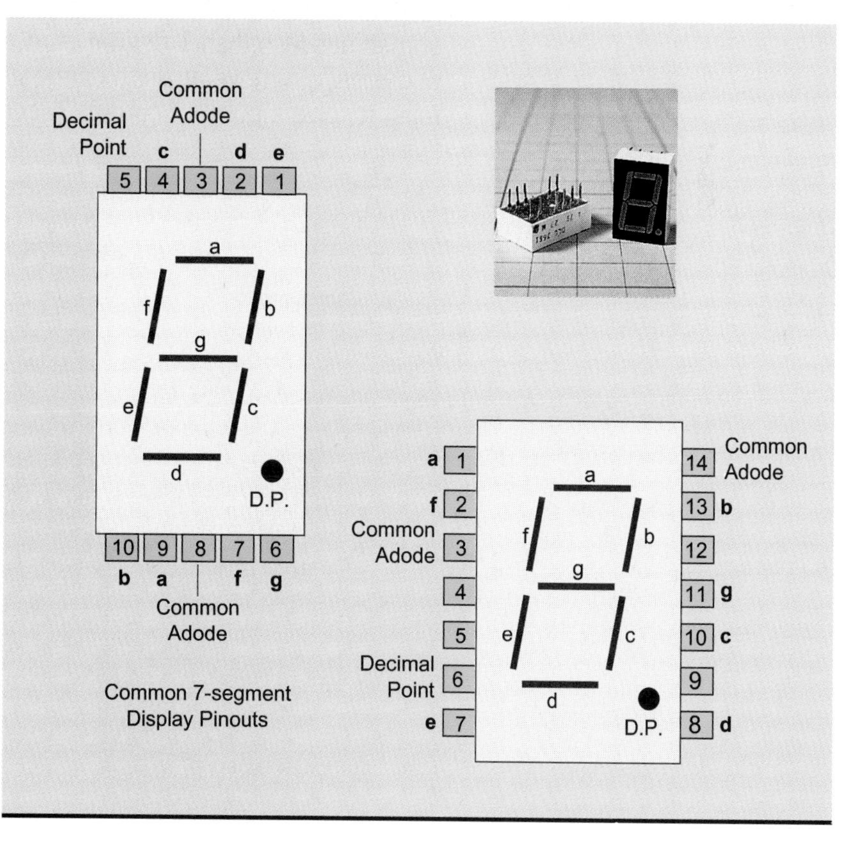

Fig. 8-16

A seven-segment LED display

Fig. 8-17

DC motors come in a variety of sizes and shapes.

DC Motors. DC motors are used to make rotary motion. They come in an almost endless variety of sizes and shapes, and can be purchased for as little as a dime. Most small DC motors contain permanent magnets and electromagnets (Figure 8-17). The DC voltage applied to the motor produces an electromagnetic field within one coil on the **armature**, which is a number of different coils wound on a rotating shaft. The magnetic field produced repels a permanant magnet fixed inside the shell of the motor, and this causes the shaft to rotate a partial turn. As the shaft turns, the electrical contacts, called the **commutator**, switch the DC voltage to another coil of wire in the armature, which causes the electromagnetic field to be produced in a slightly different location on the armature (Figure 8-18). But, because the shaft has turned a bit, this location is back where the original field was produced when the motor was first connected to the voltage source. The like magnetic fields of the electromagnet and the fixed permanent magnets now repel again, and the armature and shaft rotate a little more. Again, the electrical contacts switch the DC voltage to another coil of wire in the armature, and the whole process repeats. This repeating process causes the armature and the shaft to continue to rotate until the voltage source is disconnected.

Stepper Motors. **Stepper motors** also operate on DC voltage but use a special control circuit to feed the six or eight wires needed to function (Figure 8-19). The stepper motor moves a tiny increment, sometimes less than 1 degree of rotation, each time it receives a pulse of current. To move again, a different combination of the feed wires must receive a pulse of current. A control circuit is necessary to coordi-

Fig. 8-18

DC motor operation

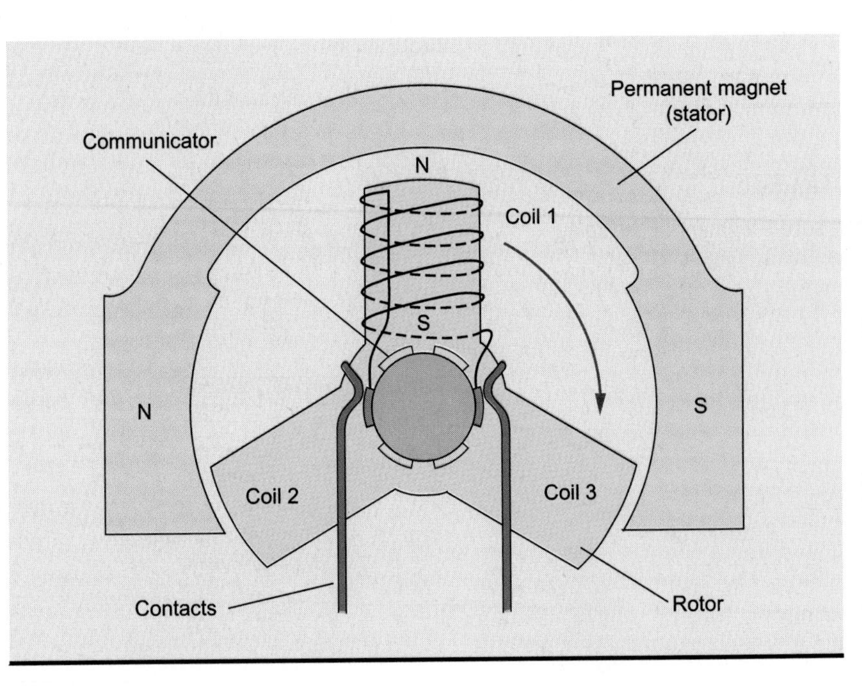

Fig. 8-19

A stepper motor provides precise rotation control.

nate the pulses to the correct input wires, and it is possible for the stepper motor either to turn continuously or to turn to a specific spot, stop, and rotate again to another location. A computer interface can be used to control a stepper motor with a simple program. In addition, there are special integrated circuit chips that are designed to control stepper motors.

These devices are ideal for moving the printing head on a dot matrix printer or positioning the table on a computer numerical control (CNC) machine. While accurate, stepper motors are not capable of much torque and are usually quite expensive when purchased new. They are, however, readily available and inexpensive as surplus.

Solenoids, Relays, and Electromagnets. **Solenoids** are devices that produce rapid linear motion over a short distance, from about ⅛ inch to several inches (Figure 8-20). They are used to engage mechanical mechanisms or gears, or to operate valves. Solenoids consist of a magnetic coil wrapped around a hollow core, with a sliding armature within the core. As the magnetic field is energized by applying a voltage across the windings, the armature moves. Some solenoids are capable of exerting a great deal of force. An example of a powerful solenoid is one that engages and disengages the starter motor gear to the flywheel gear of an automobile engine. Others are used for more precise movement and can be found in video cassette recorders and many audio cassette tape decks.

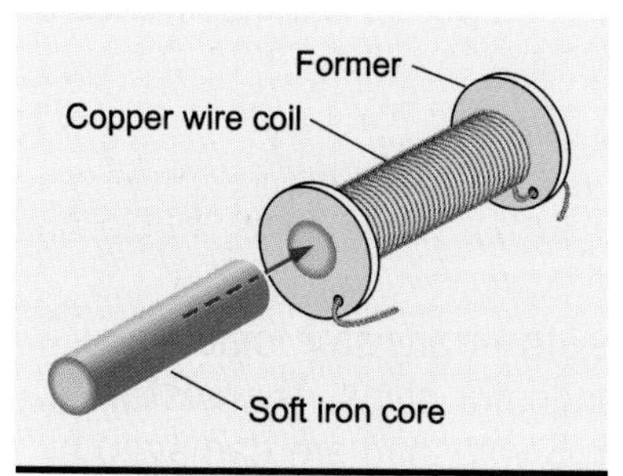

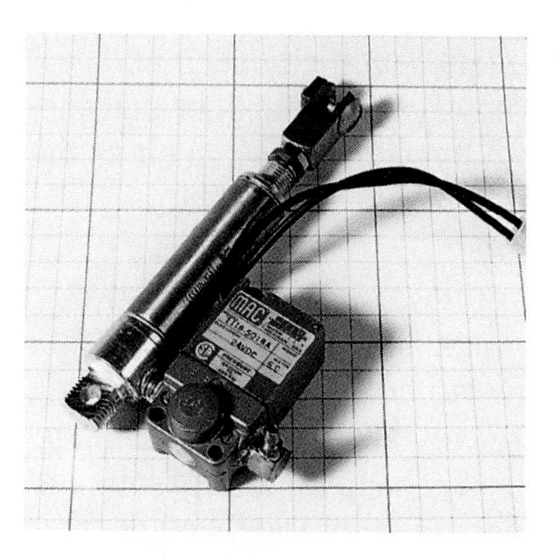

Fig. 8-20

An electric solenoid used to control the air inlet to a pneumatic cylinder

Closely related to the solenoid is the **relay**. It acts as a switch, because electrical contacts are connected to the armature (Figure 8-21). A relay is often used as a remote switch to control one circuit with another. For example, a relay would allow a large current device, such as a heavy-duty motor, to be controlled by a small switch. The starting circuit of many automobiles contain relays, because the current requirements of most starter motors is over 300 amperes. Wires that can carry this much current are usually ⅜ inch or larger in diameter, with the necessary insulating cover. To overcome the problem of switching this large current, a relay is used that is controlled by the ignition key switch.

Electromagnets are also actuators. Current running through a wire causes a magnetic field to form around the wire. Winding wire around an iron core concentrates the magnetic field into the core, and powerful electromagnets can be constructed (Figure 8-22). The strength of the magnetic field depends on essentially three factors: (1) the core material; (2) the number of turns of wire around the core; and (3) the amount of current flowing through the wire, which is limited by both the diameter of the wire and the capability of the voltage supply.

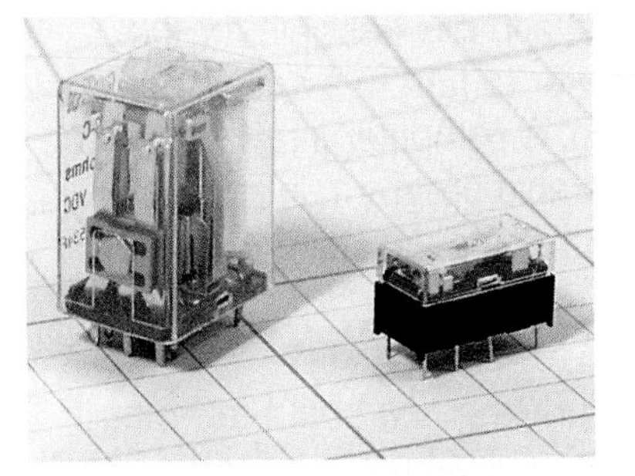

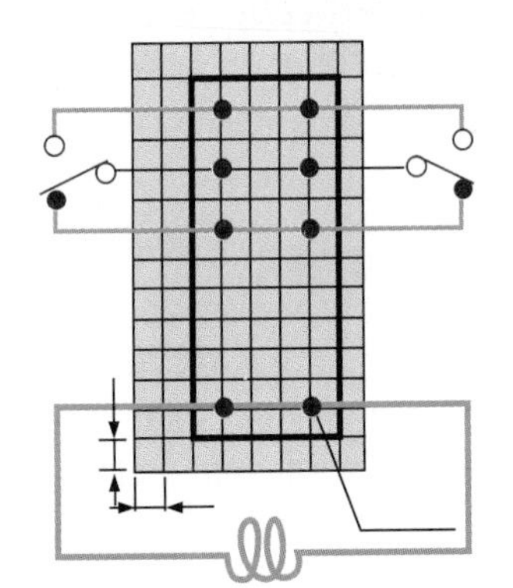

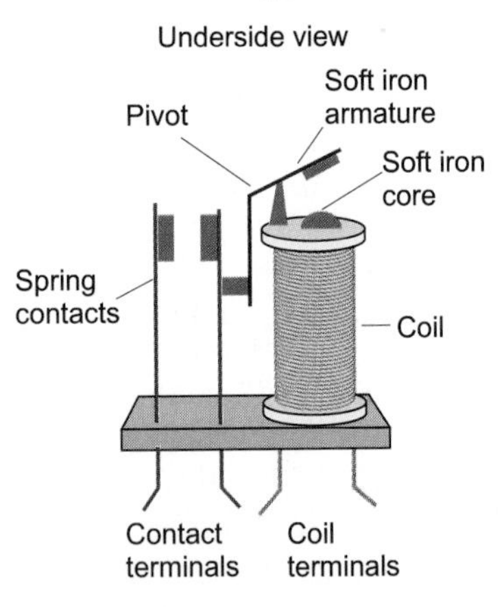

Fig. 8-21

Relays are useful switches.

Fig. 8-22

An electromagnet

Electromagnets may be used for all sorts of interesting output devices. They may be used on a robotic arm to retrieve and release metallic objects, to sort magnetic from nonmagnetic objects, to cause motion in devices by attracting a magnetic object, and many other applications.

System Processors

The part of the circuit that uses the input signals from sensors and that controls the output devices accordingly, is the control or process part of the system. These subsystems are often made up of circuits that *switch, amplify, compare,* or *time*. Complete electronic systems may combine two or more of these functions. In modern electronic systems, the transistor plays an important role in these control functions.

Three American physicists—John Bardeen, Walter Brattain, and William Shockley—invented the transistor in 1947. During the early 1960s, manufacturing techniques were developed to put many transistors onto one chip. This led to the modern **integrated circuit (IC)** (Figure 8-23). While the ICs referred to in this text have only a dozen or so transistors on one chip, ICs used in computers can have several hundred thousand transistors per chip.

Integrated Circuit Types

Two common types of IC chips are transistor–transistor logic chips and complementary metal oxide semiconductor chips. **Transistor–transistor logic (TTL)** chips operate on 5 volts and are damaged by voltages greater than 6 volts. These chips are well suited for project work, because they are not sensitive to static electricity such as found in dry, warm conditions. They do, however, have higher current requirements. Digital chips that begin with the number 74, such as the 7447 and 7490, are TTL ICs.

Complementary metal oxide semiconductor (CMOS) chips will work with voltages between 4.5 and 16 volts and consume very little current, so they are ideal for battery-powered projects. The main disadvantage of CMOS chips is that they are very susceptible to damage from static electricity. These devices must be handled with great care and stored in special conductive envelopes or pads. Digital chips that begin with the numbers 40 or 45, such as the 4001 and the 4518, are typically CMOS ICs. TTL and CMOS chips are, in most cases, not compatible, so you will need to choose the type that is most appropriate for a particular application.

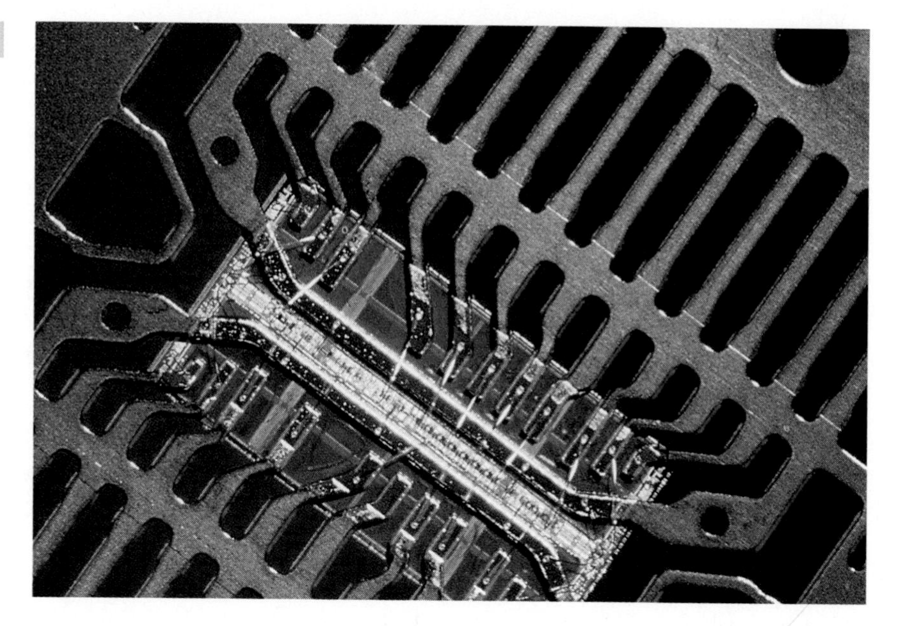

Fig. 8-23

Integrated circuit (IC) chip. Courtesy of International Business Machines Corporation

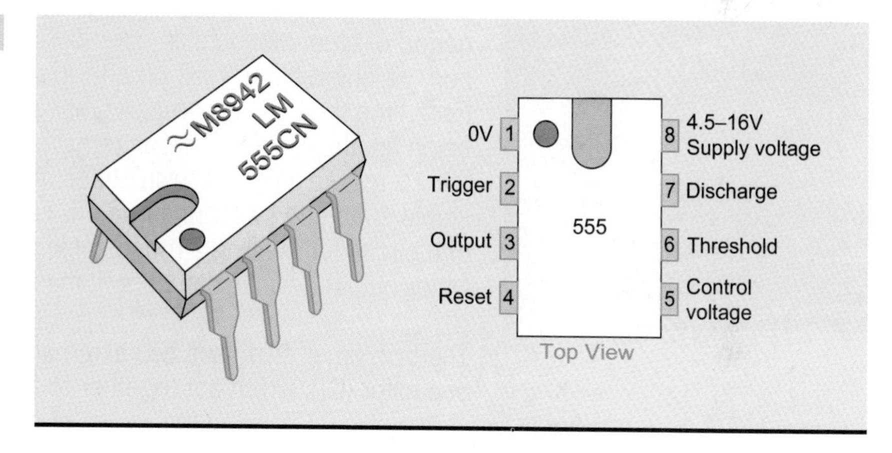

Fig. 8-24

The 555 timer chip

Simple circuits can be developed using IC chips to control various functions associated with electronic control.

The 555 Timer Chip

The 555 IC uses CMOS technology and is one of the most popular IC chips. It can be used to provide a pulse of timed duration, which means that it can be used to turn on something for a period of time. Using the chip in this mode is called "one-shot," or *monostable operation*. The chip can also be used to provide continuous timed pulses. This is called "multivibrator," or *astable operation*. The 555 timer chip is usually in an eight-pin, dual in-line pin (DIP) package (Figure 8-24).

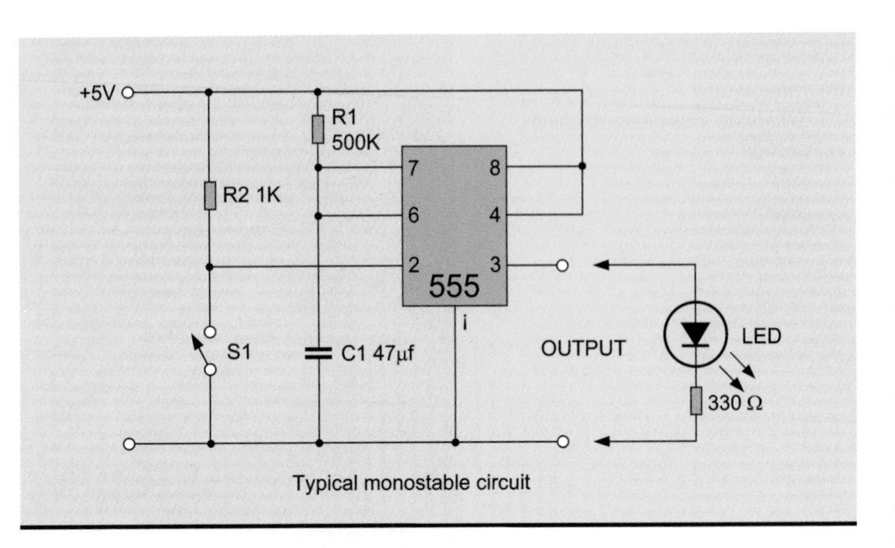

Fig. 8-25

A schematic diagram of a monostable timing circuit that will turn on an LED for a period of time

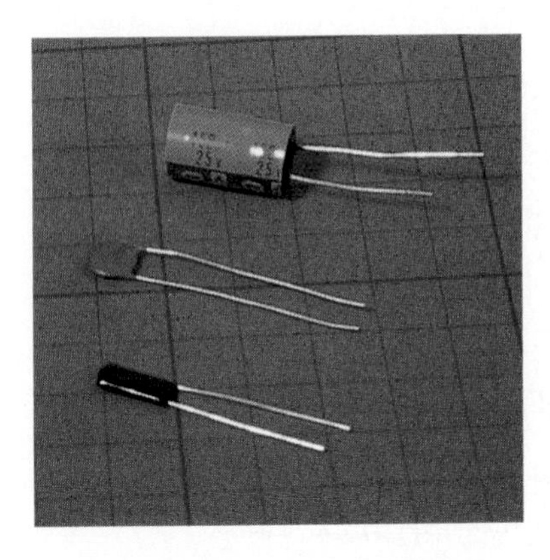

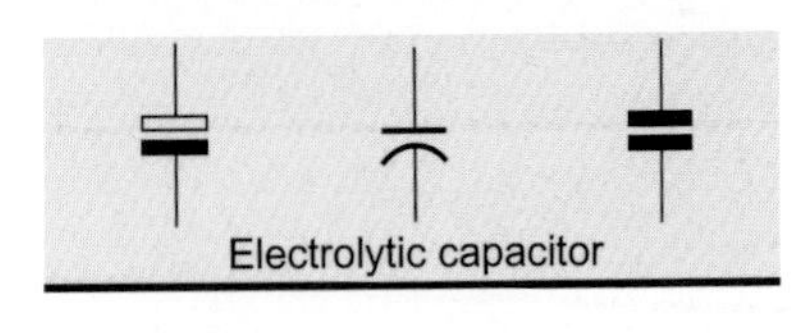

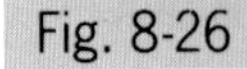

Fig. 8-26

Capacitors store an electrical charge.

Monostable Operation. A **monostable timer** is a one-stable timer; that is, it returns to its original (stable) state (off) after a certain period of time. Monostable operation is characterized by the clothes iron, which switches itself off after being unattended for 10 minutes. Each time the iron is moved, a small sensor resets the timer to zero to begin another count. The iron remains on as long as it is moved within the 10-minute limit. If it hasn't been moved, the circuit will return to its stable state and switch the iron off. Figure 8-25 is a schematic diagram of a circuit for building a monostable timer that switches on an LED. Other output devices can be switched on by this circuit.

The 555 timer chip uses two external components, a resistor (R_1) and a capacitor (C_1), which act together to form a circuit that times the monostable pulse. Changing the values of either or both of these components will change the output pulse duration. **Capacitors** are components that hold a charge of electricity (Figure 8-26). **Resistors** are components that oppose the flow of current. By using a resistor in the circuit, the time it takes to charge or discharge a capacitor can be changed.

By changing the time it takes for the capacitor to charge, you can change the timing duration of the output of the 555 timer chip. This can be done by replacing the capacitor with one of larger or smaller value, or by replacing the resistor with one of larger or smaller value. Figure 8-27 is a chart for the value of C_1 and R_1 for different timing durations.

Astable Operation. An **astable timer** is one that is constantly pulsing, never staying in one state or the other. It is constantly switching between its two states. Astable operation is characterized by the flashing light or pulsing buzzer associated with starting a car warning

Fig. 8-27

Chart of resistor and capacitor values for monostable timing durations

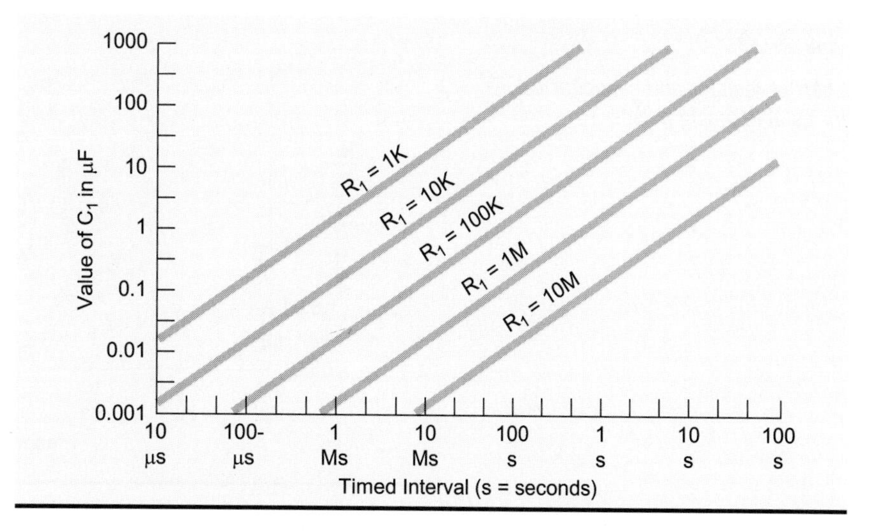

Fig. 8-28

A schematic diagram of an astable timing circuit to blink an LED

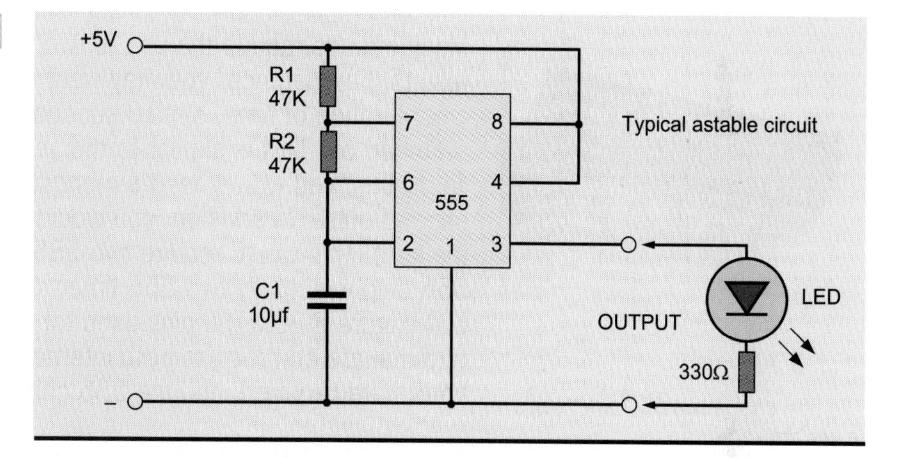

that the seat belts are not fastened. In this mode, the output pulse is fed back into the chip to trigger another pulse, and the result is a continuous pulsed signal. Figure 8-28 is a schematic diagram of a circuit for building an astable timer that blinks an LED. Other devices, such as a buzzer or a speaker, can be used in place of the LED for specific applications.

In astable operation, the 555 timer chip continuously switches between two states: zero volts and 9 volts (if 9 volts is the input voltage). Each 9-volt state is one pulse. How rapidly the pulses occur is called the **frequency**, which is measured in **hertz**. One hertz is equal to one pulse per second; 1,000 hertz is equal to 1,000 pulses per second, or 1 kilohertz (1KHz).

The 555 timer chip is capable of providing this output pulse up to 1 million times per second, or 1 megahertz, depending on the values of the components R1, R2, and C1 in Figure 8-29. Of course, for visual flashing of an LED, a much slower pulse rate would be required.

Fig. 8-29

Chart of resistor and capacitor values for astable timing pulses

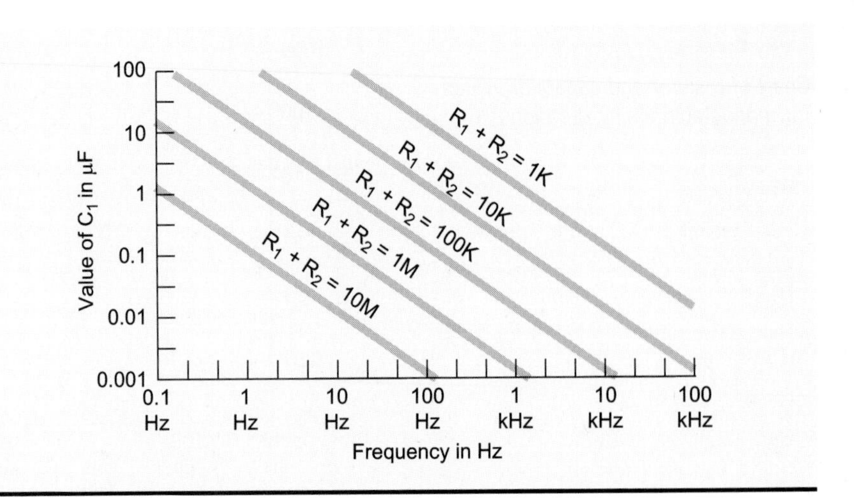

Fig. 8-30

The 556 chip is two 555 timer chips in one package.

Many circuits require the use of two timer chips to achieve the desired results. For instance, you may want to have an LED flash for a specific duration of time, say 10 seconds, each time the circuit is switched on. This is similar to the seat belt warning mentioned earlier. To do this, you would need a monostable circuit designed for one 10-second pulse. In addition, you would need an astable circuit to flash the light. This would require two 555 timer chips. Conveniently, the 556 chip is actually two 555 timer chips in one 14-pin integrated circuit (Figure 8-30). The pins used for +DC voltage in and ground (−) terminal are common to both internal circuits. In many circuits, two timers are required, and the 556 chip makes a useful package.

Counting Circuits

Counting circuits are useful in microelectronics. For example, most microwave ovens use a timing and counting circuit for displaying the time until it is shut off. First, you set the number of minutes and/or seconds that are needed to heat the food. Pressing a start button begins the cooking process. A display on the oven begins counting down from the time you have set. When the display reaches zero, the oven shuts off.

Counting circuits are developed by using IC chips made specifically for that purpose. Using a seven-segment display output, a 7447 TTL chip will light the correct segments for a common anode display when it gets a *binary-coded decimal (BCD) input.* A 4511 performs the same function in a CMOS package. Figure 8-31 illustrates a counting circuit using TTL integrated circuits and a common anode seven-segment display.

Binary Numbers

All digital electronic systems are based on the **binary number** system. The reason for this is simple: The binary system has two number digits: 0 and 1. A wire can have two conditions: off and on. Therefore, it is relatively simple to represent binary numbers in electrical circuits. In our decimal system—the one we all use to count—we have ten numerals: 0 to 9. When we get to 9 and want to count to the next higher number, we combine the numerals we have into a combination of 1 and 0 to get 10.

When we have exhausted all the combinations of two numerals to 99, we combine three numerals, 1, 0, and 0 into the number 100. Binary numbers use the the same principle, but, instead of ten different numerals, we must work with only two: 0 and 1. If we begin counting from zero, we have 0, 1, and then we run out of numerals. Using the principles of the decimal numbering system, we now need to make a combination of the two numerals we have, so the next number in our sequence is 10, a combination of 1 and 0. However, the number 10 means 2 in binary, because it comes in sequence after 1. Just as in the decimal system, each binary digit has a value according to its place in the number.

In the decimal number 234, the 4 is in the 1's place; the 3 is in the 10's place and the 2 is in the 100's place. We read the number as "two hundred thirty four." The binary number *101* is read in a similar manner, but, except for the 1's place, each numeral has a different value from the decimal system. The one at the right is in the 1's place; the zero in the center is in the 2's place; and the one on the left is in the 4's place. We read the number as "five" because 1 and 4 is equal to 5. Here is a number sequence counting from zero to seven in binary: 000, 001, 010, 011, 100, 101, 110, 111.

It is easier to understand binary numbers when we can look at a chart comparing decimal and binary numbers together. The number 0010101 in binary is twenty-one (21) in decimal because it has a 1 in the 1's place, the 4's place, and the 16's place. Each numeral place is double the one to its right, so beginning with 1, the next place to the left is 2; the next is 4; the next is 8, and so on.

Comparison of Decimal and Binary Numbers

Decimal	Binary	Decimal	Binary
0	00000	11	01011
1	00001	12	01100
2	00010	13	01101
3	00011	14	01110
4	00100	15	01111
5	00101	16	10000
6	00110	17	10001
7	00111	18	10010
8	01000	19	10011
9	01001	20	10100
10	01010	21	?

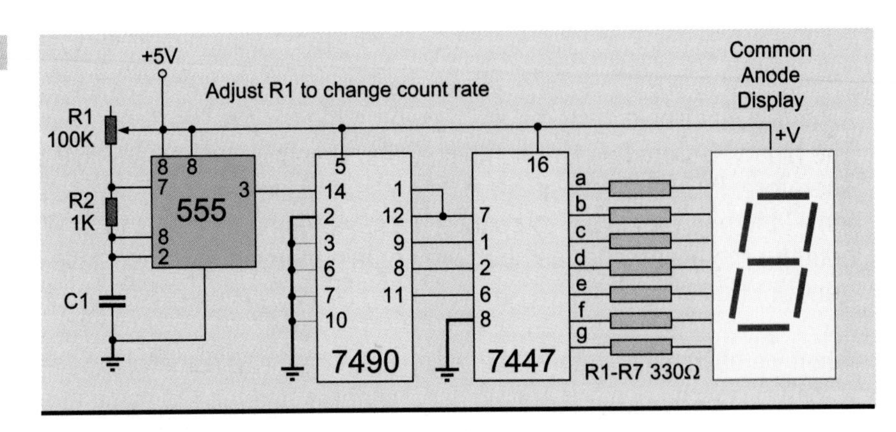

Fig. 8-31

A schematic diagram of a 0–9 counter using TTL chips

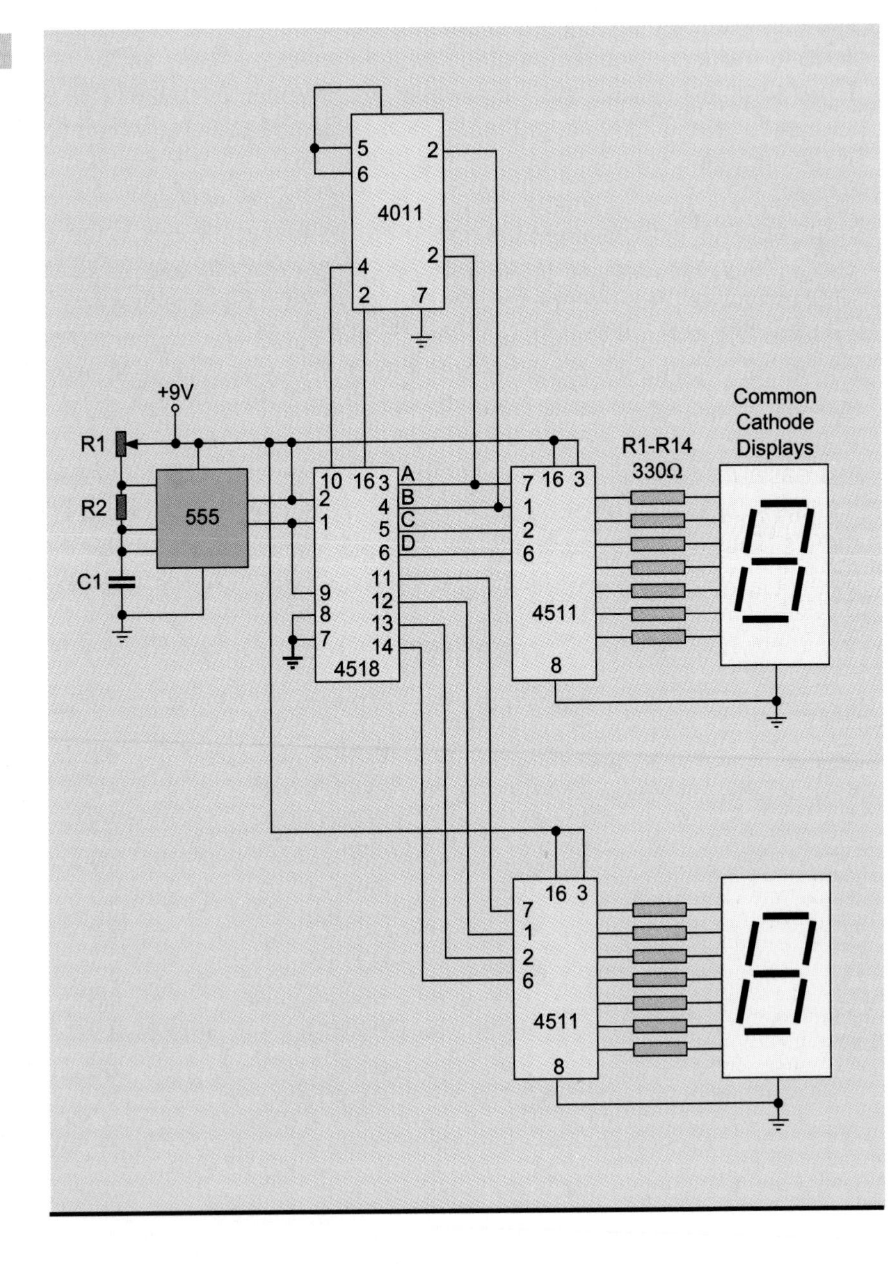

Fig. 8-32

Schematic diagram of a 0–99 counter using CMOS chips

A BCD signal in this case is four wires, each of which is either high or low, which represent the 1, 2, 4, and 8 binary bits. Another chip, the 7490, provides the binary-coded decimal output when it receives a pulse, such as that from the output of a 555 timer chip. The 4518 performs this function in CMOS. The complete circuit is a counting system that will count up to 9, in increments of 1, as each pulse is received from the 555 timer chip. By combining two or more of these circuits, you can build a system to count to 99, 999, or 9,999. Figure 8-32 illustrates a 0–99 counting system using CMOS integrated circuit chips and common cathode displays.

Switching Circuits

Transistors. The ability to switch or control one circuit with another is one of the cornerstones of electronics **Transistors** are semiconductor devices that can switch large currents on and off by using a small control current.

The control current may be from a sensor, such as the light sensor in a solar hot water system. A light-dependent resistor (LDR), which changes resistance according to the amount of light falling on it, detects the presence of light. As the level of light changes, this device causes a small change in the current entering the base of a transistor. This change in current switches the transistor on or off. The transistor is then used to control the current going to the DC pump motor in the solar hot water system. Figure 8-33 shows a simple control system that uses the LDR, a transistor, and a DC motor. If an AC motor were used, a relay would have to be added to the circuit, because the transistor cannot switch alternating current.

Fig. 8-33

Schematic diagram of a simple solar hot water heater controller

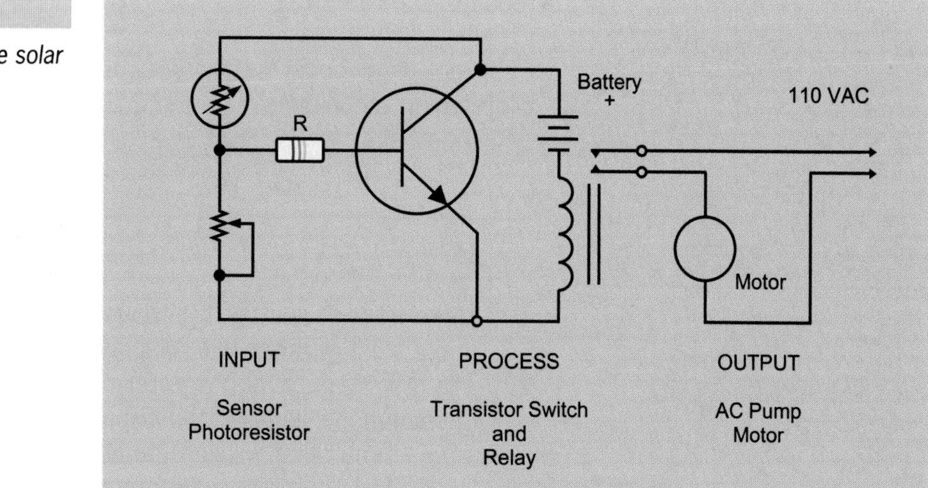

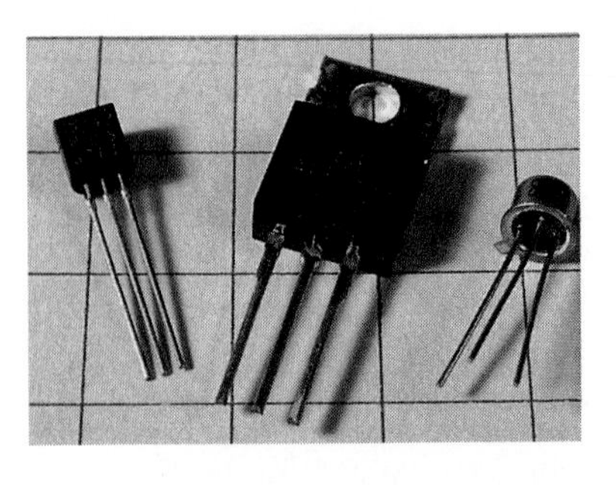

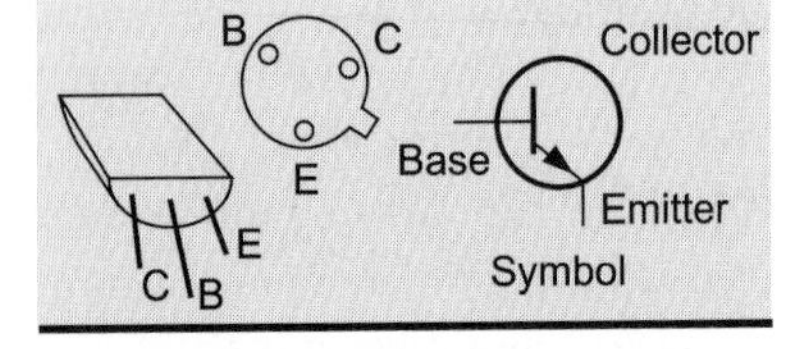

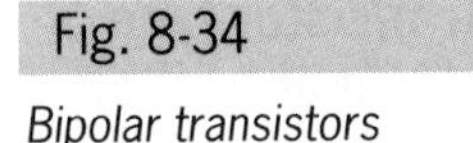

Fig. 8-34

Bipolar transistors

A **bipolar transistor** has three leads: the *emitter* (e), the *base* (b), and the *collector* (c) (Figure 8-34). The larger current flows through the emitter/collector circuit. The smaller control current flows through the base–emitter circuit. The two types of bipolar transistors are NPN and PNP, which designate how the transistor is constructed.

Darlington Pair. Some transistors are very sensitive to changes in the base current, so they will switch on with a very low input signal. Because of their construction, however, these devices cannot switch very much output current. Similarly, *power transistors* are designed to switch larger currents but are not sensitive to small input signals. In situations in which larger currents must be switched with very small sensing signals, it is common to use two transistors in combination, known as a **Darlington pair** (Figure 8-35).

In this circuit, a sensitive transistor is used to control a second transistor, which carries the main current load for the device being switched. This provides a high degree of sensitivity but with the capability of switching 5 amperes of current. Conveniently, you do not usually have to build such a circuit yourself. Darlington pair transistor packages are available, such as the TIP 120.

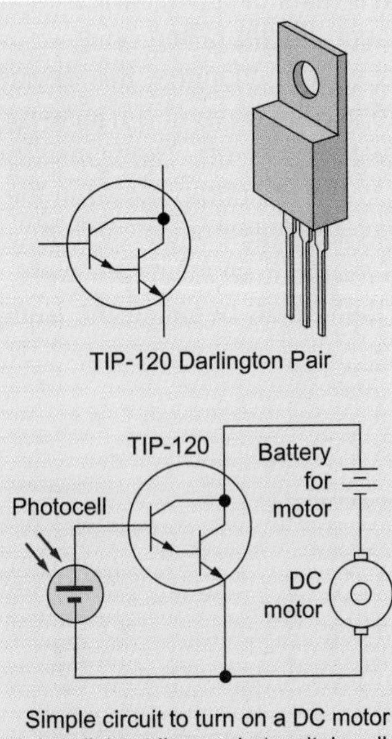

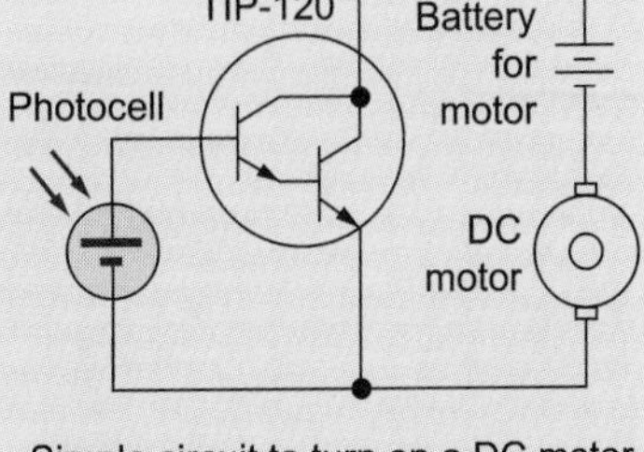

Fig. 8-35

The Darlington pair transistor

Summary

Electronic systems perform a wide variety of tasks, including timing and switching. These systems connect to the outside world through input and output devices, called sensors, transducers, actuators, and displays. The design of electronic systems to solve specific problems means that decisions will need to be made about the three system parts: sensors to gather input information; processors to take that information to control an output; and output devices to return the result of the system to the real world.

Chapter 10 will cover modeling circuits with electronic kits and prototyping techniques. Chapter 11 will cover printed circuit board development.

Observation/Analysis/Synthesis

1. Make a list of devices that have electronic timing circuits in them; make a list of those with display circuits.
2. The first stage of photographic print processing is timing the development, usually for 1½ minutes. Design and make a monostable circuit that will light an LED for exactly 1½ minutes. Redesign your circuit to light an LED after 1½ minutes have

elapsed. Using a second 555 timer chip or one 556 chip, redesign the circuit to produce an audible tone after 1½ minutes.

3. A flashing warning light is needed to attract the attention of a pilot when the plane's fuel is low. Design and make a simple astable circuit to flash an LED when a switch is closed. Using a second 555 timer chip or one 556 chip, redesign the circuit to produce the flashing pulse for 1 minute.
4. Simple electronic games are exciting for young children. Design and make a toy that will flash lights when a ball is tossed to a target.
5. Explore the possible application of the 4017 chip in conjunction with a 555 astable circuit to make a flashing light sequencer.
6. Design and make (or model from Lego™ or Fishchertechnik™) a toy vehicle powered by a small DC motor that, when turned on, will travel forward for about 2 seconds, stop for about 2 seconds, and move forward for about 2 seconds. It should do this until it is turned off. Use a transistor to switch the motor.
7. A store would like to keep track of the number of people entering during the day. It has separate entrance and exit doors. Design and make a circuit that will sense when each person enters and will display the total in seven-segment display format. Your first circuit can count a total of nine people. Redesign the circuit to count up to ninety-nine.

CHAPTER

Designing Pneumatic Systems

Introduction

Although mechanical and electronic systems are visible in many familiar consumer products, pneumatic and hydraulic systems are less well known. **Pneumatic** refers to the use of *compressed air*, which is air that has been raised to a pressure above the normal atmosphere. **Hydraulic** systems also use the pressure of fluid to do work, but, instead of a gas such as air, a liquid is used.

Both gases and liquids are fluids. The terms *fluidics* and **fluid power** are used to refer to both pneumatic and hydraulic systems. These systems are useful for applications in which linear, reciprocating, or rotary motion is needed. Electrical or mechanical devices can also produce these motions but usually at greater expense. Individual electrical components can produce linear motion, but requirements of more than a few inches of travel are impractical. With pneumatic and hydraulic cylinders, linear travel up to 10 feet is easily accomplished. High forces are easily obtained with these systems as well.

For producing rotary motion, pneumatic and hydraulic motors are more compact and can provide higher speeds than electric motors. In applications in which high heat, flammable fumes, dust and/or grit may be present, pneumatic systems are safer than electric motors.

Characteristics of Fluids

A *fluid* has the characteristic of being able to flow and to take the shape of its container. A *liquid* is a fluid with an independent volume that, for practical purposes, does not change. In other words, it cannot be compressed or expanded in volume. A *gas* is a fluid that can expand indefinitely to fill its container. It has no fixed volume, so it can be compressed or expanded.

Fig. 9-1

According to *Pascal's law*, when a pressure or a force is applied to a confined liquid, that force is transmitted to all parts of the liquid in all directions. As a result, the force is transmitted to the inside walls of the container holding it. This is called a *hydraulic system*. Let's say that you have a balloon filled with water (Figure 9-2). If you press your finger on one part of the balloon, the balloon will change shape as the force you applied is transmitted throughout the liquid water. If you were able to enclose the water balloon with your hands and squeeze, you would not be able to change the volume of the liquid, and the balloon would ooze out between your fingers.

Fig. 9-2

Squeezing a water balloon will not decrease its volume. Liquids cannot be compressed.

In addition to being essentially uncompressible, a confined liquid can be used to gain "mechanical advantage"; that is, it can *multiply force*. For example, in Figure 9-3, a simple hydraulic jack is pictured (although without the safety and other features of a practical device). A 25-pound force applied to the piston on the left can counter the

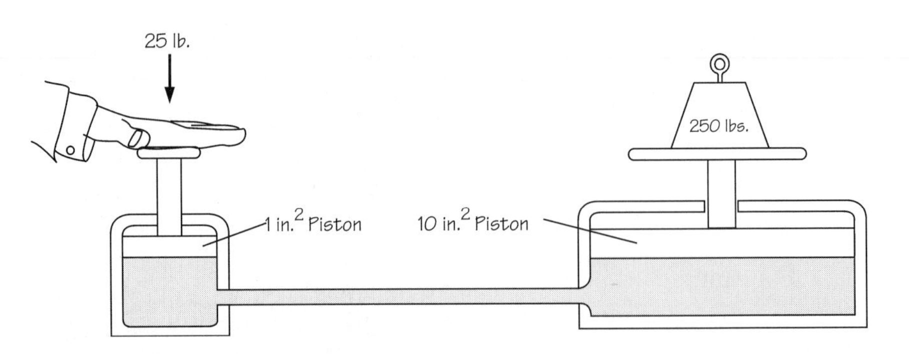

Fig. 9-3

Fluids can transmit and multiply force.

250-pound load on the right. You don't get something for nothing, so what is being sacrificed in this system?

When a force is applied to a gas in a container, the force tends to decrease the distance between the molecules. Newton's third law, which describes how each action is accompanied by an equal but opposite reaction, tells us that there must also be a force pushing the molecules away from each other. Therefore, when a force is applied to a gas, the gas exerts its own force in response. Unlike the water balloon, a balloon filled with air will decrease in volume if you can get your hands completely around it. But the smaller you manage to get the volume of the balloon, the harder you will need to squeeze it to maintain that volume. Of course, if there are spaces between your fingers, the balloon will ooze out in a similar manner to the water balloon.

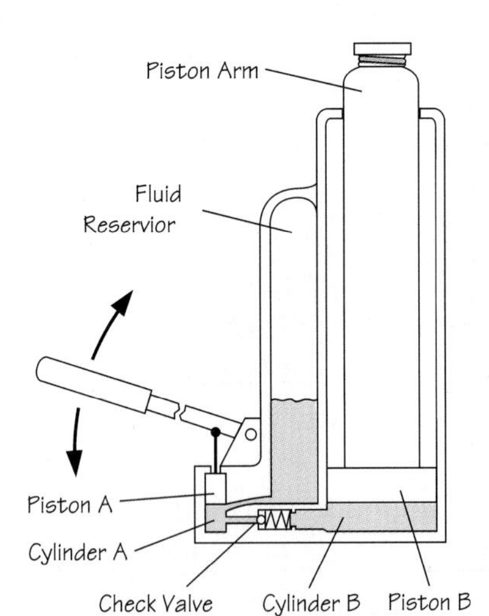

Pushing down on handle moves piston A down which closes off passage to reservoir hydraulic fluid. Fluid is forced past check valve and enters cylinder B where it displaces piston B a small distance upward. Lifting the handle raises piston A until it uncovers hole in the cylinder wall which is the passage to reservoir. Hydraluic fluid fills cylinder A by gravity. When handle is pushed down again,more fluid is forced into cylinder B displacing piston B upward again. The check valve prevents the fluid in cylinder B from pushing back into cylinder A.

Fig. 9-4

Pneumatics Versus Hydraulics

There are some applications in which one specific system is better suited. A hydraulic system allows a precise linear movement, and that movement may be stopped at any point. For example, the operator of a backhoe needs control over each axis of movement of the hoe. The blade must be able to move to a location, stop, and move again. By opening valves that allow a certain quantity of liquid to enter a cylinder, the piston will move a certain distance.

In a simple hydraulic system, such as a car jack, a person moves the handle on a pump, which forces a hydraulic oil into a cylinder (Figure 9-4). The pump can develop high pressure, but the volume of liquid it can pump is very low. Many strokes are needed to force enough liquid into the cylinder to raise the jack a few inches. Most hydraulic systems use motor- or engine-driven high-pressure pumps to force the liquid into a cylinder to extend a piston. When the piston must be withdrawn, the liquid in the cylinder must go somewhere, so return hoses or lines are used to route the fluid back to a reservoir. Liquid is recycled in hydraulic systems.

Pneumatic systems, in contrast, have valves that allow a gas under pressure, usually air, to enter a cylinder. The pressure of the gas pushes the piston, but the pressure builds up rapidly and the piston is forced from its starting location to the end of its travel. Unlike the liquid system, in which a volume of liquid would displace the piston a certain distance, gas does not have a specific volume.

Although liquid-filled systems transmit force more efficiently than gas-filled systems, there are advantages to the pneumatic system over the hydraulic system for many applications. Pneumatic systems are much cleaner and require less piping than hydraulic systems, because they do not require leak clean-up and no liquid return lines are needed. Pneumatic systems are widely used in factory automation, braking systems on large trucks, and many other applications. Because of these reasons, this text will deal only with pneumatic systems. However, many of the same principles and applications will apply to hydraulic systems (Figure 9-5).

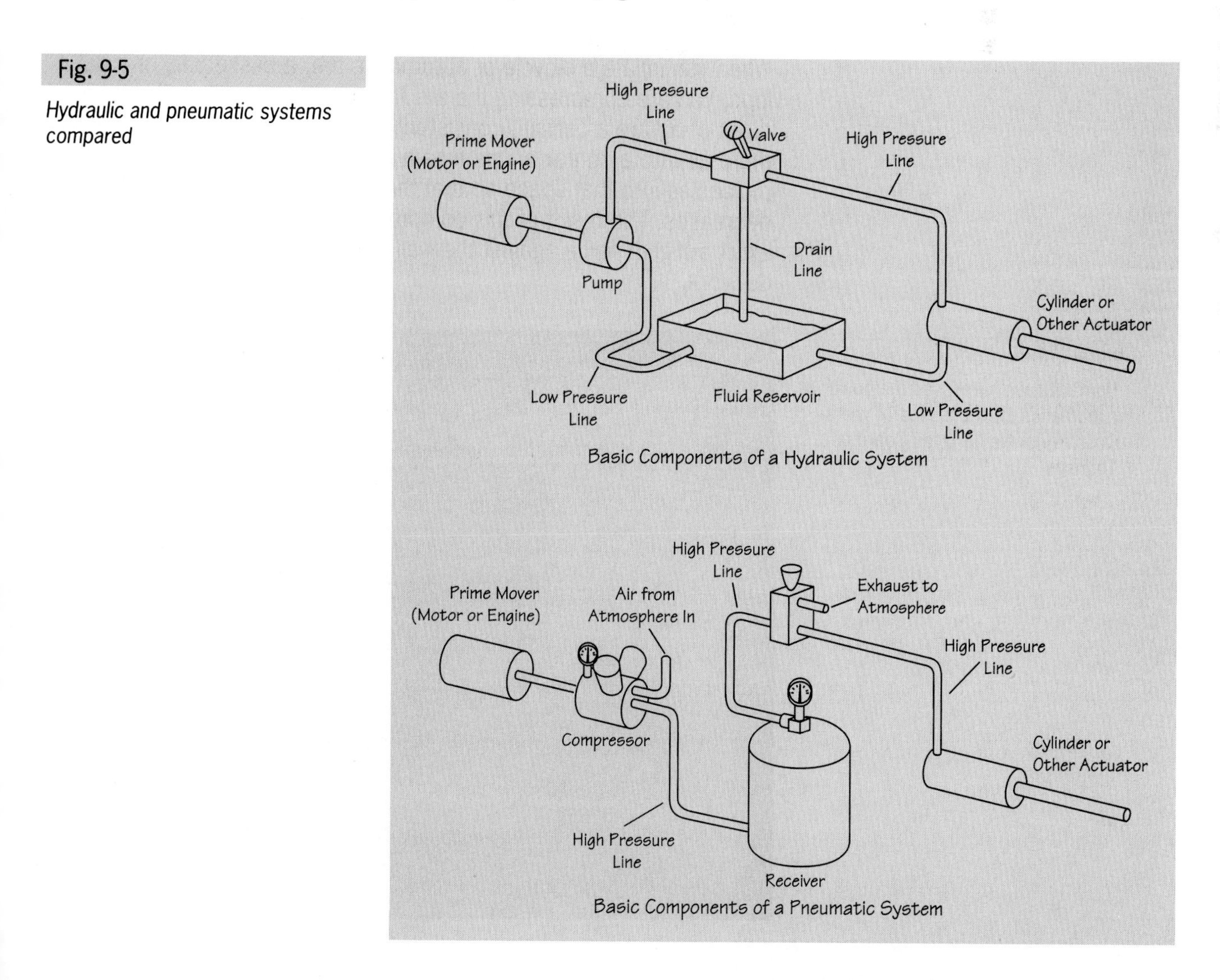

Fig. 9-5

Hydraulic and pneumatic systems compared

Principles of Pneumatics

The normal air pressure at sea level is 14.7 *pounds per square inch (psi)*. The unit of psi is only one method of describing air pressure, however. Sea level air pressure in other widely used units is 101.3 *kilopascals*, 29.92 *inches of mercury*, 1.013 *bars* and 1,013 *millibars*. The pressure of the atmosphere at sea level is a result of the weight of the molecules of air above that point. The atmosphere extends about 62 miles (100 kilometers) above the earth's surface.

Pressure differences in the atmosphere is also pneumatics. The uneven heating of the earth's atmosphere causes these pressure differences, which we see as changes in barometric pressure. Moving molecules of air, in the form of wind, try to get from high-pressure areas to low-pressure areas. Humans have used the pneumatics of the atmosphere to power sails and windmills for thousands of years (Figure 9-6).

When you inflate a bicycle or automobile tire, a basketball, or a balloon, you are compressing the air. At normal air pressure, the molecules of air are a certain, consistent distance apart. When you force more air into a container, such as a basketball, you are pushing more molecules into that space so that the distance between the molecules decreases. The result of this compression is that the molecules of air exert a higher force against the walls of the container. This force can do work.

Fig. 9-6

Humans have harnessed the power of pneumatics for thousands of years. Courtesy of Netherlands Board of Tourism

Fig. 9-7

A reciprocating-piston compressor

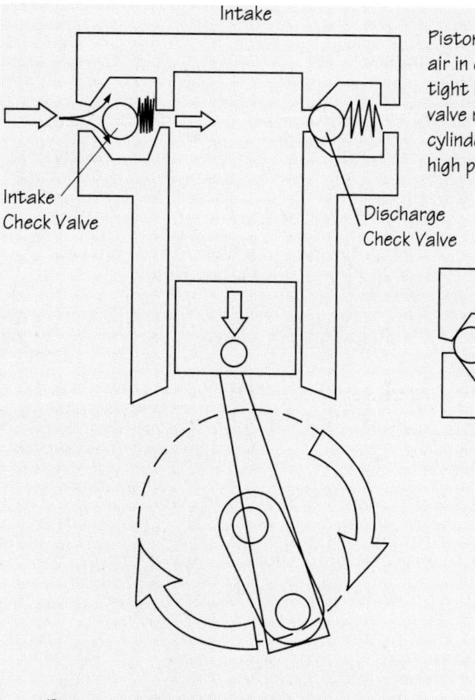

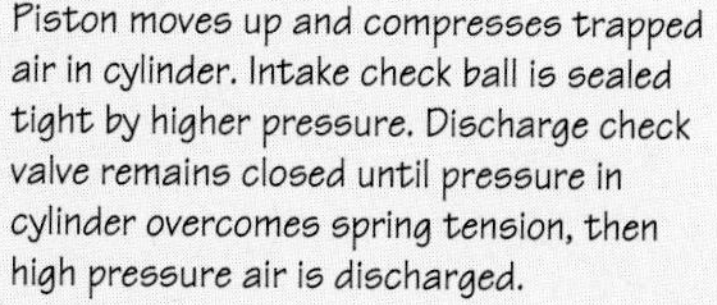

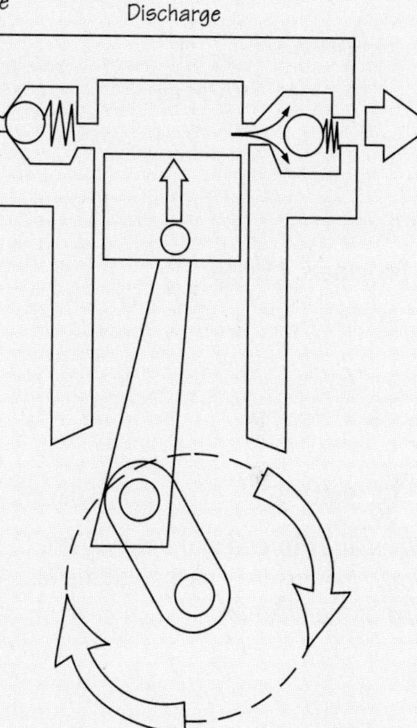

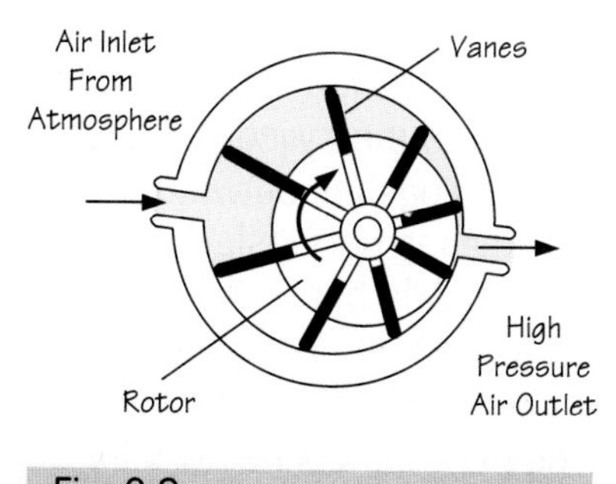

Fig. 9-8

A sliding-vane compressor squeezes air into a smaller volume as the vanes rotate.

In pneumatics, air pressure is created by means of a compressor. A compressor requires a motor or an engine to operate. Two common types are the *reciprocating-piston compressor* and the *sliding-vane compressor.* A piston compressor draws air from the atmosphere into a cylinder where it is squeezed into a small space by a moving piston (Figure 9-7). Sliding-vane compressors have movable seals or vanes on an eccentrically mounted rotor (Figure 9-8). This type of compressor is usually quieter in operation than the reciprocating-piston compressor. From the compressor output, the compressed air is fed into a storage tank, called the *receiver.*

All pneumatic systems require *pressure regulators* to limit and maintain the air pressure at a certain level. Moisture from the atmosphere also enters a pneumatic system along with the air, so a device called a *dryer* is used to remove moisture from the pneumatic line. In addition, pneumatic systems often have *lubricators* that provide a fine mist of oil droplets in the air line to lubricate the moving parts of pneumatic components. Rotating motors, such as found in impact wrenches and other air tools found in automobile repair or body shops, need this lubrication. Other pneumatic systems, such as paint-spraying equipment, would be damaged by a lubricator in the line.

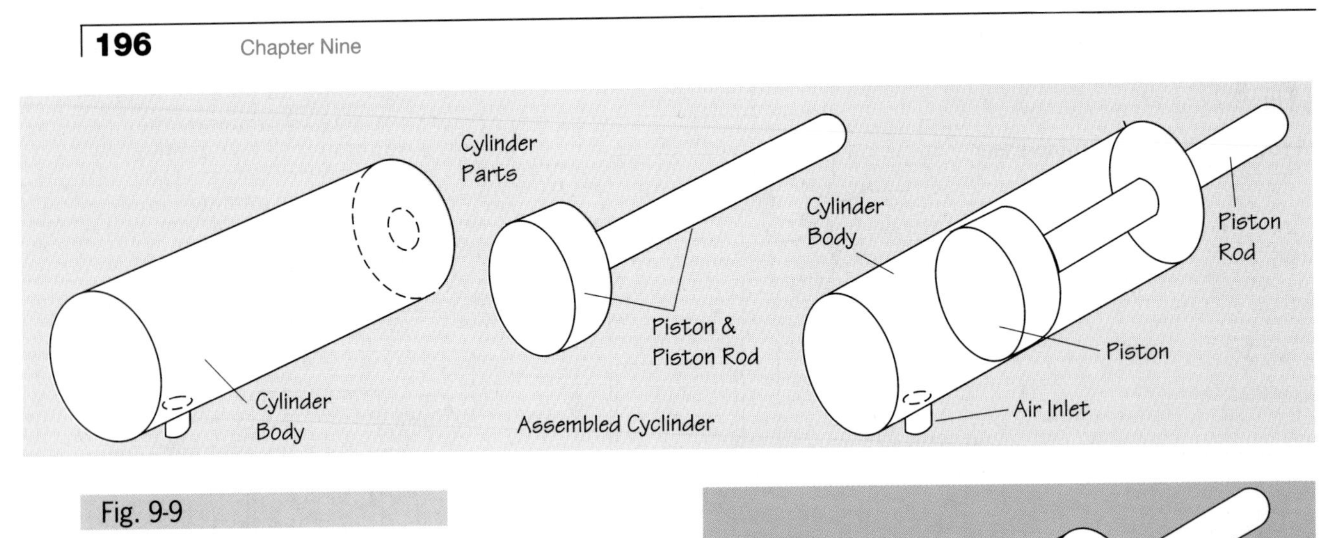

Fig. 9-9

A pneumatic cylinder

Fig. 9-10

When air is forced into the cylinder body, the piston is pushed to the far end of the chamber.

Pneumatic-System Components

A container with one movable wall is called a *pneumatic cylinder*. The cylinder is made up of the cylinder body, piston, piston rod, seals, and fittings (Figure 9-9). As air pressure is applied to the cylinder body, the force of the tightly squeezed air molecules pushes outward, equally in all directions. Since the cylinder body itself is not designed to move, the force exerted causes the piston to slide out of the cylinder body (Figure 9-10). The seals are used to prevent air loss between the piston and the cylinder walls and often to keep dirt on the piston rod from entering the cylinder body. The fittings are the connectors for the compressed air lines or hoses.

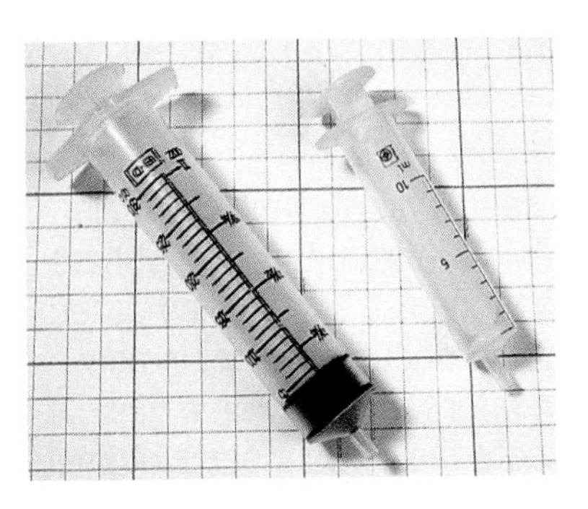

Fig. 9-11

A syringe is a simple pneumatic or hydraulic cylinder.

Cylinders and Control Valves

Single-acting Cylinders. Cylinders that are constructed so that air pressure is applied to only one side of the piston are called **single-acting cylinders**. The syringe in Figure 9-11 is a single-acting cylinder. Figure 9-12 shows a commercial single-acting cylinder. A return spring is usually installed on the opposite side of the piston to return the piston to its original position when the air pressure is removed. In most cylinders, the spring is internal and cannot be seen.

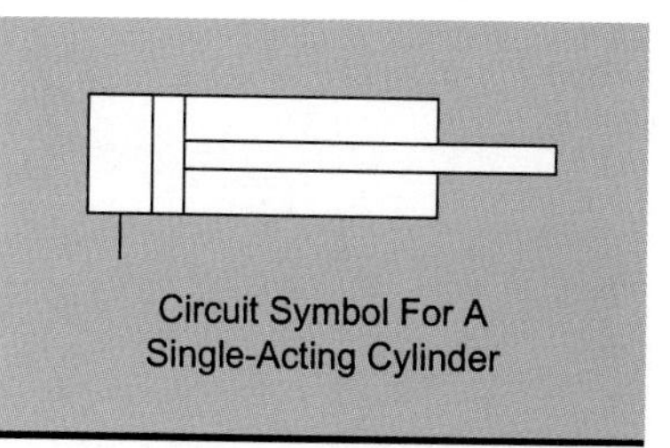

Fig. 9-12

A single-acting cylinder used in commercial systems

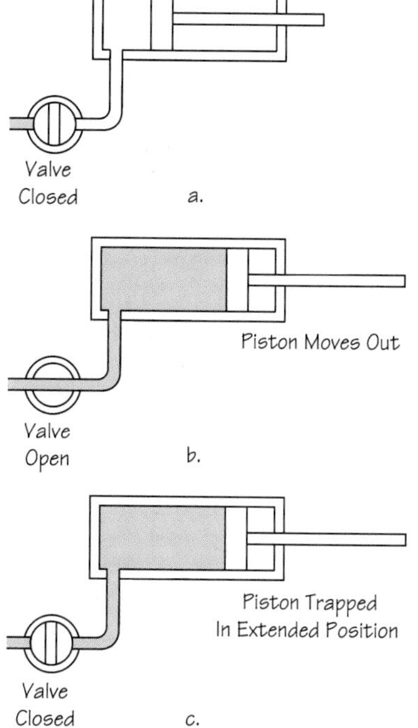

Fig. 9-13

A simple on/off valve will not operate a pneumatic cylinder.

Three-Port Valve. To control the action of a single-acting cylinder, a special valve is required. If a simple flow valve were used, the air pressure applied to the piston in the cylinder when the valve was opened (on) would remain in the cylinder even when the valve was closed (off) (Figure 9-13). To correct this problem, a **three-port valve** was created that would provide an exhaust path for the trapped air in the cylinder when the valve was off. It has one air inlet port, one cylinder port, and one exhaust port.

The three-port valve has two positions. In the "on" position, the air inlet port is connected to the cylinder port, allowing compressed air to travel through the valve and enter the cylinder. In the "off" position, the cylinder port is connected to the exhaust port, allowing air pressure trapped in the cylinder to escape through the valve into the atmosphere (Figure 9-14). These valves come in a variety of types, including push-button, toggle, plunger, and roller-trip.

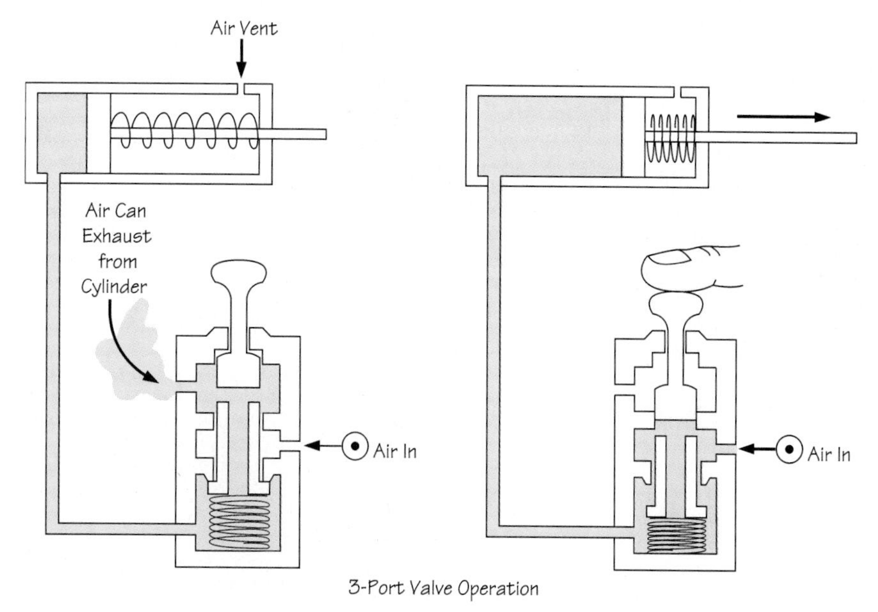

Fig. 9-14

A three-port valve is used to control a single-acting pneumatic cylinder.

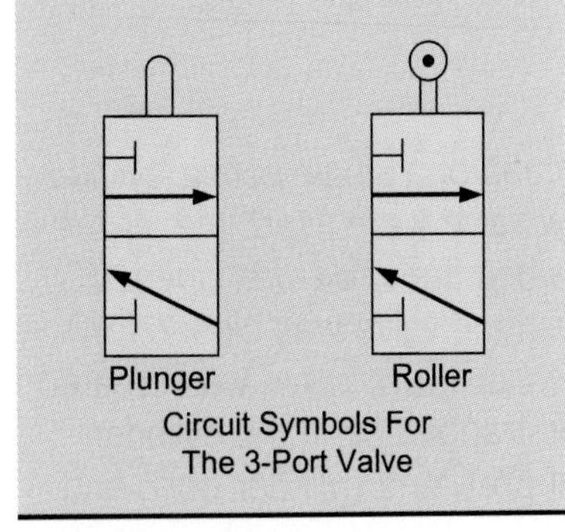

Fig. 9-15

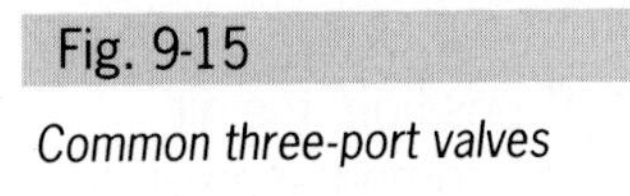

Common three-port valves

Double-acting Cylinder. Cylinders constructed so that air pressure can be applied to either side of the piston are called **double-acting cylinders** (Figure 9-16). In this way, pneumatics is used both to extend and to retract the piston, and no return spring is needed. It is possible to control a double-acting cylinder with 2 three-port valves—one controlling the extension of the piston and one controlling the retraction.

Five-Port Valve. Most applications require that the double-acting cylinder be operated by one valve. The **five-port valve** is a combination of 2 three-port valves in one package (Figure 9-17). There are two cylinder ports, two exhaust ports, and one air inlet port (so we don't call it a six-port valve). The five-port valve, like the three-port valve, has two positions. In the "on" position, the air inlet port is connected to one cylinder port. At the same time, the other cylinder port is connected to one exhaust port. In the "off" position, the air inlet port is connected to the second cylinder port. Also, while in the "off" position, the first cylinder port is connected to an exhaust port (Figure 9-18). In this way, the five-port valve allows exhaust to the side of the piston that is not under pressure. There are a variety of five-port valves, including push-button, toggle, plunger, and roller-trip.

Pressure-Operated Five-Port Valve. In addition to the valves mentioned, a very important valve in pneumatics is the pressure-operated five-port valve. This valve is just like the five-port valve just discussed, except that, instead of being controlled manually or mechanically, it is controlled by air pressure. Normally, 2 three-port valves or 1 five-port valve control the air pressure on the pressure-operated five-port valve (Figure 9-19).

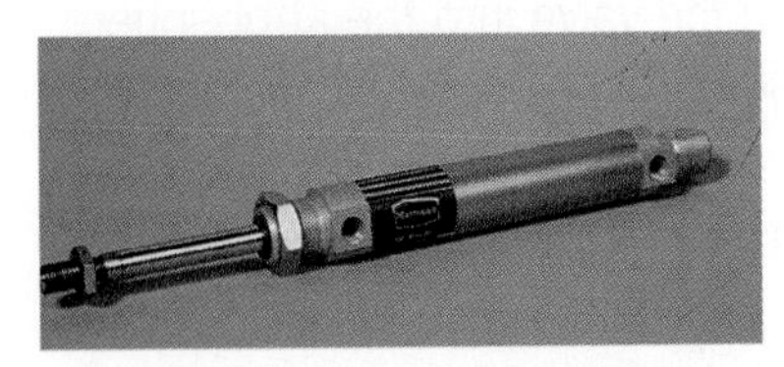

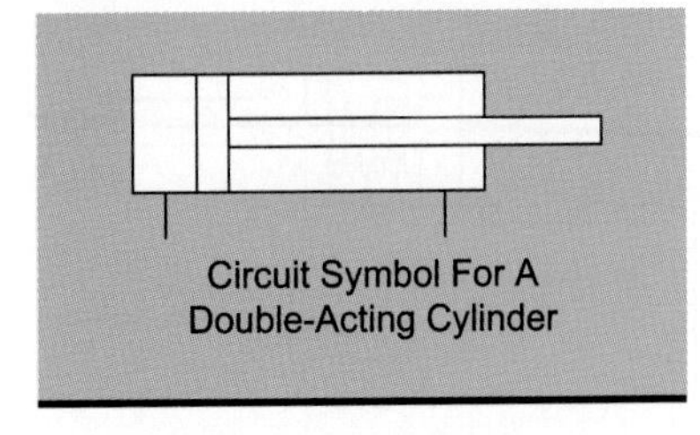

Fig. 9-16

A double-acting cylinder used in commercial systems

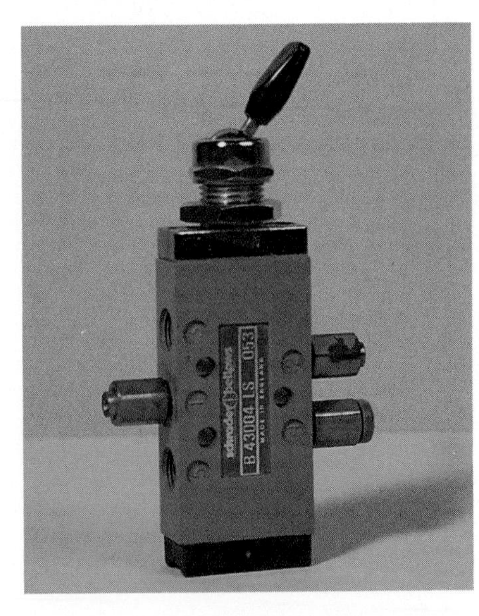

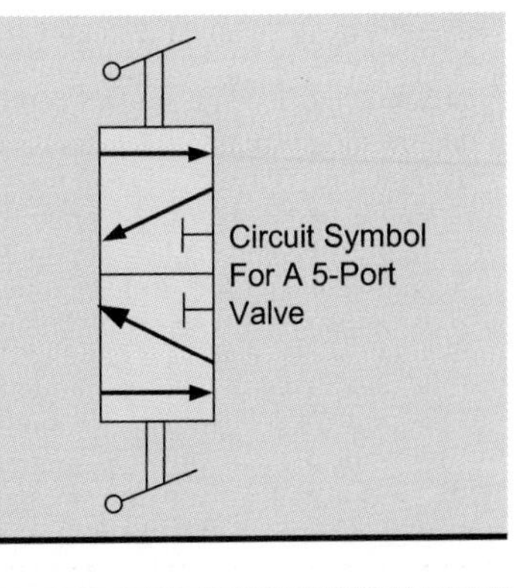

Fig. 9-17

A five-port valve is necessary to control a double-acting cylinder.

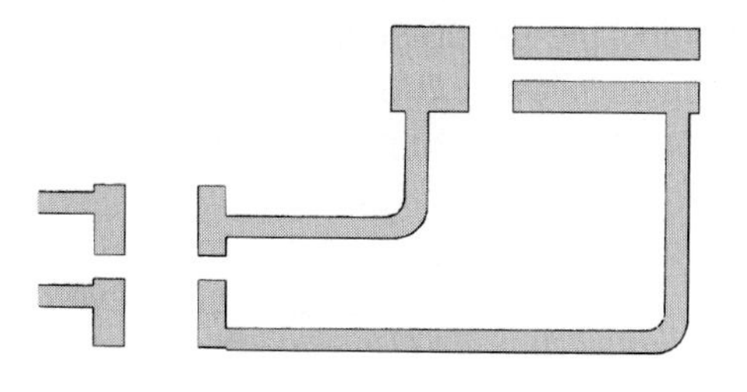

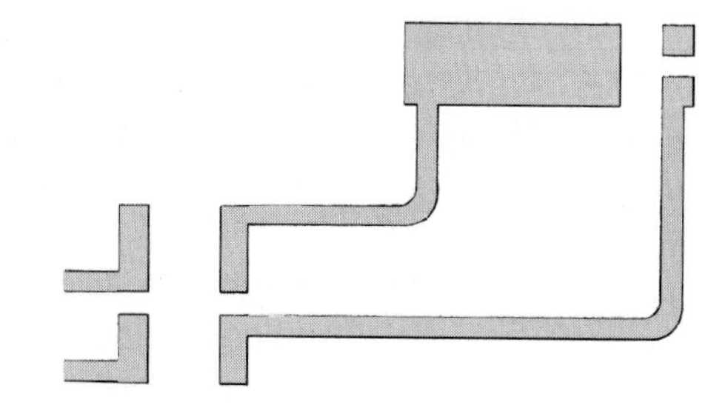

Fig. 9-18

Drawings of the two positions of a double-acting pneumatic cylinder system using a five-port valve

Other Components

The components discussed so far allow the development of simple pneumatic systems. Other components are required, however, to make these systems practical.

Flow Regulator. The flow regulator allows control of the speed of piston travel in a cylinder. It does this by restricting the flow of air in one direction only (Figure 9-20). Air flowing in the opposite direction pushes a ball or a disk inside the valve and flows freely. Flow regulators can be placed in a circuit so that either the compressed air inlet flow or the exhaust flow can be regulated. To control the motion of a double-acting cylinder, two flow regulators must be used. By adjusting the two flow regulators, the speed of extension and retraction of the piston can be controlled.

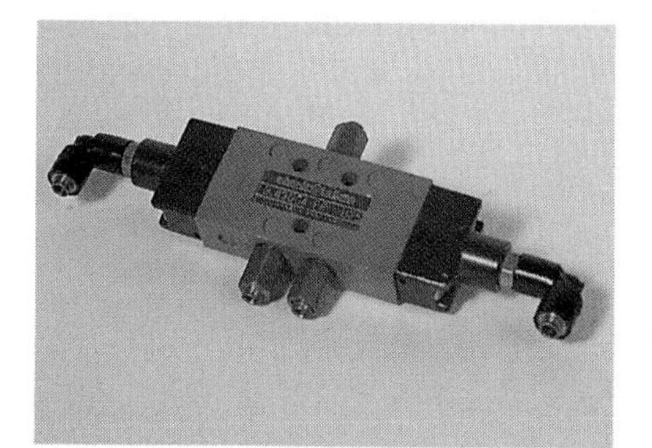

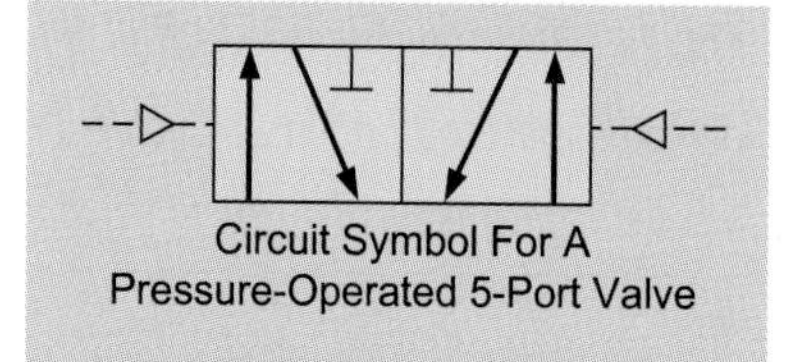

Fig. 9-19

A pressure-operated five-port valve is used when remote control of a double-acting cylinder is required.

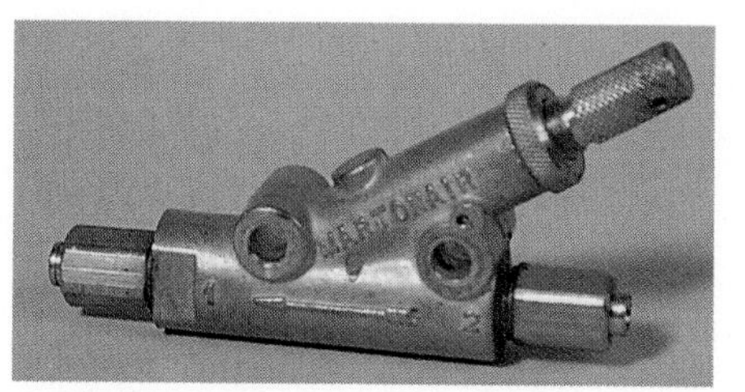

Fig. 9-20

Adjustable flow regulator with bypass

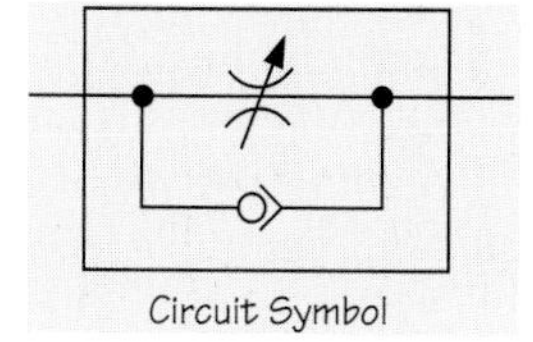

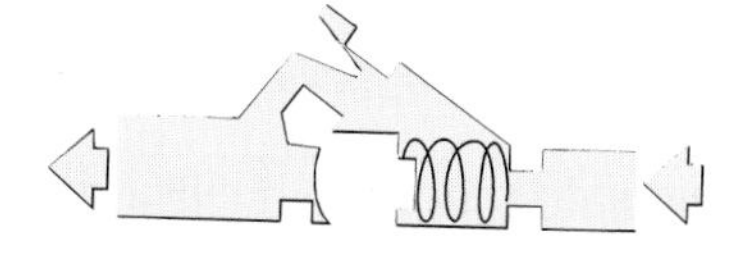

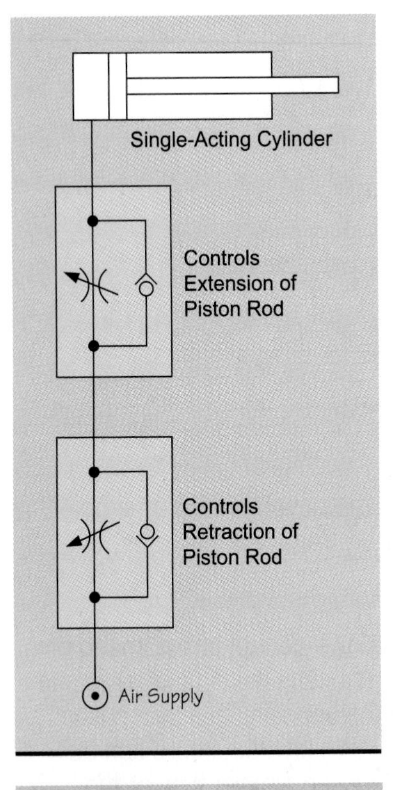

Fig. 9-21

Using two flow regulators back to back to control the extension and retraction rate of the piston rod on a single-acting cylinder

Because flow regulators only control the flow of air in one direction, one flow regulator can only control the piston speed in one direction in a single-acting cylinder. By using two flow regulators back to back, both the extension and the retraction piston speed of a single-acting cylinder can be controlled (Figure 9-21).

Shuttle Valve. A shuttle valve allows a single-acting cylinder to be controlled from two locations. It is a "T" connector with three ports, two air inlets, and one air outlet. It has a small valve inside that shifts to close off one inlet port if the other inlet port has pressure applied (Figure 9-22).

Solenoid Valve. Controlling pneumatic systems by means of electronic circuits is possible with a solenoid valve (Figure 9-23). The solenoid was described as an actuation device in Chapter 8. When electric current is connected to the solenoid coil, a plunger moves to open a valve. Both three-port and five-port valves are available.

Circuit Symbol

Fig. 9-22

Shuttle valve

Fig. 9-23

Solenoid-operated pneumatic valve

Calculating Forces in Fluidic Systems

To calculate the force a cylinder can exert, you must know the pressure of the fluid entering the cylinder and the area of the piston. The area of the piston is found by using the formula πR^2.

Situation

You are designing a simple hydraulic system to rotate the base of a robotic arm.

Problem

You need to know the distance that piston B will travel if piston A moves 1⅛ inches. You also need to know the force that piston B can exert if a force of 3 pounds is applied to piston A.

You Know

- Diameter of piston A = 1¼ inches
- Length of travel of piston A = 1⅛ inches
- Diameter of piston B = ¾ inches

You Need to Find

- Volume of liquid that the travel of piston A will displace
- How far piston B will travel if that amount of liquid enters cylinder B

Solution

Find volume displaced by travel of piston A:

Diameter = 1.25 inches Radius = 0.625 inches

Area of piston A (in^2) $= \pi R^2$
$= 3.14 \times (0.625)^2$
$= 3.14 \times 0.39$ inches
$= 1.22\ in^2$

Volume = Area × Length
$= 1.22 \times 1.125$ inches

Volume displaced = 1.37 cubic inches

Find area of piston B:

Diameter = 0.75 inches Radius = 0.375 inches

Area of piston B (in^2) $= \pi R^2$
$= 3.14 \times (0.375)^2$
$= 3.14 \times 0.14$

Area of piston B $= 0.44\ in^2$

Find piston travel of B:

Length of travel $= \dfrac{\text{Volume}}{\text{Area}}$

$= \dfrac{1.37\ in^3}{0.44\ in^2}$

Length of travel of B = 3.11 inches

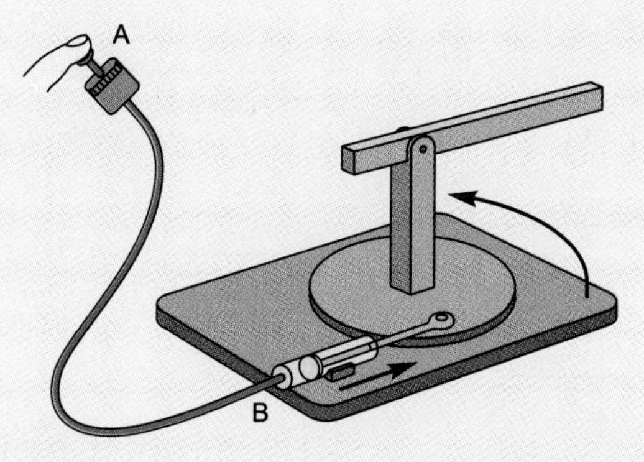

How much force does B exert if the input force on A is 3 pounds?

Find force that piston A exerts:
Force = input force × area of piston A
= 3 pounds × 1.22 inches
Force = 3.66 psi

Find force that piston B exerts:
Force = input force × area of piston
= 3.66 × 0.44

Force = 1.61 pounds

Double-acting cylinders have less piston area on one side than the other, so they cannot exert equal pressure in both directions. This is because the piston rod is connected to the piston on one side and takes up some of the piston area. When the force is calculated for movement of the piston back into the cylinder (negative direction), the figure for piston area is less, so the resulting force is less.

Safety in Pneumatic Systems

Pneumatic systems can develop extremely powerful forces. Compressed air itself can be dangerous. Here are a few simple rules that will help make work with pneumatic systems safe.

Fig. 9-24

Symbols and basic circuits

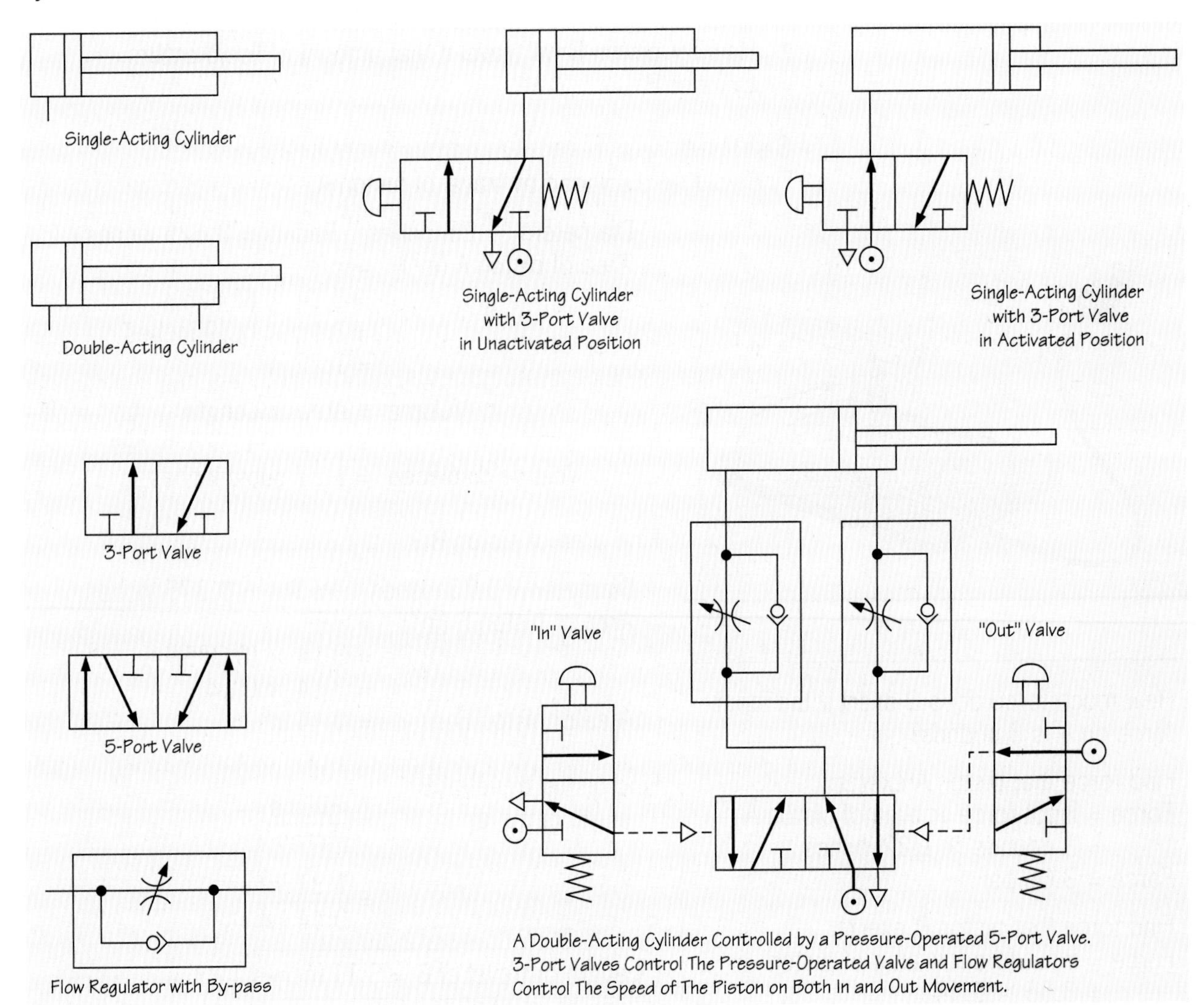

- Never blow compressed air at yourself or anyone else. Under high pressures, air can enter the bloodstream through the skin and cause serious injury or death. Eye injury can easily result from high-pressure air blown in the face.
- Components that operate on compressed air can also be hazardous. A moving piston rod will not stop if your finger or your hand is in the way. Crushing pressures are common in the moving parts of pneumatic systems, and even low air line pressures can translate into large forces.
- Always make connections and remove connections to pneumatic components with the air pressure line disconnected. Be certain that the circuit is completed and that no hoses or lines are left disconnected when air pressure is applied.
- Hoses strung across the floor or between tables invite tripping and falling accidents. Make certain all air lines are out of the way.

Summary

Pneumatic and hydraulic systems are both classified as fluid power systems. To do work, pneumatic systems use air, whereas hydraulic systems use liquids. Both systems have advantages and disadvantages.

Pneumatic systems require a source of compressed air. Through valves, the compressed air is directed to actuating components such as cylinders, which provide linear motion and can exert a great deal of force. Other pneumatic components, such as pressure regulators and flow regulators, control the amount of air pressure to the system and to specific components.

Observation/Analysis/Synthesis

1. Explain the differences between pneumatic and hydraulic systems, and provide examples of the application of these systems.
2. Design and make a hydraulic-operated remote gripper for a robotic arm using acrylic, wood, or other material and two syringes filled with water. The gripper should be able to pick up a "D" size battery when attached to the end of a 4-foot long pole.
3. What is the output force of this hydraulic system? (An input pressure of 60 pounds is applied to a 2-inch diameter piston; the output piston diameter is $4\frac{1}{4}$ inches).
4. Design a simple hydraulic system with two cylinders so that, when the piston of a $\frac{5}{8}$-inch diameter cylinder is pushed in 1.5 inches, the piston in the other cylinder moves 4 inches. You will need to determine the size of the second cylinder.

5. Using standard pneumatic component symbols, do the following:
   a. Design and draw a simple pneumatic clamp system with one double-acting cylinder, one pressure-operated five-port valve, and separate valves to apply and remove clamping pressure;
   b. Design and draw a simple pneumatic door lock system that would extend and retract a metal bolt into a hole in the edge of a door. A person must be able to operate the system on either side of the door.
6. Using kit or commercial pneumatic components simulate the following system: Design and make a system that will clamp a piece of material and drill a hole. The system should use a single-acting cylinder to clamp and a double-acting cylinder to hold the drill head. Remember to regulate the speed of the drill bit entering and exiting the material. Can you redesign the system to automatically clamp and drill in one operation?

CHAPTER 10 Generating and Developing Ideas

Introduction

In the early chapters of this book, you encountered the design process, a loop with nine steps. This design loop forms a framework for the discussion of the thinking processes and skills that go into the solution of a problem. Some of the steps have involved analytical or logical thinking, and some have more creative mind work. What should be apparent by now is that designing involves returning again and again to the various steps as the need arises. Also, there is often a need to go around the loop a number of times for any one problem. This is because a subproblem, within the overall problem, requires attention to many of the individual steps in the design process. Jumping around from step to step is common in design work. (Figure 10-1).

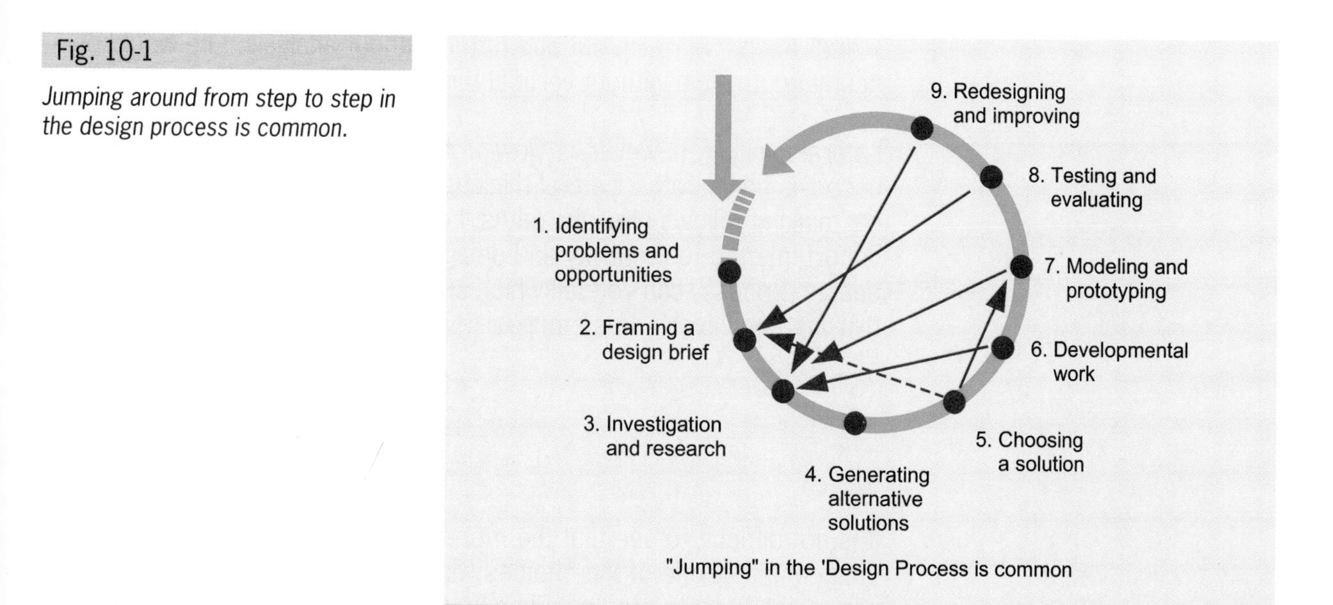

Fig. 10-1

Jumping around from step to step in the design process is common.

In the step of the design process in which you generate alternative solutions to the problem, it will be necessary to use all the skills you have learned. For example, you will probably need to revisit the investigation and research step as information is needed about a particular solution idea or direction. As ideas arise, you may need to go back to your design brief and specifications to evaluate whether there is more leeway there in light of an especially creative solution idea. Designing is not a rigid process, and that becomes apparent in this phase of the process.

Creative Thinking

Much of the process of designing involves creativity. But, without some training and practice, creative thinking does not come easily to most people. This is because most of our education experiences and decision-making situations involve convergent or deductive thinking. We are taught to look for the "right answers" to our problems, and we are taught also that not being "right" means failure. In creative thinking, there are no "right" or "wrong" answers—only ideas. Later, when we have a lot of ideas, we use logic to sift through them. Edward De Bono coined the terms **vertical thinking** and **lateral thinking** to distinguish between logical and creative thinking.

In vertical thinking, each idea rests on another idea, in logical form. De Bono describes it as "high-probability thinking," which means that we use this kind of logical thinking to function day to day. Instead of analyzing each and every action we take, vertical thinking allows us to make assumptions based on past experience, such as, when we see a door knob, we reach out and turn it without analysis. Life would be nearly impossible without vertical thinking.

Lateral thinking, however, is "low-probability thinking"; that is, it follows unconventional paths. Lateral thinking allows new ideas, whereas vertical thinking follows previous paths. Learning how to think laterally is an important step to creativity and design. As you look at the steps of the design process, can you tell which steps emphasize logical or vertical thinking and which steps emphasize creative or lateral thinking?

Generating Solution Ideas

It is not difficult to see that the more choices you have, the better the chance is that one of the choices will be a good one. Generating alternative solutions to problems is a key stage in designing. There are a number of popular techniques for developing ideas (Figure 10-2).

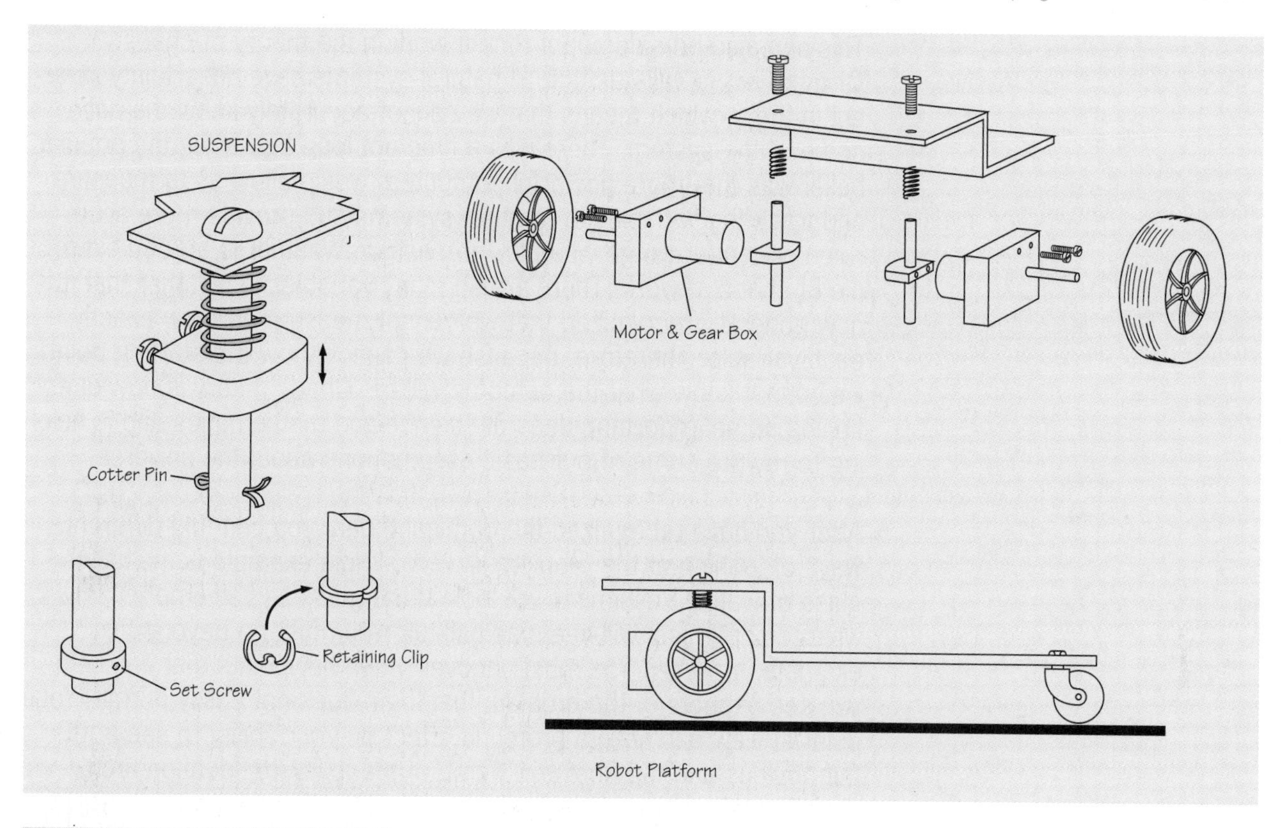

Fig. 10-2

Developing ideas for solutions to your problem is a major challenge.

Lateral Thinking

Your creative thinking potential can be developed by employing lateral thinking techniques pioneered by De Bono. One of these techniques involves the identification of the *dominant idea* in a situation in which you are attempting to find a creative solution. A dominant idea is like a hole dug in the ground. People may enlarge the hole and dig it deeper, but it is easier to stay with that hole than begin digging a new hole in another location. The often used phrase, "breaking new ground," is what digging a new hole is all about. Identifying and understanding the dominant idea allows you to step away from it to pursue other paths of reasoning.

A good example of a dominant idea can be seen in the work of Sir Issac Newton. For centuries after Newton, his ideas about the mechanics of the universe served both science and technology. There were, however, inconsistencies in his theories, but these were largely written off as a lack of understanding or error in observation. It took someone who both knew the work of Newton and could step away from it to form a new theory that fit the evidence. Einstein's willingness to question one of the most fundamental assumptions of the human mind—time—led to one of the most revolutionary leaps in our understanding of the universe.

This can also be said about recent work in the theory of chaos. **Chaos theory** tells us that all natural systems have uncertainty inherent in them, which means that we cannot accurately know the outcome of the system. There was a time not long ago when the popular belief was that more powerful computers would allow us to measure and track the variables in weather systems and provide more accurate and longer-range weather forecasting. Weather is a good example of chaos at work. There are so many variables in weather that it is now generally recognized that we cannot measure or know exactly each variable. Therefore, because we cannot know the starting point, we cannot know what the outcome will be. Chaos theory is now questioning some important ideas of science and mathematics and is helping to rewrite some of our basic assumptions about the universe.

After you have recognized the dominant idea, you can find different ways of looking at the problem. A very simple example is the half-glass of water: Pessimists will look at the glass as half-empty, whereas optimists will see it as half-full. You must go beyond this example, however. You must uncover other viewpoints that can help you see aspects of the problem that logic does not easily permit. Your point of view can have a profound effect on your ability to see both the obvious and the obscure. Some of the following ideation techniques can help you develop other ways of looking at a problem.

Analogies

An analogy is a similarity between two unlike things. In problem solving and design work, analogies can play a helpful role in generating ideas. It is said that Johann Gutenberg's invention of the printing press was, in part, a result of the analogy he saw in the coin punch and the wine press of the day. He was able to visualize a machine that combined the principles of both devices to print words on paper using individual letters.

In your problem situation, can you see similarities to other problems? Do the solutions to these problems hold useful ideas for your problem? If you are working on a problem involving structural design, can you find analogies in the structure of plants or the human skeleton? Seeing analogies may provide you with design possibilities (Figure 10-3).

Brainstorming

Brainstorming is built on the belief that creative ideas differ from conventional wisdom. In a brainstorming session, two or more people get together to exchange ideas and use these to stimulate more ideas.

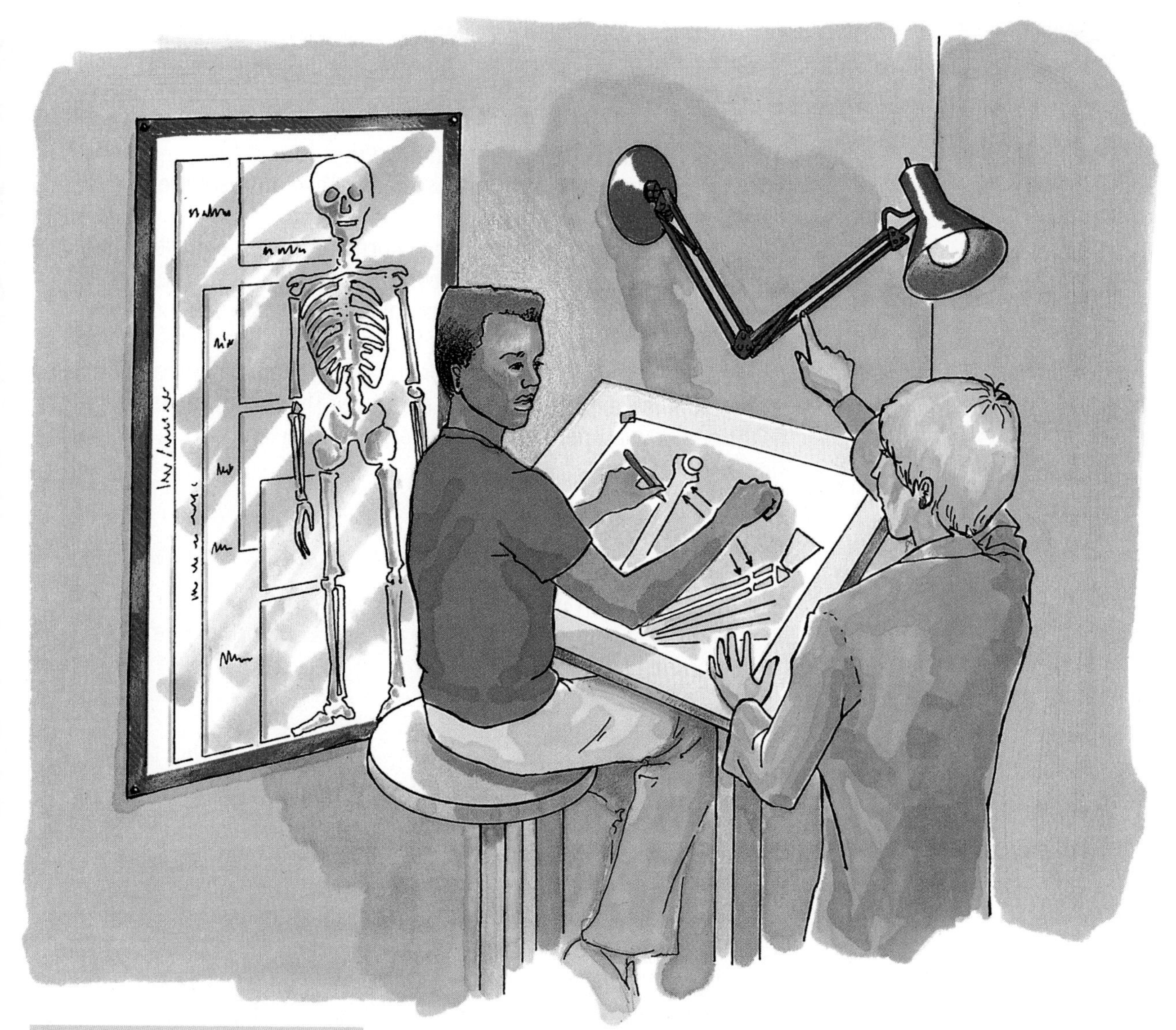

Fig. 10-3

Look for analogies in other things to help you find solutions.

They try to get away from conventional wisdom by letting their imaginations run wild and by reaching for outrageous solutions. This helps break down the "logic" of convention (vertical thinking) and provides an atmosphere that encourages creativity. In brainstorming, participants must be open and adopt a friendly atmosphere in which new ideas will be welcome (Figure 10-4).

Nothing kills creativity more effectively than criticism, so participants in a brainstorming session are not permitted to criticize. There is time for evaluation of the ideas generated after the brainstorming session. Here are some of the basic tenets of brainstorming:

- The more ideas there are generated, the better are the odds that a quality idea will emerge.

Fig. 10-4

Brainstorming is a useful technique for generating solution ideas if the rules are followed.

- Group ideation is more valuable than individual ideation, because group members stimulate one another; brainstorming is usually ineffective with one individual.
- A non-threatening atmosphere is necessary for free expression of ideas; therefore, criticism is deferred.
- A limited time frame enhances the creative atmosphere.

Often, when students attempt brainstorming, they meet with little success. This is usually because of two factors: First, they are inexperienced with the process and do not take it seriously; second, they are inhibited and are afraid of looking foolish. To be successful, you will need to practice the technique. Brainstorming has a number of rules that makes the experience more effective. The following figure describes how a brainstorming session is organized.

Organizing a Brainstorming Session

Brainstorming can be an effective strategy for generating a lot of ideas for solving a problem. Here are some important rules for brainstorming:

1. Work in a group of at least three or four people.
2. One person must take notes; recording of emerging ideas is critical for allowing revisiting of earlier inspirations.
3. Set the problem, and make sure that each person understands it.
4. Set short time limits on each problem.
5. Be spontaneous, be outrageous, be imaginative.
6. Listen to other people's ideas, and build on them.
7. Do not criticize, evaluate, or even elaborate; this is very important!
8. Go for quantity to ensure quality.
9. Evaluate only after your ideation session has ended.

Synectics

Synectics is a technique used for uncovering perspectives. The goal is to make the "familiar strange and the strange familiar." What this means is that synectics is useful for viewing a problem in a new way. In one technique, the designer role plays the part of the product or device used to solve the problem, asking "*Who affects me and whom do I affect?*" After establishing a list, the designer role plays each of the affected people or objects involved and tries to see the other person's or object's point of view.

For example, a basketball shoe is familiar to most people, and those who wear them regularly do not give them much thought. If you were involved in athletic shoe design, you might imagine yourself as the shoe, being worn by a player during a game. You could feel your layers of foam and sole materials compressing together as the wearer prepares to leap up for a jump shot, or the repeated compression and expansion as you pound your way to the other end of the court. While this may sound a bit bizarre, it is a clever and effective strategy for getting away from vertical thinking and into generating new ideas.

Sketching and Doodling

The written and spoken languages have many rules, and designers have found it more effective to use a communication technique that is more intuitive and spontaneous. Sketching and drawing are the languages of the designer.

The development of your ability to sketch will help you in your creative efforts of problem solving in technology. In Chapter 4, a number of drawing techniques were introduced that required no instruments or tools, except perhaps a pencil. Sketching helps you to image and refine devices or systems that could solve a problem (Figure 10-5). Sketching forces you to develop your ideas in terms of the relationships between components and parts of systems.

For example, if you were trying to develop solutions to a problem involving door-opening devices for people in wheelchairs, you might sketch the possible ways in which doors can open: slide, rotate, pivot, fold, and so forth. Your sketches then might include simple systems

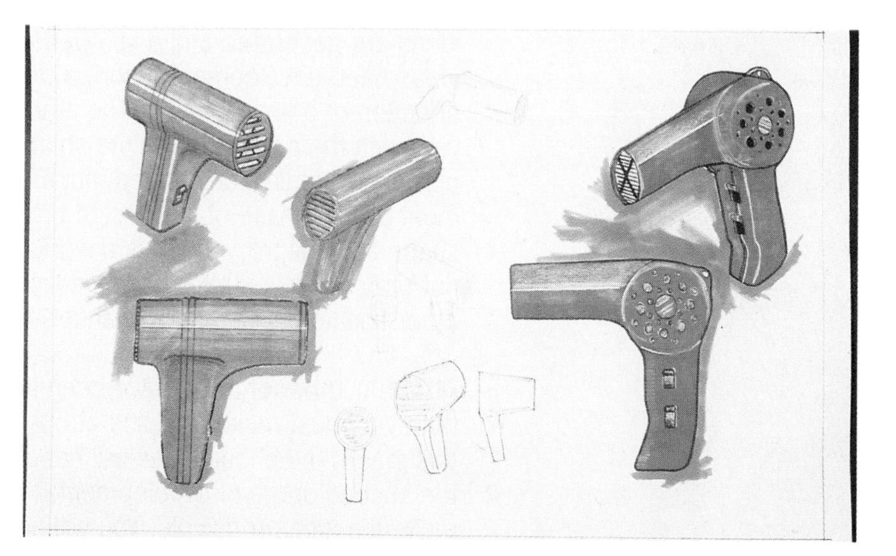

Fig. 10-5

Sketching is an excellent technique for generating and refining ideas.

for opening doors, such as motors, levers, pneumatic cylinders, and others. The use of instruments or formal drafting practices will inhibit the creative flow of your ideas. The structure imposed by these instruments and practices are just like the rules of the spoken and written word, coming between your ideas and the paper.

Incubation Period

The mind is always working, even when we are not consciously thinking about a problem. It is therefore valuable to provide a period of time during a project for ideas to "incubate" or work through in our minds. It is not uncommon for someone to struggle with a problem and go to sleep at night, frustrated that an answer hasn't been found, only to wake the following morning with a solution.

While it is unrealistic to expect that your design solutions will suddenly become clear before breakfast, it is usually beneficial to get away from the problem for a day or two and let it incubate in your mind. If you become "stuck" in some phase of the design process, go do something else. Often, when you sit down to work on the problem again, you will find new insights and ideas that you didn't see before. It is important, however, to record ideas on paper when they come to you. These "flashes of insight" can be easily lost if you do not put them on paper immediately.

Developmental Work

An idea needs development at two important stages. First, when ideas are generated and a number seem to have potential, these ideas need development before a decision can be made about the direction of a final solution. The amount of developmental effort will vary with the complexity of the solution: Simple ideas need little development; complex ideas need much development. The goal of development at this phase of the design process is to get the idea into a shape that will tell you if it is a workable solution to the problem. Do not stop when you have gotten a workable solution. Develop your other solutions in a similar manner.

After a number of ideas have been developed and deemed workable, then you must make a choice about which one will be implemented. At this stage, the chosen idea will need further development, and this is the second stage of developmental work. The goal at this stage is to make the idea producible. You will need to make decisions about size, shape, materials, fasteners, finish, and many other considerations. To do this, the final solution will need **working drawings** to enable the actual making of the product or system. In Chapter 4, working drawings, including orthographic projection, section views, and others, were briefly discussed. For more detailed explanations and direction, you may want to consult one of the many books devoted to drafting or get some help from your teacher.

Fig. 10-6

Many flexible materials are useful for developing your ideas, including paper, poster board, mat board, foam-core™ board, light plastic sheet, and others.

Developing your ideas into realistic solutions requires that you work out the *structure, function,* and *appearance* aspects of the chosen solution. This may be done with flexible materials, found objects, construction kits, or graphic techniques.

Flexible Materials

Paper, cardboard, poster board, mat board, foam-core™ board, and plastic sheet are useful modeling materials (Figure 10-6). These may be joined with a variety of tapes and glue to make useful models. Making a model of a cabinet for an electronics project from heavy paper, for example, helps you to work out size, proportion, control location, and other problems. The final cabinet may be made from acrylic or sheet metal.

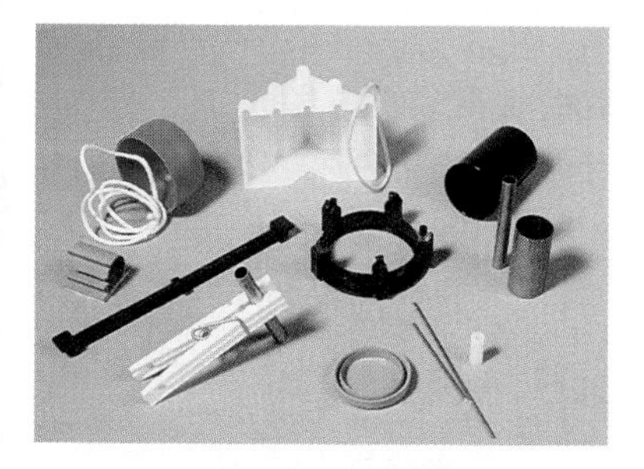

Fig. 10-7

There are many found materials that can be used for modeling.

Found Objects

Plastic bottles, aluminum cans, straws, film canisters, and other found objects make excellent raw materials for modeling development. Only your imagination will limit the possibilities (Figure 10-7).

Construction Kits

Construction kits allow complex mechanisms to be modeled. The construction kit approach allows you to easily fabricate sophisticated structural and mechanical designs that otherwise would be too difficult and time consuming to build from "scratch." You may need to design and build a gear reduction mechanism that will increase the torque of a small electric motor to perform a lifting task. Modeling from standardized parts and materials to build a mechanical system and the framework structure to hold it would take much skill and time. But this task is easily done with most construction kits, once you are familiar with the parts and joining and fastening techniques of the particular kit.

Perhaps the greatest strength of the construction kit is the ability to work out ideas in three dimensions. Drawing and sketching are used for the development and refinement of technical ideas in two dimen-

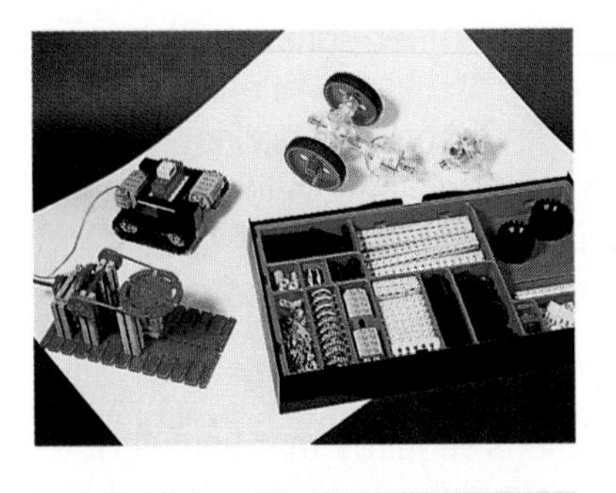

Fig. 10-8

There are many construction kits on the market that can be useful for developing your ideas.

sions, on paper. But what appears to work in a sketch may not work in actuality because of the inexperience of the designer or lack of specific detail in the design. Sketching in three dimensions is possible with construction kits and allows you to work out problems and ideas, and to get immediate feedback on the viability of the design (Figure 10-8). This is a crucial step toward the development of any "active" device and is a valuable tool for projects ultimately destined to be constructed of standard industrial materials. This is the power of kit modeling.

Graphics Techniques

Working out the shape, proportions, aesthetics, and other visual considerations of a design problem can be accomplished with graphics (Figure 10-9). Techniques like sketching, color pencil, or marker work are ideal for visualizing what a finished solution will look like. The section, "Appearance Development," found later in this chapter, outlines a number of design elements and principles that can only be realistically developed using graphic techniques such as those discussed in Chapter 4.

Other graphic techniques, such as developing orthographic projection drawings, will be required to determine the exact size, shape, and location of features, such as holes. This information is then used to actually make the finished product or prototype. It would be futile to begin making your solution if you didn't know where to cut, saw, or drill.

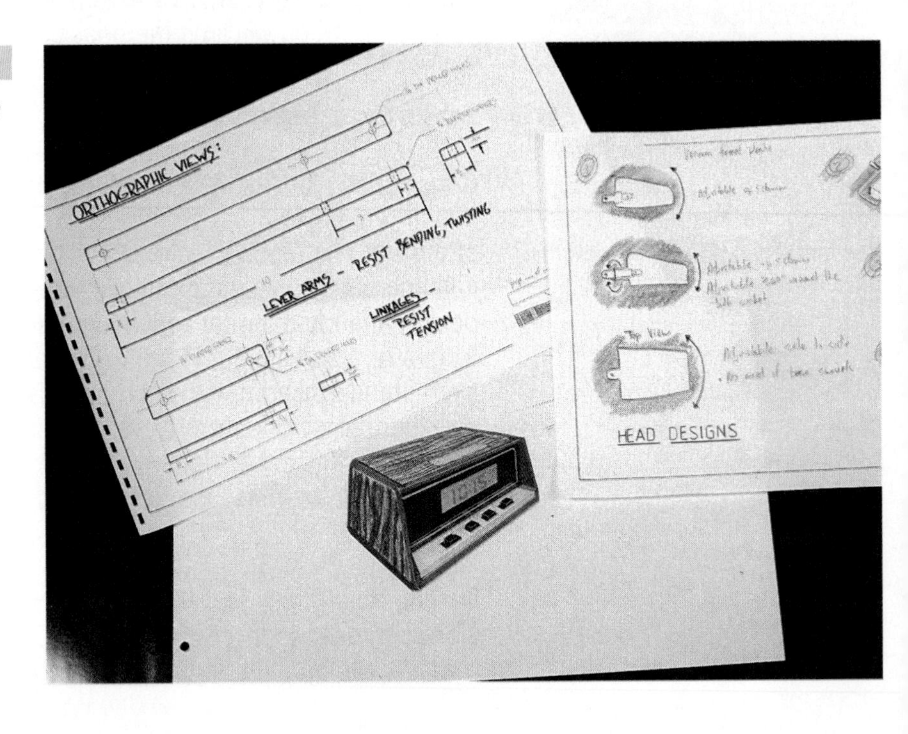

Fig. 10-9

Graphics are a necessary part of the developmental process.

Structural Development

Structural development is begun on paper, by applying the principles of structures found in Chapter 6. Through sketching and drawing, you will be able to get a general idea about how forces will act on the structure and how it may be designed to withstand those forces.

For example, you may be working on a solution for making a foldable beach chair. There are structural considerations in this design, because it must withstand both the static load of someone sitting in it and the dynamic load of that person shifting position. You may work out the initial design on paper and then construct it to scale to further develop your ideas. Your ideas would still be developing, and it should not be your intent to construct a scale model of the finished design.

Developmental modeling is an important part of the design process. Through this technique, you can see if ideas conceived on paper actually perform as intended, or you can present ideas to others for their approval or criticism. Developmental modeling is widely used in industry for these purposes.

Fig. 10-10

Mechanical systems modeled using kits

The materials used for structural development are as varied as the materials available to you. Paper, wood, plastic, metal, clay, cardboard, plaster of Paris, and many other materials may be used. Chapter 11 goes into more depth in discussing materials useful in making finished models. In addition, many construction kits are appropriate for modeling technical ideas (Figure 10-10). You are only limited by your access to materials and kits, and by your own imagination.

The Jinks' Method

A system developed for elementary school children by two British educators, David Jinks and Pat Williams, uses square dowels connected by gluing reinforcing "gussets" of index card stock (Figure 10-11). Although small children can easily use this technique, you many find it useful for quick modeling of your ideas.

Wood stock is ripped to 5/16 inches × 5/16 inches (or slightly larger) from pine or other softwood in random lengths. The only tools and materials needed are a small saw, scissors, glue, and the card stock for the gussets. This is both an inexpensive and a useful method of developmental modeling, because special equipment and facilities are not needed. But, at the same time, it can be an extremely effective technique, and elaborate models of buildings, cars and trucks, bridges,

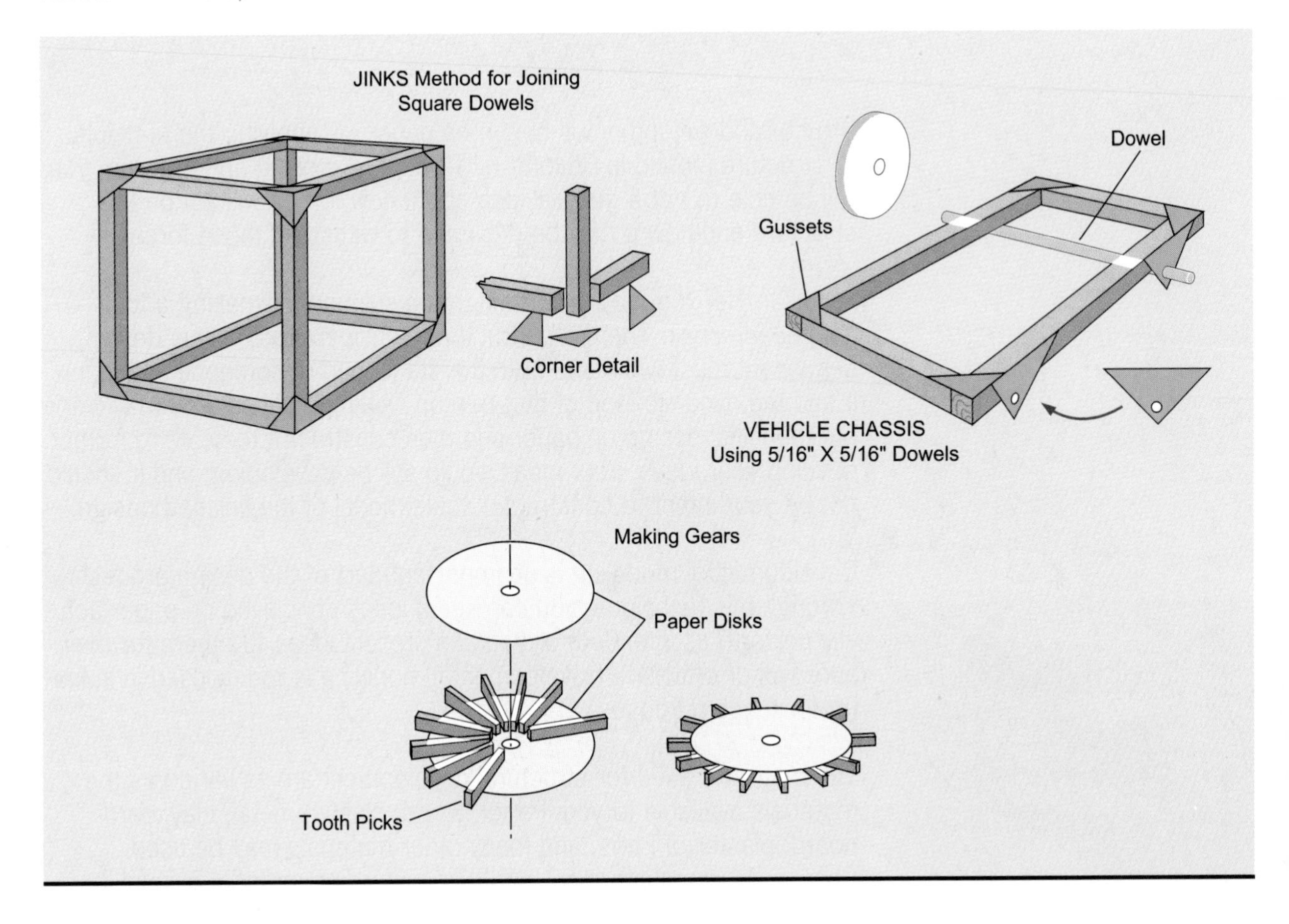

Fig. 10-11

In the Jinks' method, square dowels are connected by gluing paper gussets.

merry-go-rounds, and many other structures are possible. The modeling of the foldable beach chair could easily be done with this technique.

To make more action-oriented models, you can trace and cut out circles from the card stock and build up wheels from several layers, or use a heavier mat board for this purpose. Wheel axles can be made from round dowels and are attached to frame members with a larger gusset punched with an appropriate size hole (Figure 10-12).

Even gears can be made, using two circles of card stock with toothpicks sandwiched in between and sticking out about ½ inch. These only seem to work well when the rotating shafts of the two gears are at 90 degrees to each other, but other combinations are possible, so experiment. A handy tool for making nice circles is a compass cutter.

Covering the structural framework built from the card stock can be accomplished with paper, card, or plastic sheet. This gives the model a more finished appearance and allows for graphics and some attention to aesthetics.

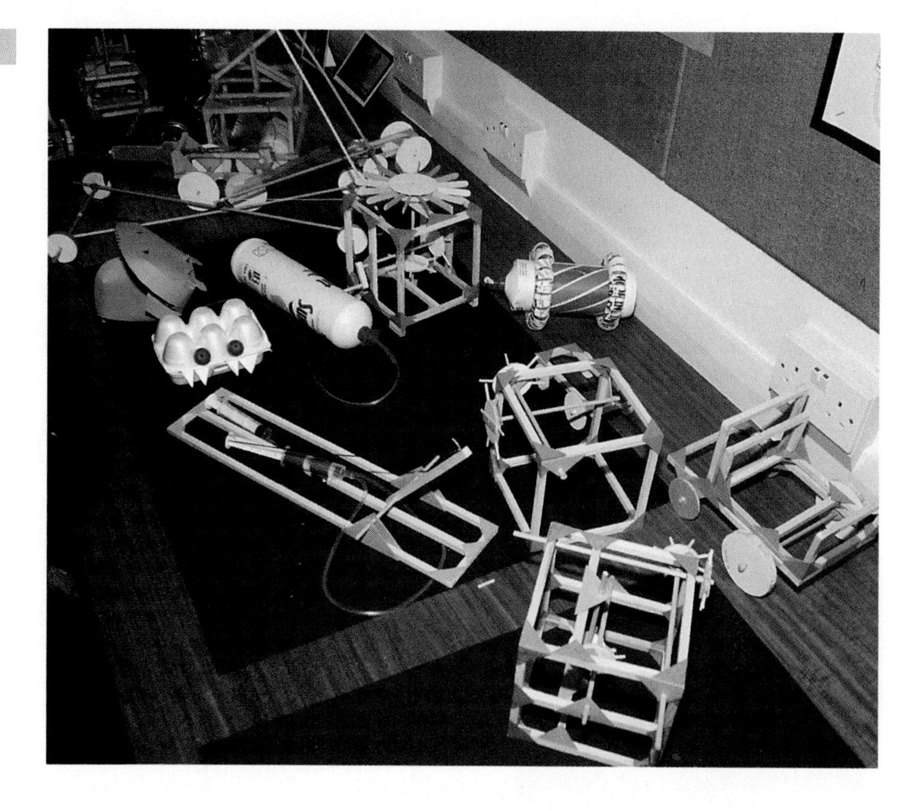

Fig. 10-12

Structures and mechanisms made using the Jinks' method

Functional Development

The functional problems of a solution will depend on the systems it will employ. Some solutions may have mechanical, electronic, and/or pneumatic and hydraulic systems, and some will not have any of these. When solutions require one of these functional systems, there are subproblems that must be solved. Here are some strategies for tackling functional problems.

Mechanical System Development

Mechanical systems were explained in Chapter 7. If a solution involves a mechanism, you may have already worked out the basic principles on paper and perhaps have calculated mechanical advantage or other needed values to determine "proof of concept." The solution now needs to be developed so that the final mechanical design can be worked out.

Low-Relief Modeling. If the solution is a device with a simple mechanical system, such as a box with a tray linked to the lid, then the length and the placement of these linkages will need to be worked out. *Low-relief modeling* is an ideal method for working out simple linkage

Fig. 10-13

Low-relief model of a mechanical system

design (Figure 10-13). In this technique, a base of heavy cardboard or homosote™ is used, and the levers and linkages are modeled full size or to scale using mat board, foam-core™ board, or poster board. The levers and linkages pivot on pushpins and are connected together with brass paper clasps or other suitable method.

Fischertechnik™. The Fischertechnik™ construction kit is a sturdy, sophisticated modeling tool. Blocks that snap together with durable nylon lugs ensure that assembled models will not fall apart easily. Because of the fine tolerances and complexity of the Fischertechnik™ components, it is more difficult to use at first. After you have had practice and some time to learn the variety of ways in which the components may be fastened together, the assembly of structures and mechanisms is much easier. Fischertechnik™ has components for building structures, mechanical mechanisms, and pneumatic-powered mechanisms.

Fig. 10-14

Mechanical model using Fischertechnik™

Lego Technik™. The Lego™ set is familiar to most students. It contains a variety of building blocks, gears, shafts, electric motors, levers, and cams. Some kits even include pneumatic cylinders. Assembly of Lego™ blocks into complex arrangements is rather easy, and vehicles, mechanisms, and entire systems may be modeled with little difficulty (Figure 10-15). The major drawback of the Lego™ system is that the assembled components come apart easily, and models that took care and time to construct may fall apart with a gentle touch. A small drop of model airplane glue can overcome this problem, but too much glue, too often can ruin parts. Talk to your instructor before using glue.

Meccano™. Meccano™ has been around for a very long time and is an excellent kit for developing structure and model mechanism. The set consists of strong, perforated metal strips that fasten together with metric machine screws and nuts. Precision brass gears, steel shafts, and a variety of cams, pulleys, and other mechanical components make the Meccano™ set an ideal choice for modeling sophisticated mechanisms (Figure 10-16). As with most construction kits, the many small parts make keeping a complete kit intact very difficult.

Fig. 10-15

Mechanical model using Lego™

Polymek™ Modeling System. The Polymek™ system falls somewhere between a kit and a supply of "stock" material. Prepunched board and flat stock, plastic and steel dowels, drill rod, plastic gears of various sizes, fastening angles, and standard metric fasteners make structure and mechanism assembly challenging but straightforward. Tools are available for heat bending the prepunched flat stock and for sawing, drilling, and taping the plastic square and round dowel stock. Because the materials are consumable, the costs are greater than standard construction kits, but students can keep their projects or disassemble the components for reuse later (Figure 10-17).

Fig. 10-16

Mechanical model using Meccano™

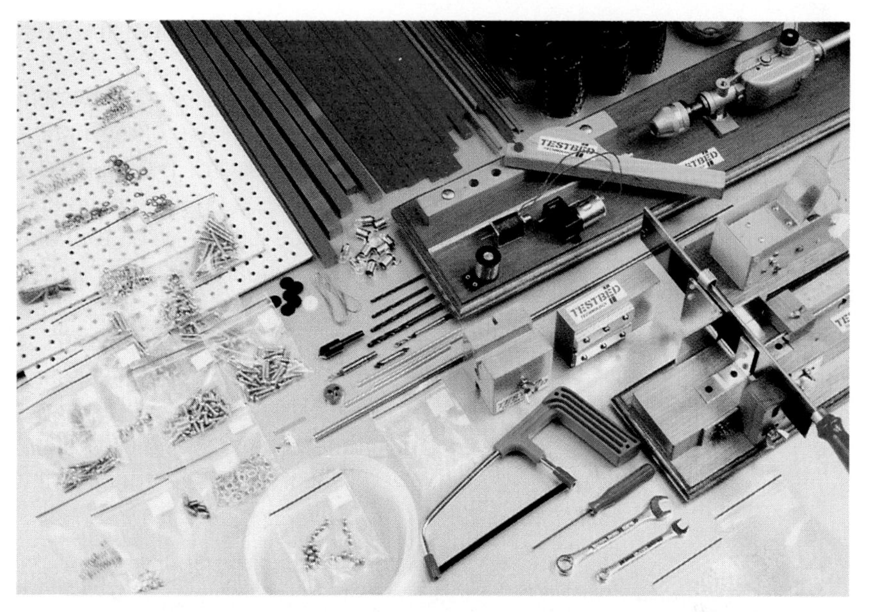

Fig. 10-17

Mechanical and structural model using Polymek™

Electronic System Development

Commercial Systems. Systems for electronics modeling most often fall into one of two categories: the multifunction board and the specific-function board. The *multifunction board* incorporates a variety of input, process, and output functions on one printed circuit board or panel, and uses jumper wires to model problem solutions (Figure 10-18). For example, a board of this type may have a switch, a light sensor, and a heat sensor on the *input* side; a buzzer, a lamp, and a motor on the *output* side; and various logic gates on the center *process* part of the board. By connecting various sensors to logic gates, such as "AND," "OR," and "NOT," and the gates to output devices, you can easily model many different sensing and control problem solutions. Many systems offer additional boards that focus on timing, binary counting, or memory.

The *specific-function board* is more sophisticated and, therefore, more useful for more elaborate problems (Figure 10-19). With this system, you will need to know more about the purpose and the operation of various electronic functions, such as amplification, flip-flop, or timing.

Boards with sensors connected to a network of "processing" function boards are then connected to boards with specific output devices to construct a complete functioning system. Each printed circuit board is a subsystem, such as a monostable timer. By connecting the boards together with wires or specially designed connectors, these systems can model elaborate electronics design solutions, from sensing and counting systems to communications systems.

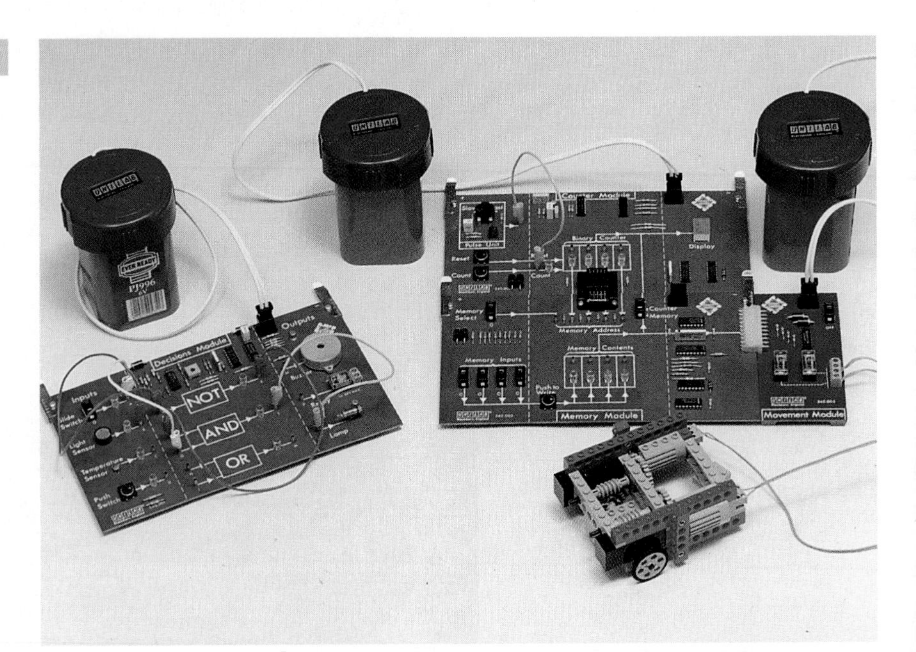

Fig. 10-18

Multifunction boards used to develop an electronic system

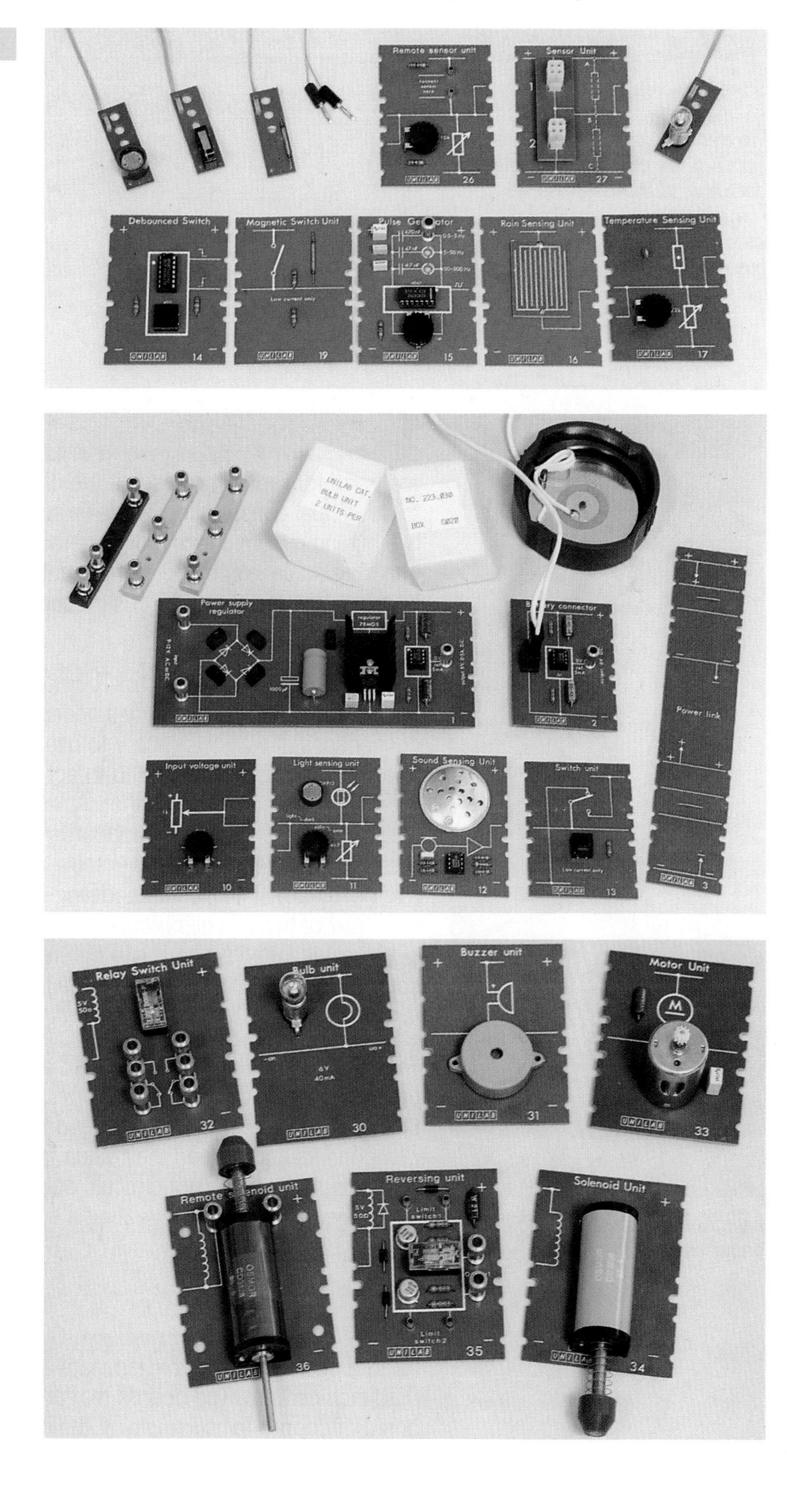

Fig. 10-19

Specific-function boards used to develop an electronic system

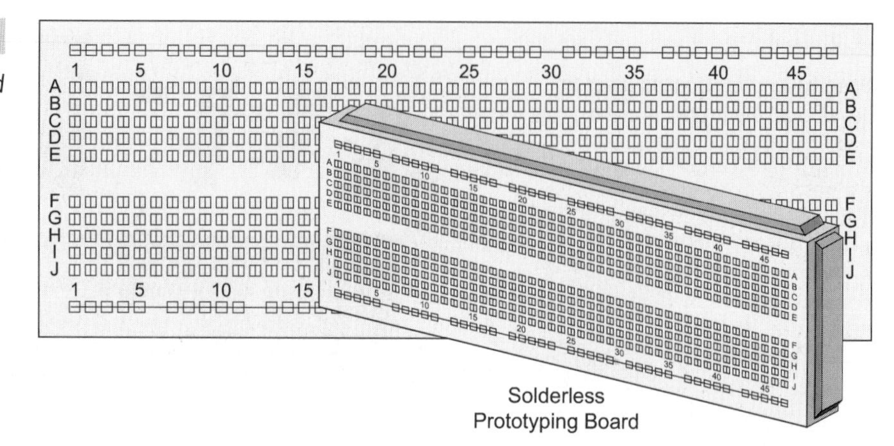

Fig. 10-20

Solderless breadboards are designed so that component leads can be inserted into common points. Terminals A–E are connected together for each row; terminals across the top row and the bottom row are all connected together and used as a bus for either "+" or "−."

Solderless Prototyping Boards. While commercial systems are good for modeling to see if the circuits will do what you want them to do, actual construction of the finished circuit for inclusion in a project is a big jump from connecting together commercial modules. Breadboarding actual electronic components before producing a finished circuit is a necessary step to ensure success of the final design.

In the early days of electronics, components were large and the current required to operate the devices necessitated large-gauge wire for connections. It was customary to use nails driven into wooden boards to anchor the components and to act as terminal posts for connecting wires and components together. Amateur electronics enthusiasts would often use kitchen breadboards for this purpose. Modern-day breadboards are a bit more versatile and a lot less cumbersome (Figure 10-20). The plastic modeling breadboards are designed so that each row of holes is electrically connected together. Component leads are inserted into the board into different rows. Connections between components are made with small-gauge solid wire.

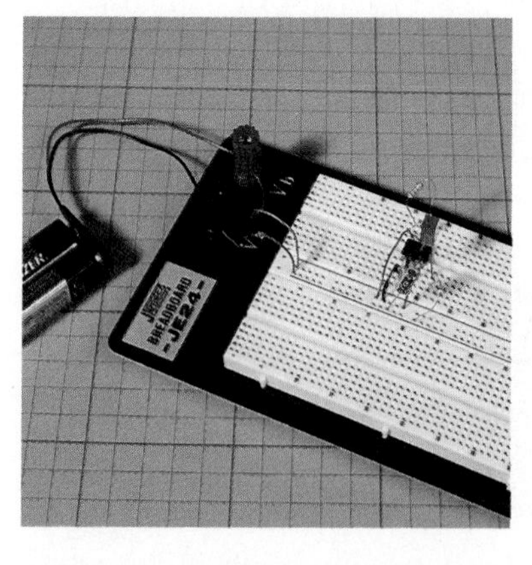

Fig. 10-21

A circuit modeled on a prototyping breadboard

To use breadboarding, you will need an assortment of electronic components, including IC chips, resistors, capacitors, potentiometers, LEDs, and others. You will also need to become familiar with these components and to read their value from the codes printed on them. Using breadboards is not difficult, but it takes practice to be able to make the transition from creating a schematic diagram to actual wiring (Figure 10-21).

Quick Wiring Techniques. Quick wiring techniques use perforated board made of thin laminate and covered with small holes on 0.10 inch centers. The boards that are the easiest to use have a small donut of copper foil surrounding the hole on one side. Component leads

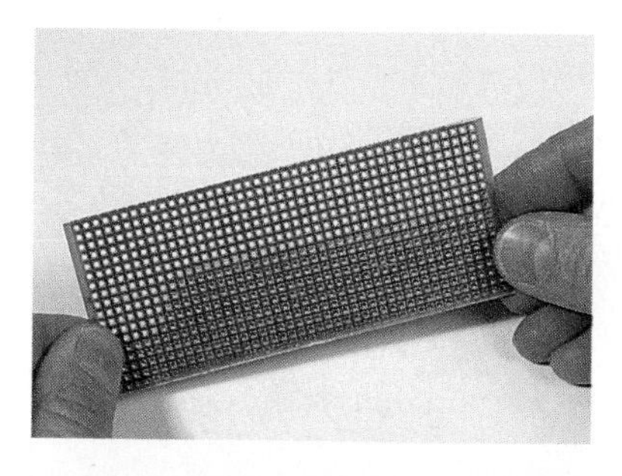

Fig. 10-22

Perfboard has holes punched at 0.1-inch centers. It comes both plain and with copper donuts used to solder component leads and wires.

Fig. 10-23

Circuit constructed on perfboard

or IC socket pins (it's not a good idea to solder directly to IC leads) are pushed through the board so that they protrude on the side with the copper foil. Small-gauge wire is then used to connect components together, and connections are soldered to the foil at appropriate locations (Figure 10-22).

Boards with no foil donuts are assembled in a similar way, except that components are not easily anchored to the board. The leads of each component protruding through a hole must be bent over to hold it mechanically in place. Component leads or insulated wire is used to make the necessary connections (Figure 10-23).

Pneumatic and Hydraulic System Development

There are three basic ways in which to develop pneumatic and hydraulic systems. The least inexpensive and most easily used technique is to use syringes and plastic hoses. These may be filled with water for a more positive hydraulic action, or the air within the lines may be used for pneumatics. The appropriate syringes are those that are used to dispense glue and other industrial liquids (Figure 10-25). The kinds of syringes used for human or animal injections are too small to be practical and are potentially very dangerous. The limitations of this technique are obvious: You have to devise methods of fastening the cylinder body and the piston arm, and there are no double-acting syringes, so lack of reverse movement of the piston must be overcome.

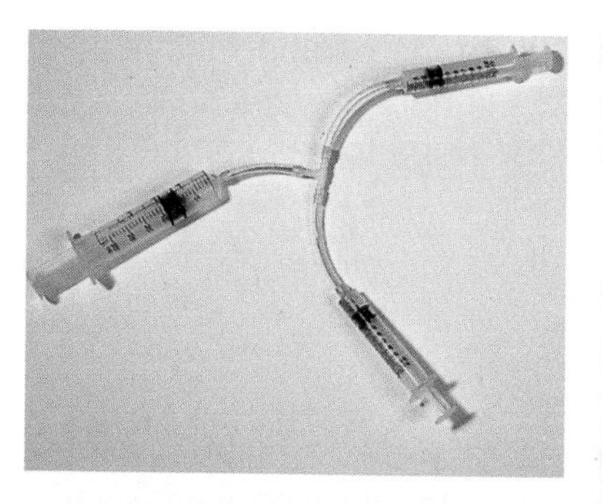

Fig. 10-24

Industrial syringes come in a variety of sizes.

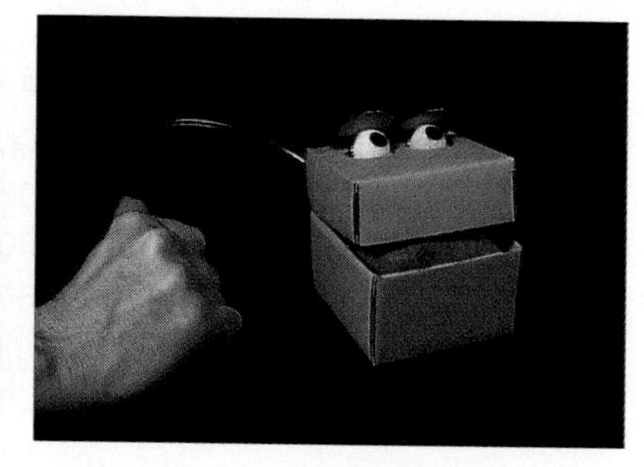

Fig. 10-25 a, b, and c

A student's pneumatic toy made using a balloon to create movement

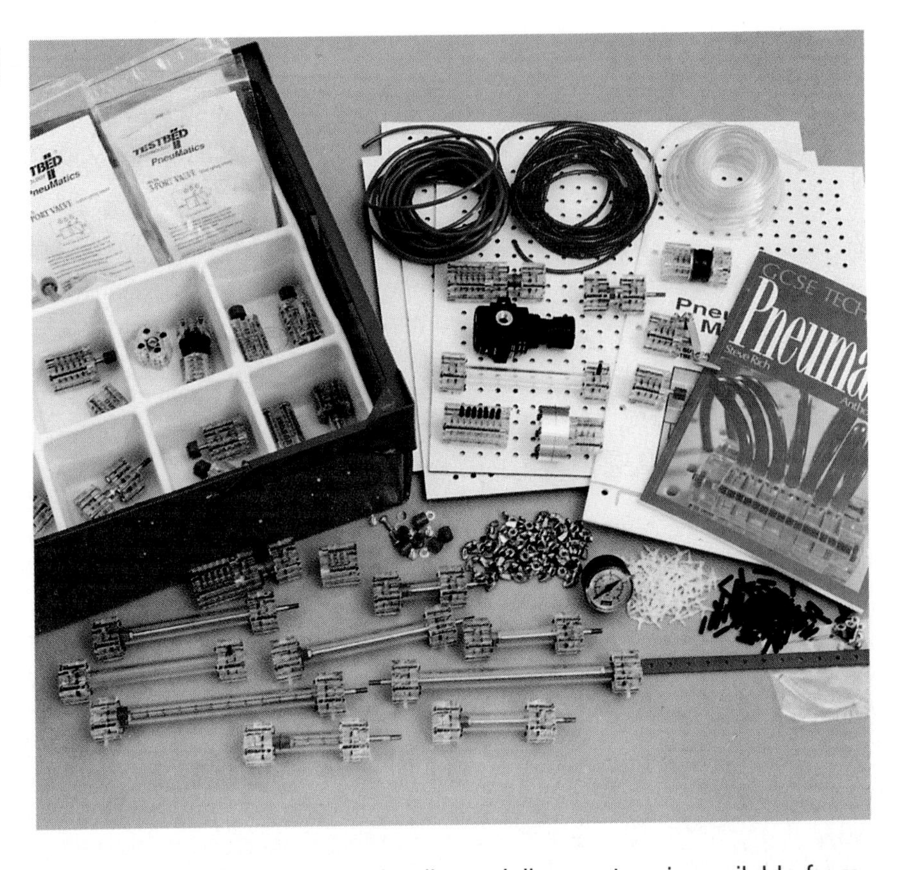

Fig. 10-26

Kit pneumatics are available from several manufacturers.

A second pneumatic- and hydraulic-modeling system is available from both Lego™ and Fischertechnik™ and Unilab. These systems are well designed and appropriate if the end product of your design work is a model (Figure 10-26). Obviously, these systems do not provide the force output of industrial systems.

A third strategy is to obtain industrial-grade pneumatic or hydraulic components with which to build your model. The cost of these components is usually quite high; they necessitate the use of special fittings, hoses, or lines; and often they require a pump or an available air line pressure. The advantage of using industrial-grade components is that they may be obtained in an almost endless variety of sizes and shapes to meet your specific requirements. Huge forces can be exerted with large-diameter cylinders.

Ergonomic Development

Fitting products to people, rather than the other way around, is called *ergonomics* or *human factors engineering*. In the developmental work you do to produce a product or a system, you will need to consider the human user (Figure 10-28). In Chapter 5, ergonomics and the use of anthropometric data were discussed in some detail.

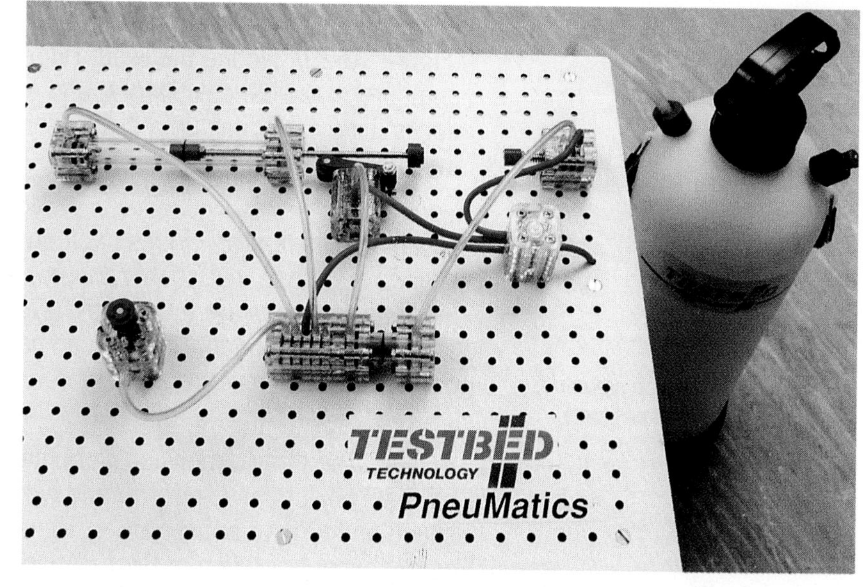

Fig. 10-27

Modeling a simple pneumatic system.

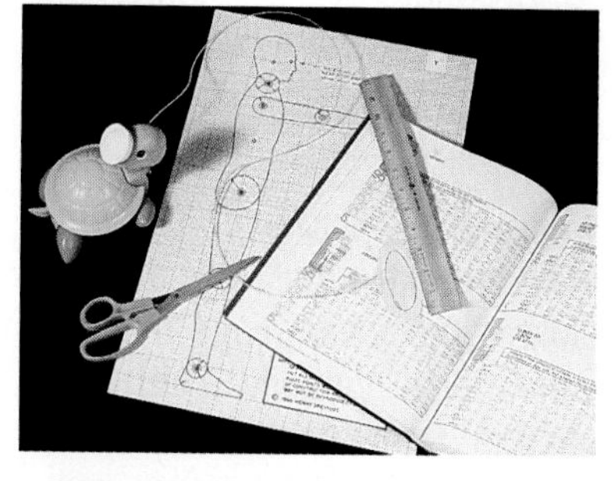

Fig. 10-28

Ergonomics must be considered when you are developing products or systems for human use.

You will need to identify who will use the product or system as a first step to employing ergonomic principles. For example, if a toy requires that a child move a knob or a lever, will the size of the child's hand fit the control? Will the child be able to exert sufficient force to move the control? The answers to these questions depend upon the age of the children in the target population. You will find the information in Chapter 5 and reference books on anthropometric data valuable for designing products for human use.

Appearance Development

The term "aesthetics" is often thought to be related to art, but aesthetics is an integral part of almost every aspect of our lives. *Aesthetic* is the word used to describe something that seems pleasing to us. Whether we are aware of it or not, aesthetics is considered in the decisions we make about what we like and don't like, and what we buy and don't buy. In the design work we do, aesthetics is an important consideration.

Design Elements

There are a number of design considerations that you may want to include in your developmental work. These elements often serve both aesthetic and functional purposes, as in the folded surface of a metal or plastic case. The folds provide a visual element of line and at the same time impose rigidity to an otherwise flexible material. Here are some design principles you might consider.

Fig. 10-29

Repeating a line is a common design element in consumer products. Courtesy of Nikon, Inc.

Line. A common way of decorating surfaces is to use *line* (Figure 10-29). Notice the linear ridges or indentations applied to computers, radios, calculators, and even the stripes on your sweat socks. Line can break up a dull or uninteresting surface. It can give the impression of motion or speed.

Shape. When a line closes back on itself, you get a shape. *Organic shapes* are those that seem random, meandering, or "natural." *Rectilinear shapes* are those that have straight sides meeting at corners and look human-made. Shapes can have hard or soft edges, be light or dark, opaque or transparent, colored or textured.

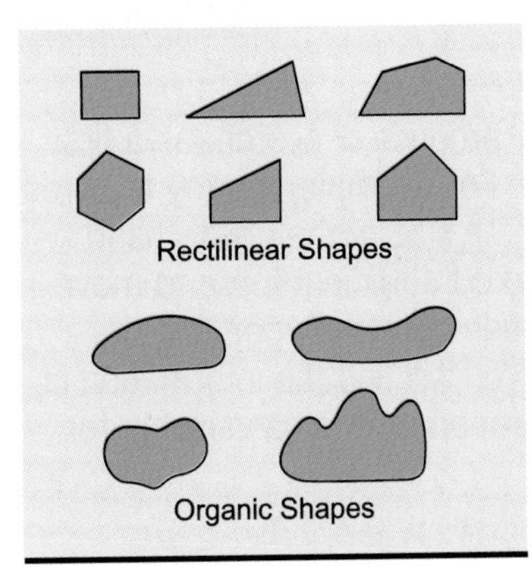

Fig. 10-30

Form. The three-dimensional counterpart of shape is called *form*. All physical objects have a form, and either they or we have to move in space and time to appreciate it. The same characteristics apply to form as to shape, but much of the functioning of products depends on form. The overall size, weight, materials, and purpose constrain the form of a car, building, or sculpture.

Value. In graphic terms, value means white, black, and everything in between. It can be more generally applied to the extremes, and the steps between them, of anything. Value always indicates things that are relative; value is seen in contrast to something else. Visually, a light or "high-value" feature makes a dark or "low-value" feature more visible.

Color. Possibly, the most accessible graphic design element is color. We react to it on emotional levels, and it directly affects our impressions. Physically, color has several dimensions, including wavelength, value, and intensity, all of which can be manipulated by the designer.

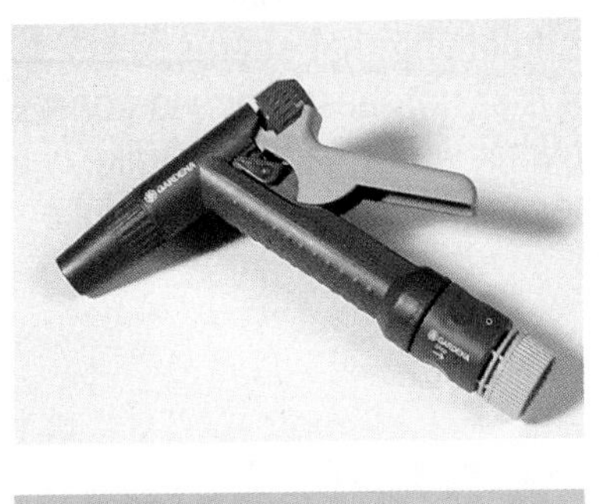

Fig. 10-31

The molded bumps on the handle of this garden hose nozzle are both functional and aesthetic.

Texture. For functional and aesthetic purposes, texture is extremely useful. As an alternative to value and color, which make surfaces interesting, texture is applied to objects that might be touched or at least might take advantage of light and shadow. Rough textures on tool handles, diving boards, and skateboards provide needed friction for grip (Figure 10-31).

Design Element Relationships

The relationships among design elements describe how those elements are arranged to achieve the designer's ends; for example, these relationships apply to the arrangement of text and photos on a printed page, as well as to how the displays and controls are arranged on a stereo. The relationships between the design elements

Fig. 10-32

Creating balance without exact symmetry makes products look interesting. Courtesy of Carver Corporation

are important aesthetic decisions, and the designer can put his or her "pieces" together in a variety of ways.

Balance. Visual and physical equilibrium is described as balance. Balance may be due to equity in area, mass, or attention within a defined area and may be *symmetrical* or *asymmetrical*. Symmetry means that balance has been achieved by one part being the mirror image of the other, such as the front of most cars. Asymmetrical balance does not rely on mirror image balance but, rather, on the careful placement of design elements within an area. It is more interesting than symmetrical balance (Figure 10-32).

Rhythm. Regularly occurring elements, such as line or shape, give a design rhythm. Rhythm is based on repetition, which can be exact (each element is the same), alternating (two or more alternating elements), or varied (recurring combinations of similar but not exact elements).

Proportion and Scale. The relationship of one part of a design to another is known as proportion. Objects have overall length, width, and height, and features within an object have other dimensions as well. The relationship between these various dimensions is proportion.

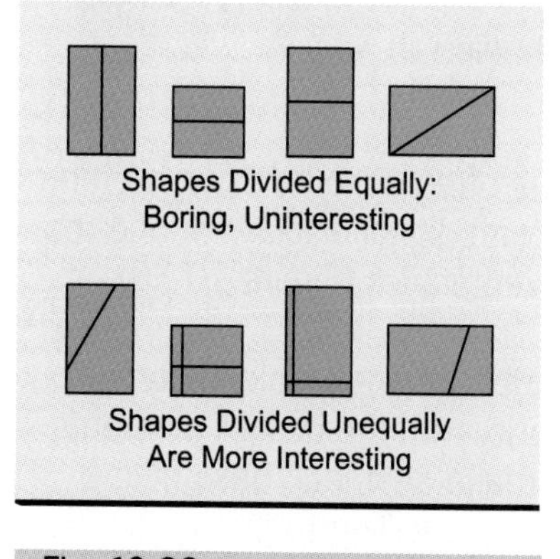

Fig. 10-33

It was recognized long ago that a shape divided into equal parts is rather dull and lifeless. More interesting shapes are suggested when areas are divided into unequal parts (Figure 10-33). The *golden section* was the proportion of a rectangle originated by the ancient Greeks based on the scale of the human body. Although this is a good starting point, do not let yourself be limited to making all objects with the golden section. Figure 10-34 shows how the golden section can be constructed.

Contrast. Contrast is achieved when two different shapes, textures, or colors are used in a design.

Harmony. When things are in agreement, they are in harmony. For example, by using similar shapes, color, or textures at several places

Fig. 10-34

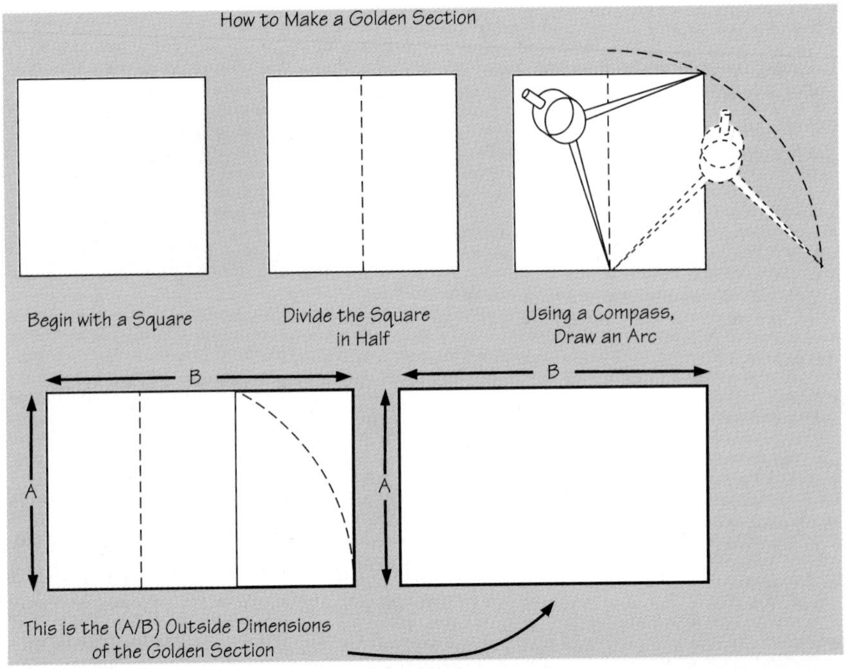

How to Make a Golden Section

Begin with a Square

Divide the Square in Half

Using a Compass, Draw an Arc

B

A

This is the (A/B) Outside Dimensions of the Golden Section

Fig. 10-35

The use of rectangles gives this product a clean, modern appearance. Courtesy of Yamaha Electronics Corporation

on an object, harmony is achieved. This is often done by using rectangles on consumer electronics products, with digital displays in one rectangle and controls grouped into similar rectangles (Figure 10-35).

These and other common design concepts are applicable in architecture, music, and product design. Familiarity with the design elements and sensitivity to possible relationships make aesthetic designing a reliable process, accessible to anyone who is willing to pursue an idea.

Design Styles

Because aesthetics is associated with taste and opinion, you may find it helpful to investigate a number of recognized visual design styles to see if any of them are to your liking. Although there are probably too

Fig. 10-36

The metalwork in this Paris Metro station entrance is a beautiful example of Art Nouveau style.

many styles to list here, there are a number of major aesthetic styles that have had an impact on product design and architecture since the late 1800s. They represented a departure from the traditional design of the time, and many are still visible today.

Art Nouveau. The Art Nouveau design style is based on natural shapes. In this style, patterns and designs are often based on curving, intertwined plant and animal shapes, such as climbing vines and flowers, leaves, birds and insects, and women with flowing hair. The stained-glass windows and lamps by Louis Comfort Tiffany are a good example of the Art Nouveau style. Begun in the early 1880s, Art Nouveau lasted until the beginning of the 1900s (Figure 10-36).

Arts and Crafts Movement. The concern for function and honesty in design was one of the basic tenets of the Arts and Crafts Movement, formed by William Morris in 1883. Until this time, much of design was concerned with adding features and decorations to an object in order to make its appearance pleasing. Morris, and the followers of the Arts and Crafts Movement, however, emphasized that the appearance of an object should clearly reflect its function. Adding decorations was out (Figure 10-37).

Fig. 10-37

An example of furniture from the Arts and Crafts Movement. Courtesy of This End Up Furniture Company

Bauhaus. A school of design in Weimar, Germany, called Bauhaus, was founded in 1919. One of the basic aims of this school of design was the design of products suitable for machine and mass production. The phrase "form follows function" has come to be associated with the Bauhaus, because the designs that emerged were very functional and reflected the materials from which they were made. Materials were not disguised but instead became a statement about the product itself.

Fig. 10-38

The Barcelona chair, 1929, is in the Bauhaus "form follows function" design style. Mies Van der Rohe's Barcelona chair courtesy of Knoll Studio, a division of the Knoll Group

Modernism. The modern design movement in Europe reached the United States in the 1930s, just after the Great Depression. It became part of the country's effort to get the economy moving again. Modernism had a futuristic style that looked toward tomorrow (Figure 10-39). The Chrysler building in New York City and such things as streamlined trains, the jukebox, and the classic American diner grew out of the futuristic style of Modernism.

An important influence on Modernism was the introduction of the many new materials in the early part of this century. Materials such as Bakelite, Plexiglas, vinyl, plywood, urea formaldehyde, and nylon had a tremendous impact on design. At the same time, fluorescent lighting was introduced and neon became popular. Both fluorescent and neon lighting influenced many designs for decades.

Design in the 1950s and 1960s. It is difficult to characterize design during the 1950s and 1960s because of the many contradictions. At the time that Henry Dreyfuss was developing anthropometic data for the emerging field of human factors engineering, Detroit automobile manufacturers were designing cars with huge fins and engines. While there was still a significant functionalist movement in design, American society was beginning to gear up for an economy that demanded relentless production and selling.

Functionalism is founded on design that minimizes and optimizes. The affluent society of the 1950s, 1960s, and even the 1970s demanded new products every year. The consuming public was convinced that a new-model refrigerator or other appliance should be purchased, even if the old one was still working. "New, improved!" "The latest model!" and other slogans emerged.

a. Courtesy of Eastman Kodak Company

b. Courtesy of California State Railroad Museum

Fig. 10-39a, b, and c

Streamlined products of all kinds are characteristic of Modernism, which emerged in the early 1930s.

c. Courtesy of Henry Ford Museum and Greenfield Village

Fig. 10-40

An example of high-tech design. Courtesy of Canon

Contemporary High-Tech Design. A recent trend in design since the 1970s is a slick, clean look with flowing curves and sharp edges. This design style seems to have no formal name, but many have referred to it as *high-tech* style. You can see this design style in the latest automobile model, consumer electronics, and many other products (Figure 10-40).

The convenience products of the 1980s are giving way to the environment friendly products of the 1990s. Designers are becoming more concerned about using both less material (conserving natural resources and reduction of waste) and environmentally safe materials. The new "Green Revolution" has encouraged smaller products and the development of new materials, such as soy ink for printed products.

Choosing the Best Solution

Once you have generated a number of workable ideas, you will need to evaluate them to find the "best" one. This is Step 5 in the design loop, and the more complex the solutions are, the more difficult this step becomes. It is often necessary to develop some of the solutions further to see if they are workable and should be considered. The nature of the design process is not linear, and, in this step (choosing the best solution), this is very clear.

For example, you may need to revisit Step 3 (research and investigation) to find out the characteristics of certain materials and if they are available to you in order to consider a particular solution seriously. Or, you may need to jump ahead to Step 6 (developmental work) to model a solution in cardboard or Lego™ in order to determine if the size, shape, or mechanism itself will fit your needs. Designing is not a linear process, and jumping around to various steps is not only necessary but also encouraged.

You will need to look at a number of factors in this selection process. First, what are the important things you should consider in making the decision about which solution is best? Second, once you have these criteria, how will you use them to reach a final decision?

Criteria Selection

The idea of a "best" solution can mean many things. It is important to base this choice on realistic and well-defined criteria. Some solutions, although better in concept, may not be realistically tackled in the time permitted for a school project; others may require equipment or expertise that is not available to you; still others may require expensive materials or materials that are difficult to obtain. Therefore, your judgment of the "best" solution needs to take into account all of these factors and more. In a commercial venture, similar limitations must be taken into account.

The first step you must take to decide on a final solution is to establish how well each solution being considered meets the design brief and specifications. If a solution does not achieve the desired results, do not discard the solution, but, rather, put it away in case you need it for later reference. Many solutions that at first seem unworkable turn out later to be valuable ideas. But there are always "shades of gray" in design, so it is probable that some solutions will more closely meet the design brief than others. Go through the specifications, and rate each solution on each point.

Specifications established early in the design process do not cover every consideration. Additional criteria that are important but that may not have been included before are *aesthetics, ergonomics, cost, available resources, personal skills, ethics, time,* and *safety.*

Here are some of the questions you will want to ask for each of the solutions you are considering:

Aesthetics. Is the solution pleasing to look at? Is it pleasing to a sample market population?

Ergonomics. Does the solution fit the human user?

Cost. What will it cost to make the solution? How would mass production affect the cost?

Available Resources. Are the materials, tools, and machines, as well as other necessary resources, readily available?

Personal Skills. Do I have the skills, or can I develop the skills, necessary to make the solution?

Ethics. Are the materials and processes that will be used to make the solution environmentally friendly?

Time. Will there be enough time to develop and to implement this solution?

Safety. Will the solution be safe to make, operate, use, store, and dispose of?

Rating each solution with regard to its potential to meet each of these criteria will provide you with additional information, based upon which you can reach the decision as to which is the "best" solution.

Attribute Matrix

Making sense of the information that you have generated about each solution can be simplified by developing a chart. The *attribute matrix* is a chart that has on one axis the different solutions being considered and on the other axis, the criteria being used to evaluate the appropriateness of each solution (Figure 10-41).

Fig. 10-41

An attribute matrix

Criteria	Solution Idea I	Solution Idea II	Solution Idea III
Appearance	3	5	4
Cost	3	2	4
Ergonomics	5	3	3
Available Resources	2	2	4
Ethics	4	5	5
Time Requirements	5	1	4
Safety	5	5	5
TOTALS	27	23	29

5 = meets criteria totally
4 = meets criteria mostly
3 = meets criteria somewhat
2 = meets criteria very minimally
1 = does not meet criteria

Develop a simple rating system, such as the numbers 1–5, that describes how well a particular solution achieves a specification or a design criteria. You will need to develop a rating system in order to assign a point value (e.g., 5 points if the solution meets the specification completely, 4 points if it meets it very well, 3 points if it mostly meets it, an so forth). Adding up the points for each solution will give you a more objective answer to the "best" question.

Summary

Generating and developing ideas are important steps in the design and problem solving process. The need to generate as many ideas as possible cannot be overemphasized. It must be understood that the greater the number of ideas generated, the greater the probability of finding the best possible solution to the problem. In technology, problem solvers are challenged to develop a variety of thinking skills. DeBono's horizontal model was used to show how new and novel solutions can be found by using traditional and nontraditional thinking skills. By using techniques such as the dominant idea, analogies, brainstorming, Synectics, and sketching or doodling, additional ideas may be found. All ideas need a waiting period to be fully developed.

After the problem-solver decides that sufficient ideas must be generated, developmental work on each promising idea must begin. This time is given to see which ideas will work. Usually the structure, function, and appearance for each idea must be developed. To do this, the designer uses a variety of materials and techniques. Some commercial kits and materials are available to aid in the development of the structure and function of the proposed solutions. The human user and the environment must always be considered during the development stage. The final appearance of each idea is also developed at this time. Both aesthetics and style are major elements of appearance.

At this point, the designer will select the best solution from the many ideas. The selection process should be based on established criteria that are usually placed in an attribute matrix for final evaluation. While much effort and time has been devoted to preparing possible solutions, considerable energy will still need to be invested in developing the best solution into a successful product. If the process of generating and developing ideas has been done well, then there will be confidence that the idea selected will be a successful solution to the identified problem.

Observation/Analysis/Synthesis

1. List the steps in the design process in which lateral thinking is employed, and give examples.
2. Using a group of three or four people, set up a brainstorming session, and come up with as many solutions as you can to one of the following problems:
   a. How can a new product be displayed?
   b. How can we raise money for a class trip?

3. Using synectics, role play a coffee maker, a piece of luggage, and a bicycle tire.
4. Use the low-relief modeling technique to model the working action of an architect's lamp, a folding chair, a production clamp, a backhoe and a robotic arm.
5. Investigate the kinds of rides found in an amusement park, and speculate on the mechanical systems that cause the motion of a specific ride. Using the Jinks' method, design and build an action model of the ride of your choice.
6. Using a construction kit, model a mechanism that will wind thread on a spool. The thread should be wound evenly with from thirty to forty turns per layer.
7. Using a solderless prototype board, model an electronic system that will count to 9; modify the system to be triggered by a light-dependent resistor.
8. Sketch a number of designs for a clock face, using the elements of line and color, and a computer mouse, using the elements of form and texture.
9. Obtain a child's toy, and analyze it in terms of aesthetics, ergonomics, cost, resources, ethics, and safety. Begin by developing a list of criteria by which you might determine at least three levels in each category.
10. Obtain at least three different brands of the same product, and develop an attribute matrix using criteria you have identified: a disposable razor, a retractable-tip ball-point pen, and a stapler.

CHAPTER

11 Materials and Modeling

Introduction

You have learned that technology is the human quest for solutions. It is interesting to observe that the early history of the human race is organized by the materials used in that quest for solutions. During the

Fig. 11-1

Early people used available natural materials to adapt to their environment.

Stone Age (prior to 3000 B.C.), people used naturally occurring materials to solve basic problems. Tools to extend human capability were shaped from stone, wood, and animal bones. Other natural materials were used to make clothing and to provide shelter (Figure 11-1). During the period of history known as *ancient times* (3000 B.C. to 400 A.D.), problem solvers found new materials to use.

The ancient times, divided into the Copper Age, the Bronze Age, and the Iron Age, reflected the development and use of new materials with broader capabilities. It was during this time that clay-making techniques were developed and that wood was used to make boats, carts, windmills, and waterwheels. Also during this period, aesthetics became more important. Furniture designs included extensive shaping of the material, and inlays and decorative metal parts. The importance of protective finishes was recognized, and shellac resins (a naturally occurring polymer) began to be used in dyes and varnishes.

Tools evolved from crude, basic tools to remove bark or to cut small timber. Over time, more complex tools were developed that allowed more sophisticated shaping of materials, such as the lathe, which was used to make round wooden objects, including chair and table legs (Figure 11-2).

Naturally occurring materials were modified to give them desired qualities. Alloys were developed by combining various metals. Bronze, for example, is the result of combining copper and tin. Iron, developed later, was shaped by beating plant fiber into the surface to make a stronger, harder metal. Today we know that the process combined carbon from organic plant material to create a crude form of steel.

Fig. 11-2

Eighteenth-century wood-turning lathe. Courtesy of Smithsonian Institution

Prior to the twentieth century, most material development was done with limited scientific knowledge. Today, material design and material science are closely allied.

Materials

Materials are necessary to produce models of design solutions and later, during production, the actual product itself. The technologist is limited, not only by his or her creative ideas, but also by the availability of appropriate materials. In 1832, Charles Babbage developed an early type of computer (Figure 11-3). Because Babbage had only clumsy mechanisms, rather than modern electronic components, his automatic calculator design was extremely limited in its capability when compared with today's microelectronic calculators. Computers, skyscrapers, jet engines, and satellite communications would not be possible without special materials and production processes. Appropriate materials and production processes are an essential part of technological design and problem solving. The need for new materi-

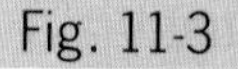

Fig. 11-3

In 1832, Charles Babbage (1791–1871) developed this "analytical engine," a precursor to the modern computer. Although never fully completed, it remains today a marvel of technological innovation. Courtesy of International Business Machines Corporation

als and production processes to meet specific requirements results in new problems for people to solve.

Material Selection

When designers consider material selection, they usually think about the characteristics of the material they plan to use. In Chapter 10, you learned that evaluating materials was part of choosing the best design solution. Materials have both physical and aesthetic qualities.

Physical Properties

Most materials must support a load or resist an outside force, meet specific design specifications, and have the ability to be processed into a finished product. Following are some characteristics of materials that should be considered when you are choosing the appropriate material for a specific job:

- **Strength**, also known as *stiffness*, is a measure of the material's ability to withstand a change in shape (strain) while under a load (stress). See Young's Modulus of Elasticity, which was discussed in Chapter 6. The strength of a material determines its ability to resist mechanical deformation from tension, compression, or torsion forces.
- **Ductility** describes how well the material can be shaped without fracturing. The ductility of sheet metal is important and allows the material to be pressed into new shapes, such as car body parts or a kitchen sink (Figure 11-4).
- **Brittleness** describes how quickly a material fractures as it is deformed. Window glass is extremely brittle.
- **Hardness** is the measure of the material's ability to withstand scratches, dents, and wear. A diamond is extremely hard, whereas balsa wood falls at the other end of the hardness scale.
- **Elasticity** describes how much the material can be deformed and still return to its original shape, an important factor in the material of an archer's bow.
- **Dimensional stability** is used to describe any change in the material when the material is exposed to changes in temperature, pressure, and humidity. Many plastics, such as acrylic, change size significantly with changes in temperature.
- **Electrical conductivity** is a measure of the material's ability to conduct or to transmit electrical energy. Copper is the most widely used conductor of electricity, but other excellent conductors include silver and gold.

Fig. 11-4

Many materials with different characteristics have been shaped to form products in this kitchen, including the stainless steel sink. Courtesy Northern Homes, Incorporated

- **Thermal conductivity** is a measure of the material's ability to conduct heat. Insulation, such as those materials found in the walls and ceilings of your home, have very low levels of thermal conductivity.
- **Other properties** can describe materials. For example, materials can have optical and magnetic properties.

Understanding material processing is extremely important when considering if the design solution is producible. Knowing the material properties needed to make a design allows the designer or engineer to select the most appropriate material to make a successful product. Materials, however, must be selected not only for their physical qualities but also for their cost, aesthetic value, and environmental impact. Charts will be used in this chapter to list commonly used materials and to compare important qualities. Materials used in modeling will be presented later in the chapter.

Material Classification

Materials can be classified as either *natural* or *synthetic*. In the beginning of this chapter, it was shown that early humans used natural materials to modify their environment. Materials such as stone, wool, clay, plant fiber, leather, wood, and metal are all found in nature. Synthetic materials are those invented and produced for broad use by industry. For the purposes of this book, both natural and synthetic materials will be divided into the following five categories:

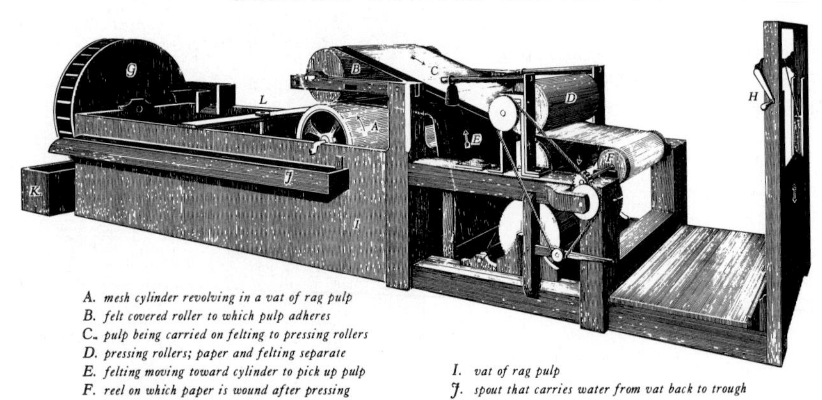

Fig. 11-5

The first American papermaking machine, constructed by Thomas Gilpin and put into operation in Gilpin Mills, 1817. Courtesy of The Hagley Museum and Library

- Paper and paperboard
- Metals
- Ceramics
- Polymers
- Composites

Paper and Paperboard

Paper is produced from wood and rag fibers. The word *paper* comes from *papyrus*, a reed used in early Egypt to make a writing material. True paper from wood cellulose fiber was invented in China around 100 A.D. The Chinese later found that paper could be made from other fibers. Papermaking spread to Europe by the thirteenth century. A ready supply of paper was one factor that influenced the invention of printing by Johann Gutenberg in 1450 (Figure 11-5).

In 1803, the Fourdrinier brothers developed a machine for making a continuous roll of paper. Their name is still associated with papermaking machines and the wire used to form the paper in the machine. By the mid 1800s, machines had been developed to grind logs into small chips and to separate (called "digesting") the cellulose fiber from the wood (Figure 11-6).

Paper was important for the early development of the publishing industry and newspapers, in particular. Paper also began to be used for packaging products. Many commodities, such as sugar and flour, were shipped to merchants in bulk containers. If a customer forgot to bring a container to the store, merchants would take paper and twist it into a cornucopia or funnel shape. For convenience, merchants began making up containers in advance. These packages developed into the modern paper bag. Again, it can be seen how a need led to the invention of a new product.

Paper and paperboard materials come in a variety of sizes, thicknesses, textures, and colors (Figure 11-7). Fine papers are used to print everything from stationery to corporation year-end reports. These papers are usually relatively thin in weight. Paperboard products are thicker and are commonly found in packaging applications. Posterboard, railroad board, and blanks are paperboard products ranging in thickness from three ply (0.015 in.) to ten ply (0.036 in.). These cardboard sheets are laminated with a thin sheet of white or colored paper that is capable of carrying a high-quality printed image (Figure 11-8).

Recycling of paper products is receiving greater attention. Presently, between 40 percent and 50 percent of all trash in the waste stream represents paper products. As new materials are being developed, greater attention is being paid to both the use of recycled materials in manufacturing and the recyclability of new products entering the marketplace. *Source reduction*, or the design of new products using less material, is also an important part of new package design work.

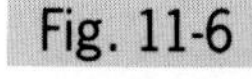
Fig. 11-6

The papermaking process. From Karsnitz, Graphic Communication Technology, *2nd edition, Copyright 1992 by Delmar Publishers Inc.*

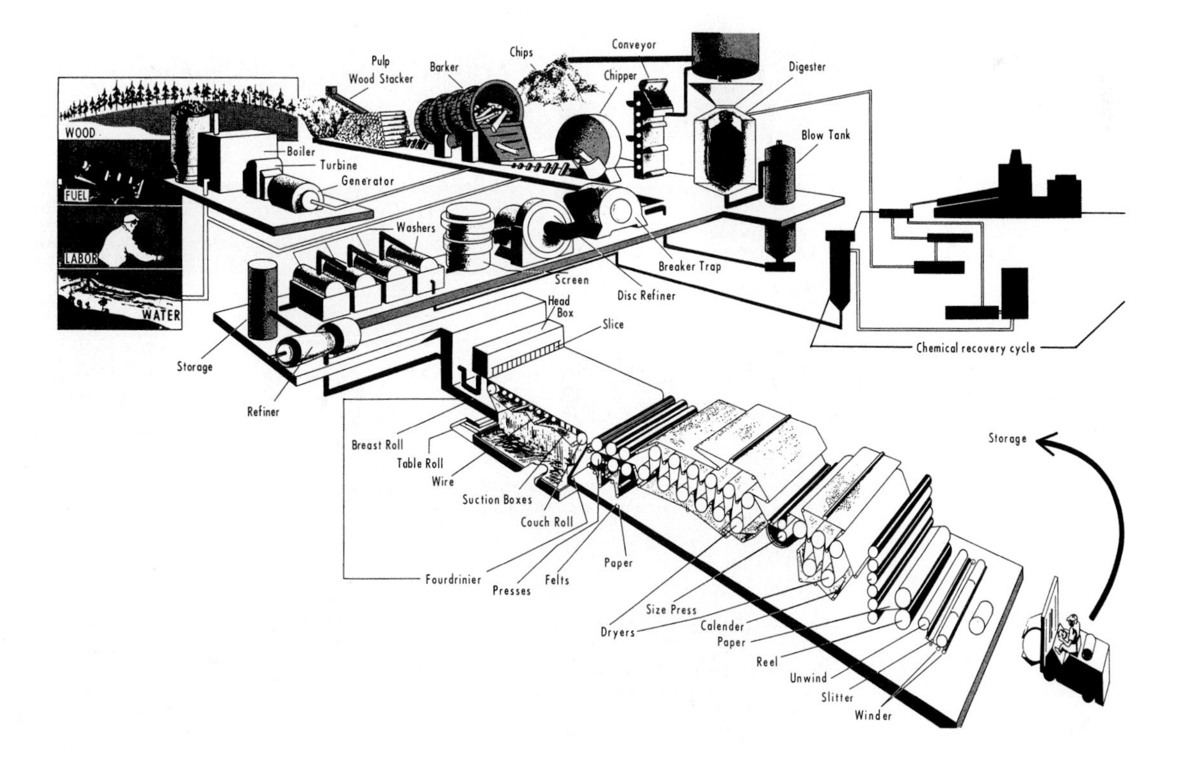

Some products rely completely on the use of recycled materials, such as *Homosote*™, a board made from recycled newspaper. For many years, it was used only in the construction industries as an insulating material, but today it is finding new applications in packaging and shipping. Do you know how paper egg cartons are made?

Metals

Metals are classified as ferrous or nonferrous (Figure 11-9). *Ferrous metals* contain iron. Iron ore is one of most common elements in the crust of the earth. During the Industrial Revolution, iron production

Fig. 11-7

Paper samples

Fig. 11-8

Printed packaging samples on laminated paperboard materials

Ferrous Metals

Name/Grade	Properties and Working Characteristics	Uses	Tensile Strength
Cast Iron	Brittle metal with a hard surface and a soft core. Strong under compression but cannot be bent or forged. Self-lubricating.	Heavy equipment and machinery frames and components. Brake drums and discs for transportation industry. Tools such as vices and small machine parts.	20,000–80,000 psi
Carbon Steel—1000 Series *See Appendix C for more information.* 1010/1018/1020	Low carbon series of "mild steel," containing between 0.10%–0.25% carbon. Can be cold or hot formed and welded. Relatively soft with low strength. Medium grades contain 0.25%–0.55% carbon. High carbon steel contains between 0.55%–1.70% carbon. Carbon generally increases the strength of the metal.	Fasteners such as rivets and nails. Also used to make wire, pipe, chains, machine parts, and structural components. Medium carbon steel is used for gears, shafts, tool shanks, and heat-treated machine parts. High carbon steel is used for screwdrivers, cutting tools, blades, drills, springs, knives, and striking dies.	40,000–190,000 psi
Carbon Steel—4000 Series *Alloy with molybdenum, chromium, and nickel in various combinations* 4130/4130/4340/4620	Good workability, fabrication, and welding characteristics. May be heat-treated. Good impact toughness, stress, and fatigue resistance.	Aircraft components, "chro-moly" bicycle frames.	100,000–165,000 psi
Carbon Steel—5000 Series *Alloy with chromium* 5160/52100	High tensile strength, tough, and ductile. Not easily machined or welded.	Springs and bearing applications.	280,000 psi
Stainless Steel *Alloy with chromium, nickel, and magnesium*	Hard and tough. Very good corrosion and wear resistance. Difficult to cut and fabricate.	Sinks, cutlery, kitchen utensils. Medical and surgical tools and equipment.	190,000 psi

Fig. 11-9

increased to meet the demand for wrought iron. *Wrought iron* was refined from pig iron and had good ductility. It was easily welded to other parts. Wrought iron was necessary for the production of steam engines, railroads, bridges, and large buildings of the Industrial Age.

Fig. 11-10

Courtesy of Corning, Incorporated

Today, around 90 percent of all iron is produced as mild steel. *Mild steel* contains between 0.10 percent and 0.30 percent carbon. Mild steel is difficult to machine and cannot be hardened or tempered. Mild steel with less than 0.20 percent carbon, however, has excellent ductility and can be easily drawn into needed shapes, such as rods, bars, and sheets. To increase its hardness and strength, more carbon is added to steel. Higher carbon steel material, however, becomes less ductile. High carbon steel contains between 0.60 percent and 1.70 percent carbon. High carbon steel is used to make cutting tools, such as drill bits. Stainless steel, an alloy with iron, is used for making specialized products such as kitchen utensils and medical equipment (Figure 11-10).

Nonferrous Metals. The most common nonferrous metal is aluminum (Figure 11-11). Nonferrous metals contain no iron. A magnet is often used to determine whether a metal is ferrous or nonferrous. Other metals in this category include titanium, copper, zinc, antimony, magnesium, tin, and lead. Tin cans are really made of steel with a thin tin coating to provide a corrosion-free container for foods. Lead, which was used extensively in paints, printing ink, and as a gasoline additive, has been found to be environmentally unsafe, and its use has been discontinued in these and many other products.

Most metals used today are alloyed, which means that they are combinations of two or more metals. Alloys have physical characteristics that are different from the original materials from which they were made. For example, bronze is stronger and has a higher melting point than either tin or copper from which it is made. Metals are very useful materials for design solutions in which high strength and stiffness are required. Metals also have good ductility, conductivity, and toughness.

Nonferrous Metals

Name/Grade	Properties and Working Characteristics	Uses	Tensile Strength
Aluminum—1000 Series (Min. 99% pure metal) *See Appendix C for more information.*	Pure-grade aluminum. It is soft and ductile, with excellent working characteristics. Nonheat treatable but the most weldable of all aluminum alloys. Excellent resistance to corrosion.	Good for intricate forms. Takes decorative finishes well. Used in chemical and food processing industries.	#1010 13,000–24,000 psi
Aluminum—2000 Series *Alloy with copper* 2011/2014/2017/2024	Alloy with excellent machining characteristics and high strength. Good strength-to-weight ratio. Machines to a high finish. May be heat treated. Not weldable. Usually anodized to improve corrosion resistance.	Parts such as fasteners, heavy-duty structural members, or items needing extensive machining. 2024 is used in aircraft structural components, hardware, truck wheels, and parts for the transportation industry.	#2024 27,000–70,000 psi
Aluminum—3000 Series *Alloy with manganese* 3003	Excellent working characteristics, with 20% increase in strength through addition of manganese. May be deep drawn, spun, brazed, and welded.	The most used aluminum alloy. Good for cooking utensils, decorative trim, awnings, siding, storage tanks, chemical equipment.	#3003 16,000–30,000 psi
Aluminum—5000 Series *Alloy with magnesium* 5005/5052/5083/5086	Similar to 3000 series, with very good resistance to marine environment and salt water corrosion. 5083/5086 offer better welding applications.	Marine and transportation parts, heavy duty cooking utensils, and bulk food processing. 5083/5086 are used in missile containers, truck and trailer assemblies, boat hulls, and superstructures.	#5052 28,000 psi
Aluminum—6000 Series *Alloy with magnesium and silicon* 6061/6063	The least expensive heat treatable aluminum. Good mechanical properties and corrosion resistance. If annealed, it has good workability. Well suited for anodizing in a variety of colors.	Truck bodies, screws, and structural components where better corrosion resistance is desirable, such as in window and door frames, store fronts, and trim.	#6061 18,000–45,000 psi
Aluminum—7000 Series *Alloy with zinc* 7075	Highest strength aluminum alloy. Heat treatable. Can be spot welded, but arc welding is not recommended.	High-stress parts.	#7075 33,000–83,000 psi

Fig. 11-11

Nonferrous Metals

Name/Grade	Properties and Working Characteristics	Uses	Tensile Strength
Aluminum *Casting alloy with copper and silicon*	Casts well both by sand and die methods. Good machinability, and tougher, harder, and greater fluidity over other aluminum alloys.	Transportation industry; engine blocks and components.	23,000–38,000 psi
Copper (CU)	Good working characteristics, ductile, tough. Suitable for hot and cold working. Excellent conductor of heat and electricity. Corrosion resistant, easily joined by soldering or brazing. Finishes well but is relatively expensive.	Domestic and heating water pipes. Electrical wire and electronic circuit boards.	25,000 psi
Brass *Copper alloy with zinc*	Stronger and harder than pure copper. Golden color. Corrosion resistant, good for casting and good machinability. Easily joined by brazing. Good conductor. Finishes well by polishing and enameling.	Jewelry, water valves, boat fittings, and other marine hardware, electrical sockets, rifle shells, ornaments.	37,000 psi
Bronze *Copper alloy with tin*	Strong, tough, and good wear characteristics. Good corrosion resistance. High tensile and fatigue strength. Also alloyed with silicon, aluminum, and beryllium.	Pipe fittings, statues, coins, and bearings.	37,000 psi
Tin Plate *Steel plated with tin*	Nontoxic coating on a malleable steel core.	Tin cans.	

Fig. 11-11 Continued

Ceramics

The word *ceramic* comes from the Greek word *keramos,* meaning "potter's clay." *Ceramics* are synthetic materials made from a variety of elements, called silicates (Figure 11-12). Minerals such as bauxite, clay, silica, feldspar, and talc are used to produce ceramic products. Glass, an **amorphous** material, is produced from sand. These materials form most of the earth's crust and could provide a nearly limitless supply of new raw material. Ceramics are highly resistant to chemicals and high temperatures. Most people think of ceramics as being used to make pottery, glass, and bricks. There are, however, other important uses for ceramic materials.

New developments in ceramics have led to other very promising applications. For example, ceramic materials are typically very brittle, but, if the air is removed, the material gains considerable elasticity. To

Ceramics

Name/Grade	Properties and Working Characteristics	Uses
Abrasives	Hard crystalline structures. Natural and synthetic examples include flint, garnet, diamond, aluminum oxide, silicon carbide, emery, pumice. Synthetic materials are most frequently used.	Belts, paper, grinding wheels, and sand blasting.
Clay Brick and tile	Brittle, lacking ductility and malleability. Low thermal conductivity. Kaolin, stoneware, and slip clays commonly used. Can be glazed.	Construction facing and interior decoration. Pottery and chemical stoneware.
Glass Soda-lime Lime-alumina Borosilicate	Brittle, low thermal conductivity, clear. Amorphous. 90% of glass is soda-lime. It is the cheapest form of glass and is made from silica (SiO_2) and soda (Na_2O).	Window glazing, optical glass, bottles, jars. Specially treated glass used to make cookware (Pyrex™). High-quality glass used in fiber.
Porcelain	Similar characteristics to clay. Very low electrical conductivity. Usually glazed.	Bathroom fixtures, insulators for spark plugs, and other electrical applications. Used as coating on metal to make household products.
Refractories	Excellent thermal qualities. Can be heated to over 1,500° C.	Oven, kiln, and furnace linings. Engine parts.

Fig. 11-12

demonstrate this characteristic, a spring has been produced from this new ceramic material. Another characteristic of ceramic that can be developed is a *piezoelectric* property. Mentioned briefly in Chapter 8, this is the property of a material that develops an electric charge when pressure is applied. Ceramic microphones are made of a ceramic material that converts the pressure of sound waves into an electrical signal.

Many abrasives are made from ceramic materials. *Abrasives* are hard, sharp-edged materials that are used for cutting, grinding, and polishing various materials. Alumina, carborundum, garnet, pumice, and silicon carbide are common abrasives. Bricks, tile, concrete, gypsum (plaster), porcelain, and glass are common ceramic materials used by the construction industry. A number of consumer products, including dinnerware, pots and pans, and glassware are made from ceramics.

An important area of ceramics is the design of materials to withstand high temperatures. Refractories are ceramic materials that can be used in temperatures above 1,700 degrees C. These materials are used in furnaces, as well as on vehicles that travel at very high speed, such as the space shuttle. Ceramics are also used in medicine and the computer industry. Ceramics will continue to gain importance as a source of new materials for design consideration.

Polymers

Naturally occurring polymers include shellac, cellulose, and caseins. A **polymer** is a type of molecule formed when two or more molecules, called "monomers," combine in a process called *polymerization*. Rubber, plastic, and adhesives are examples of polymers. The first synthetic polymer, called "Bakelite," was invented by Leo Baekeland in 1906. Bakelite was extensively used to make telephones and pot handles. Today, most synthetic polymers are produced from fossilized sources, such as natural gas, coal, or petroleum. Polymers are lightweight and have good corrosion resistance but have relatively low strength and do not hold their shape at high temperatures. Polymers are classified as thermoplastic, thermosetting, and elastomer materials.

Thermoplastic. Thermoplastic materials include acrylic, cellulose acetate, nylon, polyethylene, polypropylene, polystyrene, and polyvinyl

Thermoplastics

Name	Properties and Working Characteristics	Uses	Tensile Strength
Acrylonitrile-Butadiene-Styrene (ABS)	High-impact strength and toughness. Scratch and chemical resistant. Translucent. Easily fabricated, light and durable, high surface finish.	Kitchen ware (food processors/mixers), camera cases, appliance and tool housings, instrument panels, luggage, sporting goods, safety helmets, telephones, battery cases.	4,000–8,000 psi
Cellulose Acetate (CA)	Hard, tough, and stiff. Lightweight, transparent, and nonflammable. Easily machined. Poor chemical and electrical resistance. Absorbs moisture.	Photographic films, magnetic recording films, handles and knobs, sheet packaging for vegetables and display materials. Brush and combs. Lacquers in coating industry. Textiles.	3,000–8,000 psi

Fig. 11-13

Thermoplastics (continued)

Name	Properties and Working Characteristics	Uses	Tensile Strength
Polyamide (PA) Nylon	Hard, tough. Creamy white color. High temperature and solvent resistance. Resilient to wear, self-lubricating. Machines well. Difficult to join except by mechanical means. Absorbs water.	Bearings, gears, casings for power tools. Packaging for foods such as bacon and cheese. Cook-&-serve bags. Hot-melt glues. Textiles, including lightweight tents, fishing lines, and carpets. Rollers and bearings.	8,000–10,000 psi
Polyethylene (PE) High and low density	Most popular plastic. Tough, inexpensive, light. Has good electrical and chemical resistance. Difficult to bond, high thermal expansion. Attracts dust. Wide range of colors.	Food and liquid containers, squeeze-type bottles, toys. Films for textile goods and dry cleaners. Coatings for papermaking. Garbage cans, milk crates, buckets, bowls.	4,000 psi
Polypropylene (PP)	Similar to high-density polyethylene. Light, hard, impact resistant. Good chemical resistance. Can be sterilized, welded, bent, and hinged. Difficult to bond.	Medical equipment, syringes. Containers with hinges. Automobile ducts and trim, outdoor carpet. String, rope, nets, crates, and kitchenware.	3,400–5,300 psi
Polymethyl Methacrylate Acrylic	Hard and very durable. Transparent. Scratches easily. Good light qualities (fiber optics). Safe for food. Easily machined and colored. Polishes well. High moisture absorbing.	Models, automobile taillight lenses, aircraft canopies, windows, furniture, sanitary ware, illuminated signs.	5,500–10,000 psi
Polystyrene (PS)	Hard, stiff, brittle, low-impact strength. Transparent. Safe for foods and good water resistance. Expanded foam is buoyant, lightweight, crumbles easily, has good sound and heat insulating qualities.	Model kits, blister packaging, lenses, disposable plates, cups, and utensils. Foam 2lb/ft^3, sound and heat insulation, marine applications.	5,000–9,000 psi
Polyvinyl Chloride (PVC)	Soft, flexible, and lightweight. Low thermal resistance. Good chemical and weather resistance. Wide range of colors and can be easily made with a variety of patterns. Must be stabilized for exterior use.	Film for meat packaging. Pipes, gutters, bottles, soles of shoes, roofing material, and window frames. Soft goods include raincoats, dresses, handbags, shoes, shower curtains, and wall coverings.	1,000–5,900 psi

Fig. 11-13 Continued

Fig. 11-14

Thermoplastic products. Courtesy of Dow Chemical Company

chloride (PVC) (Figure 11-13). Thermoplastics can be formed at high temperatures, cooled, reheated, and reformed many times without affecting the material. Thermoplastics can be easily formed into shapes including fibers, bottles, pipe, records, chairs, and films (Figure 11-14).

Thermosetting. Thermosetting materials include alkyd, epoxy, phenolic, polyester, urethane, and silicone (Figure 11-15). Thermosetting polymers are heat formed as strong, cross-linked polymer chain bonds. Because thermosetting materials are formed in a heated environment, they cannot be heat worked as are thermoplastics. Once heated, thermosetting materials are permanently formed. Because they can resist high temperatures, they are commonly used in cookware and electrical applications. Many adhesives are made from thermosetting polymers.

Elastomers. Elastomers are natural and synthetic rubber materials. Elastomers are polymers that are capable of stretching over 200 percent without permanent deformation. Sulphur is added to rubber in a process called vulcanization. Vulcanized rubber is harder, more rigid, and has better wear characteristics. Elastomers are used to make belt drives, tires, inner tubes, and many other products.

Composites

Composites are materials made from two or more different materials (Figure 11-16). Reinforced concrete, plywood, and fiberglass are commonly recognized composites. These materials are designed to increase the strength-to-weight ratio, as well as the temperature and shock resistance of the material. *Graphite Reinforced Plastic (GRP)* is a new composite material that exhibits extraordinary strength and flexibility. It is used for fishing rods, pole-vaulting poles, antennas, and many other applications. Glass cloth reinforced resin, commonly known as fiberglass, has been in use for decades for everything from bathtubs to automobile bodies.

Thermosetting Plastics

Name	Properties and Working Characteristics	Uses	Tensile Strength
Elastomers Styrene-Butadiene-Rubber (SBR)	Soft, flexible, and capable of considerable elongation. Vulcanization increases stiffness and wear.	Resilient floors, sealants, adhesives, vibration mounting materials, footwear, tires, hoses, and latex paints.	3,000 psi
Epoxy Resin (ER)	Two-part material with low shrinkage. Strong, good chemical, corrosion, and wear resistance. Good bonding to other materials. Relatively expensive. Can be reinforced for increased strength. Resists heat up to 250° C.	Adhesives, pressure vessels, tool-forming dies, aircraft components, surface coatings, castings, electronic components.	4,000–13,000 psi
Melamine-Formaldehyde (MF)	Harder and more resistant to heat, water, scratches, and chemicals than urea-formaldehyde. Self-extinguishing, odorless, and stain resistant. Made in a variety of colors. Does oxidize over time.	Additives in papermaking, coatings (Formica™ counter tops), decorative laminates, electrical insulation, marine plywood, and textiles.	5,000–8,000 psi
Polyester Resins (PR)	Hard, stiff, and brittle. Good heat and chemical resistance. An electrical insulator. Can be colored.	Gears, pump impellers, lawn sprinklers, textile fibers, laminated panels, boat and car bodies, shell chairs, and containers.	4–10,000. Molding 50,000 psi
Silicones (SI)	High-service temperature, tough, and flexible. Available in liquid, compounds, lubricants, resins, and elastomers. High water resistance but low strength. High cost.	Molds, finishes for glass and fabrics. Adhesives, sealants, and gaskets. Silly Putty™. Medical implants. Electrical insulation on wire and components.	3,000–5,000 psi
Urea-Formaldehyde (UF)	Scratch, heat, and solvent resistant. Brittle, hard, and strong. Wide range of colors. Cheapest of the thermosetting plastics. Oxidizes over time.	Electrical components, plywood adhesive, appliance parts such as pot handles and knobs, textiles, insulating foam (0.7 lb/ft^3).	4,000–8,500 psi

Fig. 11-15

Composites

Name	Properties and Working Characteristics	Uses	Tensile Strength
Fiber-Reinforced-Plastics (FRP) Glass-reinforced-plastic (GRP)	Improved strength, stiffness, weight, and cost ratios over materials with comparable characteristics. Fibers used include glass, graphite, and plastics. Fillers include epoxy, polymer resins, and aluminum.	Automobile industry (1953 Corvette). Truck parts. Aerospace industry. Sporting goods such as archery bows, golf clubs, skis, bicycles, tennis rackets, and pole-vaulting poles. Circuit boards for the electronics industry.	8,0000–250,000 psi
Plywood	Increased strength and stiffness. Laminated wood with thermosetting resins.	Construction sheathing and furniture.	
Sandwich Boards	High strength-to-weight ratio. Cores consist of honey comb, waffle, corrugated, foam, and tube construction. Can be laminated with a variety of materials, including aluminum, wood veneers, vinyl and paper (foam-core™ board).	Hollow-core doors, room dividers, display models, and furniture.	

Fig. 11-16

Wood

Wood is a naturally occurring fiber-reinforced polymer composite. *Wood* consists of cellulose fiber bound together with organic lignin resins. Hardwoods from deciduous trees have a thicker tubular cell wall, which gives the wood more density or hardness. Many dimensional and manufactured wood materials are available (Figure 11-17). Wood is a renewable resource and is easily worked. The natural beauty of wood has been recognized for centuries. Wood has good strength but is not dimensionally stable and is subject to decay. Wood is often used to create design models.

Adhesives

Adhesives hold things together. Lignins in woods and animal glues are examples of natural adhesives. Today, most adhesives are synthetic and can be made for a general or specific application (Figure 11-18). Adhesives have been developed to a point where they can be used to hold many aircraft components together, eliminating the need to use rivets in many applications. The scientific principles of cohesion or adhesion, or a combination of both forces, are applied in modern adhesives.

Hardwood (Deciduous Trees)

Name/Color	Properties and Working Characteristics	Uses	Density
Ash Pale cream to light brown	Open-grained, tough, and flexible. Somewhat difficult to work. Good finishing characteristics. Good elastic qualities.	Tool handles, ladders, laminating, and sports equipment.	0.60
Beech White to pinkish brown	Close-grained, hard, tough, and strong. Somewhat difficult to work. Good finishing characteristics. Will warp.	Chairs, toys, tools, turned work, steam bending.	0.64
Elm Light reddish brown	Cross-grained, tough, and durable. Somewhat difficult to work. Does not split easily. Good in wet conditions. Has a tendency to warp.	General furniture construction, garden furniture, and turned items.	0.50
Oak White or Red	Open grain. Very strong, durable, hard, tough, and heavy. Somewhat difficult to work. Will stain around metal fittings.	General furniture construction and veneer plywood. Boat building, garden furniture, and posts.	0.63
Poplar Greenish White	Closed-grained. Softer than most hard woods. Easiest to work. Relatively inexpensive. Takes stains and finishes well.	General furniture and model construction.	0.63
Walnut Yellowish brown with dark lines	Cross-grained. Attractive, durable. Relatively easy to work. Finishes well.	Veneer and solid wood for furniture, handles such as gun stocks.	0.55
Basswood Cream-white	Tight, strong, straight-grained. Carves and machines easily.	Commercial model making and pattern making.	0.37

Fig. 11-17a

Cohesion. The atomic or molecular bonding of two similar materials is known as **cohesion**. An adhesive of this type employs *chemical bonding*, and it is the strongest form of bonding. Metals, ceramics, and most thermoplastics can be bonded together using chemical adhesives of this type. *Covalent bonds* formed between two similar atoms with free electrons is a common principle applied in chemical-type adhesives. In covalent bonds, the free electrons are shared, linking the two atoms together. Covalent bonding occurs when two metals are welded together. The welding process causes the joining point of the materials to be melted (made liquid) and the atoms of

Softwoods (Coniferous Trees)

Name/Color	Properties and Working Characteristics	Uses	Density
Balsa Creamy white	Straight-grained. Very soft and light. Relatively weak.	Model making.	0.14
Cedar (Red) Dark reddish brown	Straight-grained. Weak, soft, and light. Generally knot-free. Fragrant with natural oils that makes the wood resistant to weather, insects, and rot. Easy to work.	Outdoor furniture and siding, posts.	0.47
Fir/ Hemlock Greenish white	Cross-grained, tough, and durable. Somewhat difficult to work. Does not split easily. Good in wet conditions. Has a tendency to warp.	Residential construction.	0.39
Pine White	Straight-grained, relatively strong. Can have considerable knots. Easy to work. Reasonable cost and availability. Takes paint well.	Finish construction work, trim, furniture.	0.45

Fig. 11-17b

Adhesives / Materials	China/Ceramics	Concrete	Cork	Fabric	Glass	Leather	Paper	Plastics	Rubber	Metal	Wood
Contact Cement				B		B		B	B	B	B
Duco™ Cement	B/W				B/W	G/W		G/W		G/W	G/W
Epoxy	B				G					B	B
Glue Stick™							B				
Hot Melt						B	B				B
Resorcinol		B/W	B/W	B/W		B/W		B/W			B/W
Silicon Adhesive/Sealant	G/W				B/W					B/W	
Super Glue	B				G			G	B	B	G
Wall Paper Paste				B			B				
White Glue (Polyvinyl Acetate)											B

Fig. 11-18

Note: Check each specific application. Read and follow all package directions carefully. Most products must be used in well-ventilated areas with proper eye and skin protection.

Code: B = Best Application, G = Good Application, W = Waterproof

Fig. 11-19

Product with all finishing operations completed. Courtesy of Steelcase, Incorporated

both materials to link together as covalent bonds. In many thermoplastic processes, a solvent is utilized to weld the two materials together. Similar to welding metals by heating them to their melting temperature, the solvent softens (makes liquid) the adjacent areas of the two materials without heat. The result is a covalent bond.

Adhesion. When a physical or mechanical bond is made, the principle of adhesion is being applied. **Adhesion** is the fastening together of two materials by having molecules interlock through *absorption* (taking molecules from the material into the adhesive) and *diffusion* (molecules of the adhesive moving into the material). All thermosetting materials, as well as wood and paper, must be glued by adhesives.

Most modern adhesives are toxic and must be handled carefully. (*Safety Note:* Before using any adhesive, read and understand the directions. Ask your teacher if you do not fully understand the instructions. Follow all safety precautions and special handling instructions.) Most adhesives are safe after they cure. All adhesives are packaged with instructions that include the type of adhesive it is, its application, the material surface preparation, the curing time and temperature, and a safety notice. Read all the instructions before you use the adhesive.

Finishing

A quality finish will have a tremendous effect on the success of a product (Figure 11-19). Finishing is usually done to achieve one or more of the following outcomes:

- To protect
- To color
- To give texture
- To inform

The finishing stage should be carefully planned, and adequate time should be allowed for completion. Finishing often requires lengthy time periods for proper drying.

Surface Preparation. The material to be finished must be properly prepared. The first area of preparation is the removal of unwanted materials. Some materials, including most metals, may have a thin film of oil or wax on their surface. Plastic acrylic sheets have a protective paper coating. The removal of all foreign material must be done prior to smoothing the surfaces.

Surface Smoothness. The condition of the surface is critical to the final appearance of the product. Finish will accentuate surface irreg-

Abrasives

Name	Properties and Working Characteristics	Uses
Aluminum Oxide	Synthetic material. Not as hard as SiC but more impact resistant.	Floor sanding, used on harder metals, grinding wheels.
Diamond	Natural material. Harder than other materials by 10–100 times. Low-impact resistance. Very expensive abrasive.	Abrasive for brick- and concrete-cutting tools such as saws and drills. Used on metal-cutting tools.
Flint /Garnet	Natural material.	Abrasive paper.
Pumice	Natural material.	Finishing wood, cleaning rollers.
Silicon Carbide (SiC)	Synthetic material manufactured from sand, coke, and sawdust at 4,500° F. Next to diamond in hard-ness (trade name, Carborundum).	General-purpose abrasive. Refractory, heating elements, and grinding wheels.

Fig. 11-20

Fillers

Name	Properties and Working Characteristics	Uses
Plastic Filler Hobby Putty™ Stuff Putty™	Minimum shrinkage and cracking. Fill with multiple layers. Wet sand to a glass-like finish.	Model making.
Spackle	Some shrinkage. Water-based filter. Dry sand.	Drywall residential construction.
Putty	Flexible, easily formed. Weather resistant. Dries to a hard surface.	Window glazing.

Fig. 11-21

ularities; therefore, all irregularities must be smoothed. Usually, an abrasive material can be used to smooth the surface (Figure 11-20). When the irregularity is too extensive to be smoothed with an abrasive, then a *filler* will be needed (Figure 11-21). Fillers must be compatible with the material being smoothed. Putty and spackle are common fillers that are ideal for many modeling applications. The best fillers are those that do not shrink or crack, dry hard, and can be sanded. If the product is to be painted or finished with some other material, the filler must be compatible with that finish.

Coatings. Coatings are usually designed to protect, as well as to add color or texture to the material. Treating wood with *creosote* protects wood used for external application, such as fence posts. This is not to be confused with pressure-treated lumber, which is produced by forcing a chemical into the fibers of the wood under pressure. Many paints and varnishes are designed for applications on wood, metal, and other materials for both interior and exterior applications (Figure 11-22).

For some materials, a *primer* (first) coat of finish is recommended. On many ferrous metals, this coating consists of a zinc-chromate-based primer. Nonferrous metals, such as copper and brass, are often protected from oxidation by applying a thin coat of clear lacquer finish.

Fig. 11-22

Finishes \ Materials	China/Ceramics	Concrete	Cork	Fabric	Glass	Leather	Paper	Plastics	Rubber	Metal	Wood					
Anodizing										P/I						
Creosote											P					
Glazing	P/I	P/I														
Ink			I	I	I		I	I		I	I					
Laquer										P	P/I					
Latex Paint		P/I	I					I	I	P/I	P/I					
Oil											I					
Oil-Based Paint		P/I			I		I	I		P/I	P/I					
Plastic Coating										I						
Shellac											P/I					
Stain						I					I					
Polyurethane											P/I					
Varnish											P/I					
Wood Preservative											P					

Note: Check each specific application. Read and follow all package directions carefully. Most products must be used in well-ventilated areas with proper eye and skin protection.

Code: P = Protection, I = Improves Appearance

Fig. 11-23

PVC nonslip grip on a tool handle

On wood, a sealer and primer coat is often recommended. Always read the instructions on the container before applying finishing materials, and follow all safety precautions. *Polyurethane-based* finishes are both heatproof and waterproof, and dry to a very hard, durable finish. *Epoxy-based* materials are the hardest of all finishes. These are excellent for products or components that are subject to chemicals or hard wear. The DeLorean automobile, which was designed to have a long life, had a steel frame that was dipped in epoxy to prevent rust and corrosion.

Many new finishing materials are water or *latex* based. These finishes dry to a water resistant or waterproof finish but allow for the convenience of water wash up and have a reduced environmental impact over petrochemical-based materials. Latex, alkyds, and acrylics are common water-based paints. It is important to match the paint with the material to be finished. When you are finishing a material such as a plastic foam, a nonsolvent-based paint must be used. Solvents in the paint can dissolve the material being finished. When the compatibility between the finish and the material is unknown, always test the finish on a scrap piece of material.

Finishing materials do more than color and protect. *Texture* can be achieved by adding sand or some other material to the finish. A spackle-type paint uses two materials with different coherent qualities to create a texture. *Dip coating* with PVC materials give a thick, nonslip grip surface to tool handles (Figure 11-23). When the material is to be laminated with another material, a vinyl covering is often used. Vinyl coverings are used to cover frames such as model airplane wings and fuselage parts, as well as for other applications. Some inexpensive, simulated wood products use a laminated covering material over a less expensive board material. Coverings are often used in modeling because they can be used to simulate plastic, metal, or other types of surfaces without going to the expense of machining the actual part (Figure 11-24).

Graphics

No discussion on finishing would be complete without presenting information on the wide array of graphic materials available to the designer (Figure 11-25). One of the important aspects of finishing is being able to communicate not only feeling and mood through color and texture but also messages. Graphic materials are available in a variety of colors in both tape and sheet form. Most are self-adhesive. *Transfer type* and press type are names for preprinted sheets of the alphabet available in a variety of type styles, sizes, and colors. By pressing on the sheet with a pen or a stylus, these materials can be

FORM•X•FILM Application Instructions

Dry application of FORM•X•FILM

Use the Dry Application Method on all surfaces which cannot be sprayed with water, and for areas of less than 12 × 12 inches.

In all applications, be sure the surface to which FORM•X•FILM is to be applied is free of dust, dirt and grease. Clean plexiglas and other plastic surfaces with mild detergent, rinse with clean water, and dry with a lint-free cloth.

1. Cut a section of film slightly larger than required. Use the guidelines on the FORM•X•FILM backing sheet for accurate cutting and economy.

2a. Remove backing paper from section of FORM•X•FILM cut in Step 1. Starting along the narrowest edge, gradually lower film onto the work surface while smoothing it into place with your hand.

2b. Use the FORM•X•FILM applicator to gently work out any wrinkles or air bubbles, using overlapping strokes and working from the center to the outside edges. Trim away excess material and squeegee again.

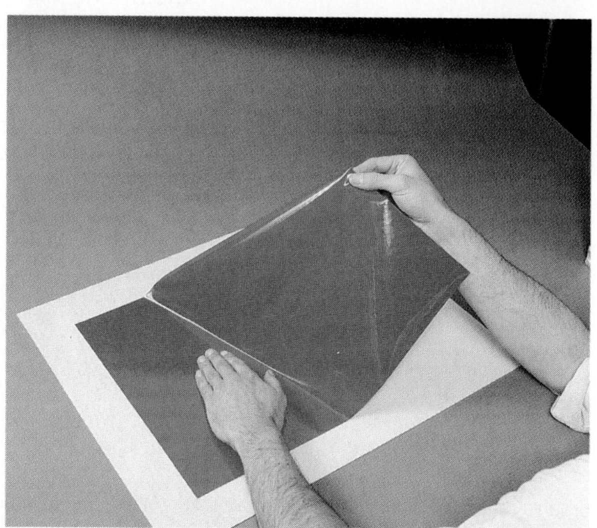

Fig. 11-24

Courtesy of Graphic Products Corporation

Fig. 11-25

Graphic materials. Courtesy of Chartpak

transferred directly onto the product letter by letter, but careful planning is necessary to ensure proper placement and alignment (Figure 11-26). The Letraset™ color-imaging system is similar but allows the designer to custom make complete type, logo, or other graphic images for transfer onto a product (Figure 11-27).

Preprinted materials also include a wide variety of images, including borders and other graphic forms and motifs. To complement transfer type, rule (line) tape, screen, and line patterns are available in transfer form. Most of the graphic materials are intended for use in modeling. They are used to represent the images that will be commercially printed on the product during the final manufacturing stage. Preprinted materials allow the designer to create a model that closely resembles the finished product. In this way, decision makers, including market analysts and potential consumers, can make a more informed judgment as to the possible success of the design.

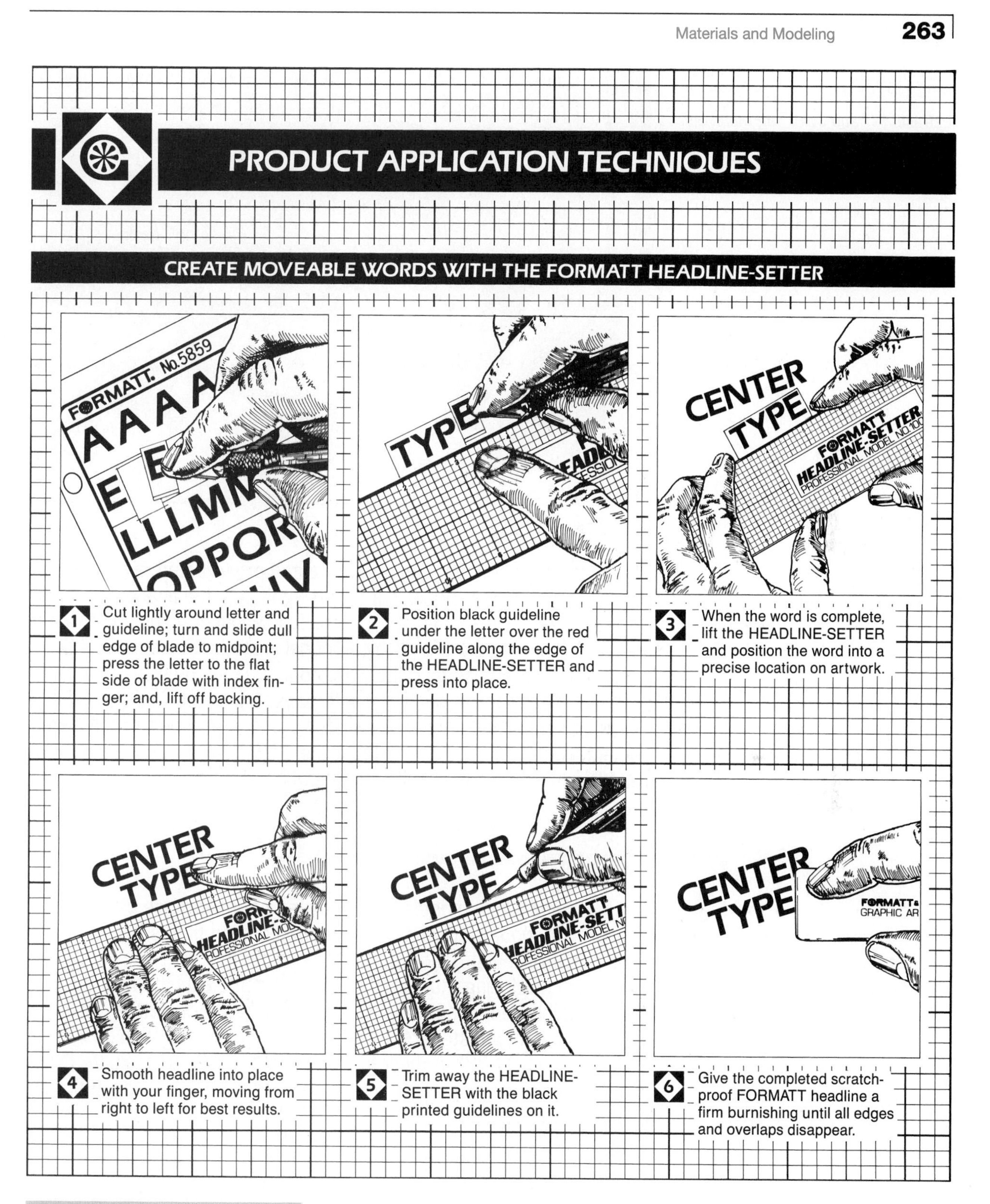

Fig. 11-26

Applying preprinted type. Courtesy of Graphics Products Corporation

Fig. 11-27

Letraset INT and Color Tag materials. Courtesy of Letraset USA

Modeling Processes

Before a design concept can be fully tested or evaluated, a model must be produced. A *model* is a representation of the final product and may be made from a variety of materials. *Computer simulation models* are often created before a physical model is made. *Scale models* are made to a smaller size than the final product. In the previous chapter, some simple modeling was done with paper (low-relief modeling) and modeling kits. These concept or sketch models were part of the developmental activity necessary to bring the design into clearer focus. What is now needed is a final representation of the design.

Appearance Models

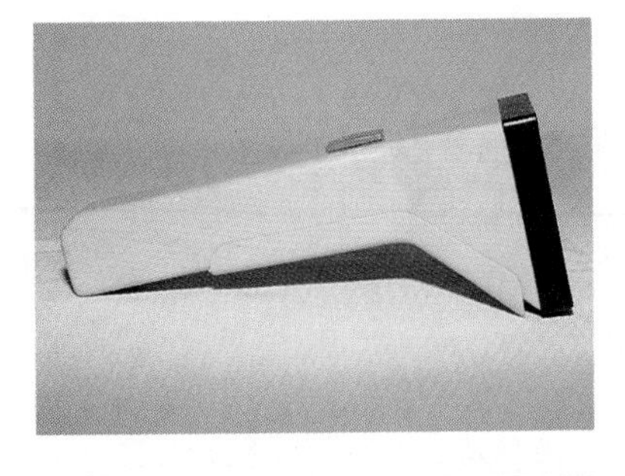

Fig. 11-28

Appearance model of a flashlight design. Courtesy of Beth Deene

Mock-ups and *block models* are two types of appearance models. *Appearance models* consist of any representation that is intended to "look like" the final product. Because considerable cost can be incurred in producing models, they are not made until the final design is well established. In appearance models, all surface details, including texture, screw fasteners, paint, and graphics are meticulously crafted (Figure 11-28). Appearance models are used to evaluate the aesthetics of the design and to make photographs for advanced promotional advertisements.

Prototype Models

Sometimes these models are called working or operational models. *Prototype models* are exact representations of the final product in all

Fig. 11-29

Camera prototype models.
Courtesy of the University of the Arts, Philadelphia

functional and appearance aspects, with one exception—they are custom produced or fabricated and may have some simulated materials and parts (Figure 11-29). As with the appearance model, prototype models should not be made until all design and operational questions have been resolved. Prototype models are expensive to produce. These models not only must look like the final product but also must function like the final product. The prototype model may have some simulated parts. This would be necessary, for example, if a complex plastic or metal part were needed. Because expensive machining will be required to produce the part in actual production, this expense cannot be justified before final approval of the project has been given. Under these conditions, simulated parts will be created. However, these parts must look and function like the final part.

Modeling requires extensive knowledge of materials and processes. Model making is interesting because it involves doing a variety of operations, from planning how to create the model, to actually producing the final model. Making models requires great care and craftsmanship, and good model makers are in high demand. Completing a well-made model is very rewarding.

Modeling with Paper

Fig. 11-30

Truck modeled with Checkcard™ paper material

A wide range of paper-based materials is available for modeling. Modeling with paper is relatively easy, because little specialized equipment is needed. Most students are familiar with 20# bond (notebook) paper. This paper, although sufficiently heavy for writing reports, is not substantial enough for modeling. *Cover, index, bristol,* and *tag papers* are better suited for some models. A 80# cover or 110# index paper is approximately equal to the thickness of three sheets of 20# bond paper. Cover papers come in a variety of textures and colors, including leather grain and metallic finishes. Index, bristol, and tag papers have smooth, hard surfaces. These papers are excellent for applying graphic images, including press type, tapes, and patterns. They can be colored with markers and paint.

Checkcard™ is a poster-board-type material, produced specifically for modeling. It comes in (25¼ in. × 17¾ in.) (64 cm × 45 cm) sheets in a variety of colors and with a 1 cm grid pattern printed on one side. The grid makes it easy to lay out, cut, and fold complex shapes, as well as to scale up or down (Figure 11-30). Modeling with Checkcard™ gives neat and sharp-looking results.

Fig. 11-31

A variety of materials suitable for modeling

Paperboard Materials. When the product, such as a package design, is to be modeled, *cardboard, illustration board,* or *railroad board* is typically used (Figure 11-31).

Packaging for large appliances or large consumer items usually requires a corrugated cardboard or plastic-based packaging material. Many packages are made by laminating a preprinted paper sheet onto a cardboard or corrugated sheet prior to forming and assembling the package.

Other Boards. A number of plastic and plastic-laminated boards are available for modeling. Many of these materials are used for making display and architectural models (Figure 11-32). *Foam-core™ boards* are constructed of a Styrofoam center sheet sandwiched between two high-quality paper sheets. These boards are available in ⅛-inch, ¼-inch, and ½-inch thicknesses. They are cut with a hobby or utility knife, take an excellent finish, and can be assembled by gluing or taping. Unfortunately, the material is relatively expensive and must be handled carefully, because it is easily dented.

Styrofoam™ brand art board is a multicellular board of extruded polystyrene. It is manufactured in 48″ × 96″ boards in thicknesses of ¾ inch, 1 inch, and 2 inches. It is white in color with a high-quality flat, smooth surface. It is used for signs, exhibits, displays, and architectural models. Dow art board is combustible and should not be exposed to flames or intense heat. Alucobond Technologies, Inc., produces a polyvinyl chloride (PVC) foam sheet material called Sintra®. This material is available in several colors, a variety of stock sheet sizes, and in thicknesses ranging from 1mm to 6mm, 10mm, 13mm, and 19mm. Sintra material is easy to work with and can be thermoformed. It can be assembled with mechanical fasteners, glued or welded. It is often used for exhibits, signs, and displays (Figure 11-33).

Fig. 11-32

Architectural model

Fig. 11-33

Display made with Sintra® materials. Courtesy of Alucobond Technologies, Incorporated

Fig. 11-34

Camera models. Courtesy of the University of the Arts, Philadelphia

Fig. 11-35

Environmentally friendly bag. Courtesy of Eco-Tote Company

Modeling with Wood and Metal

Wood and metal are the materials of choice for making appearance and operational models when the final product is to be manufactured using these materials. Modeling with wood and metal can require extensive laboratory equipment and experience. Very often, a combination of materials, including wood or plastic parts, is used to make a model (Figure 11-34). Because wood is relatively easy to work with, it can be used to make appearance models. However, making wood look like plastic or metal can require time-consuming finishing operations. Wood typically is too weak and lacks the dimensional stability needed for prototype models.

For wood modeling, the choice of the type of wood is particularly important. Some woods lend themselves to shaping and finishing, whereas some do not. Many woods with a soft, fine, tight grain, such as basswood, are well suited to the development of models. Woods with a coarser grain are not suited, as they tend to splinter and separate. Wood with knots is not acceptable.

Modeling with Textiles

Textiles and other nontraditional materials should be considered for modeling. If the design is for a tote bag, then a canvas-type material may be cut and assembled by sewing (Figure 11-35). For problems such as a new tent design or a Windbreaker™, nylon may be the material of choice.

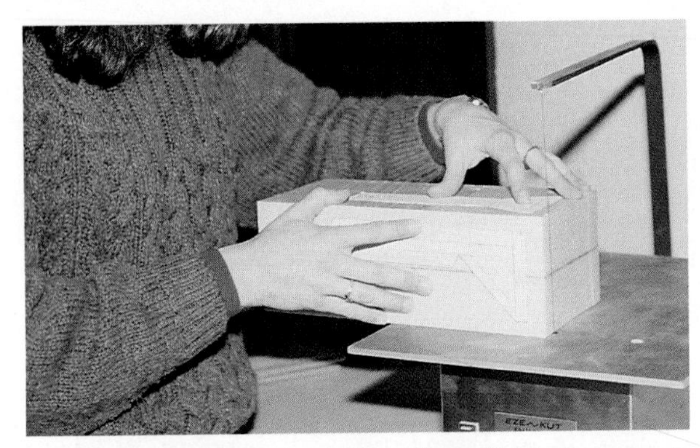

a. Glue the pattern on a Styrofoam™ block and cut shape with a hot wire or mechanical cutter. (note: Use a hot wire cutter only in a well-ventilated area)

b. Complete final contour by sanding or filling. Use a non-solvent based filler to develop fillets, add raised features, or repair damaged areas.

c. Carefully fine sand product. If the product is to be painted, a non-solvent based sealer should be used.

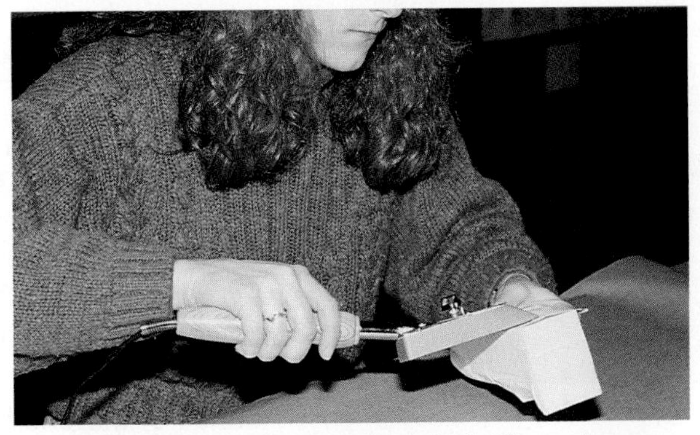

d. Apply Monokote™ or other plasic film using a heated iron. Paint can also be applied over the film if desired.

e. Plastic trim, such as the lens frame in this model, can be shaped using a strip heater. The trim can be fastened with an adhesive. Test the adhesive on a scrap piece of styrofoam™ to ensure that it will not melt the material.

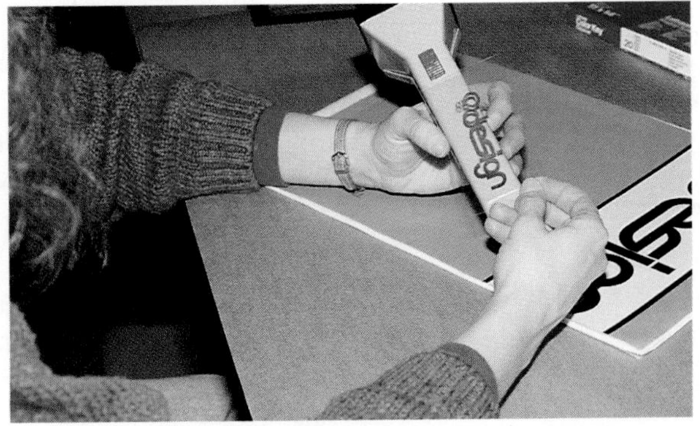

f. Graphics can be applied to the final model to give it a finished appearance. Model development courtesy of Beth Deene.

Fig. 11-36

Modeling with Styrofoam™

Polystyrene can be an excellent material for making appearance models. It is easily shaped using a hot wire, saw, router, drill, sharp knife, file, or abrasive papers. The use of a hot-wire cutter causes noxious fumes to be released, which have been known to cause headaches in some individuals. Only use a hot-wire cutter where there is adequate ventilation, and discontinue using it if anyone in the lab complains of a headache or other reaction.

Styrofoam™ has a smooth surface and can be finished with nonpetroleum-based paints. However, some precautions are necessary, since the material cannot be used near open flames or high temperatures. In addition, some polystyrenes are manufactured using chlorofluorocarbons (CFCs), which are harmful to the environment (Figure 11-36).

Materials manufactured with CFCs should not be used for modeling, because cutting and shaping them releases the gases into the atmosphere, where they damage the ozone layer. Fortunately, Styrofoam™ brand plastic foam made by Dow is free of CFCs since July 1990. The material is now manufactured using Isotron 142b. In addition, Dow is actively involved in polystyrene recycling of both preconsumer and postconsumer plastic waste. Styrofoam™ is used primarily in the construction industry. Dow, the world's largest manufacturer of extruded polystyrene foam insulation, estimates that, since 1970, over 5 billion gallons of fuel oil have been conserved because of polystyrene foam insulation. Styrofoam™ is being used to replace wood, steel, corrugated board, cork, fiberglass, and other materials in construction and packaging applications. It comes in compressive strengths, ranging from 15 to 115 lb/in^2, densities ranging from 1.6 to 3.5 lb/ft^3, and will not promote bacteria or fungus growth. Styrofoam™ has a smooth surface and is resistant to bases, salts, alcohol, and many acids but not to petroleum-based solvents.

Modeling with Plastics

Plastics can be easily formed by basic molding and fabricating techniques. In model making, simple plastic shapes can be made by mechanical forming (Figure 11-37). Mechanical forming requires a heat source, suitable thermoplastic such as acrylic, and a simple form around which to shape the plastic. For simple mechanical forming, such as bending sheet plastic, a strip heater is used (Figure 11-38).

Thermoforming. Thermoforming is an excellent process for making models when a free form or shape in plastic is to be simulated (Figure 11-39). *Vacuum forming, blow molding, drape forming,*

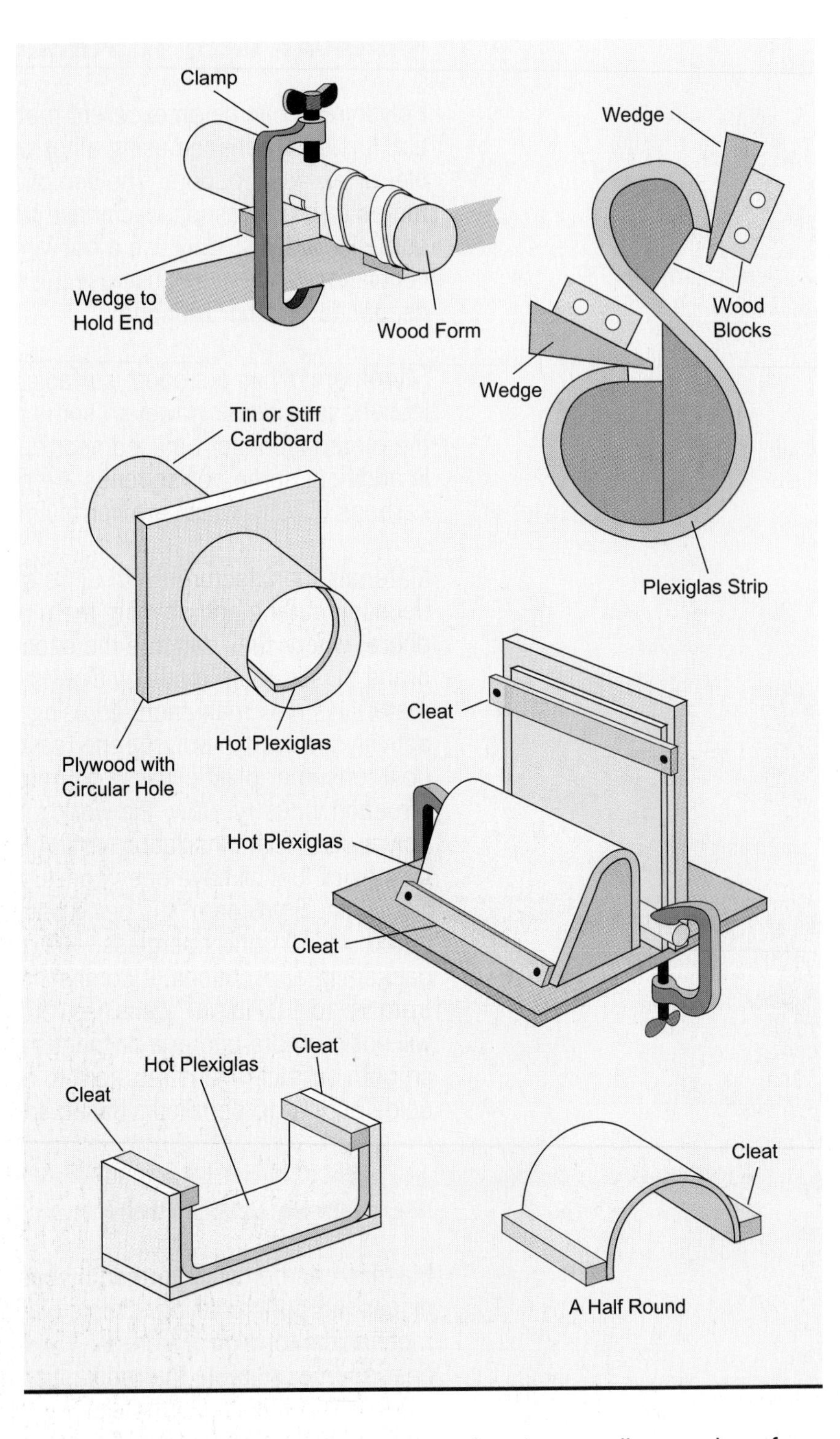

Fig. 11-37

Simple shapes made in plastic by mechanical forming

matched-mold forming, and *plug-assist forming* are all examples of thermoforming processes. One of the advantages of thermoforming is that low-cost plaster or wood molds can be used to form the plastic. Many products such as light fixtures, ice cube trays, kitchen ware, and instrument panels can be thermoformed.

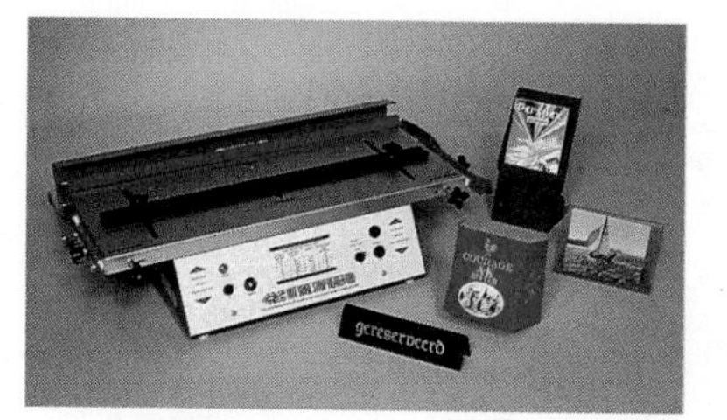

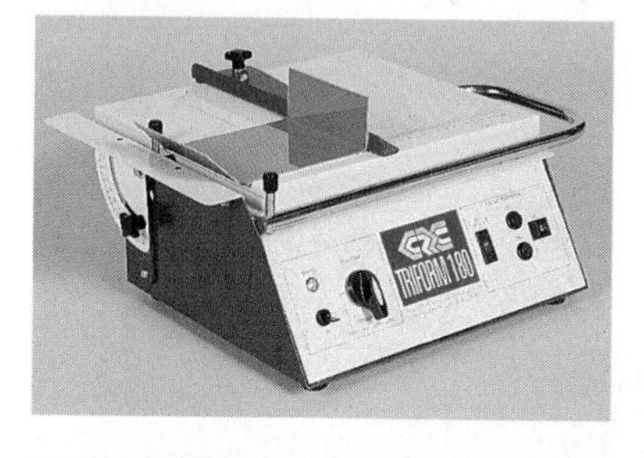

Fig. 11-38

11-38a Hot wire strip heater. Courtesy of CR Clarke & Paxton Patterson
11-38b Triform 180 hot wire bender/cutter. Courtesy of CR Clarke & Paxton Patterson

Vacuum Forming Model Parts. The process of vacuum forming is shown in Figure 11-40. The first step is to make a mold from the design drawings. The mold must have slots or holes distributed over all low or unconnected parts. These holes allow air to be evacuated from the mold. The holes should be no larger than 0.025 inches (0.65mm) in size. In addition to holes, the mold must have from 2 to 7 degrees of draft. *Draft* is the slope on the sides of the mold needed to allow the molded part to be released from the mold.

A number of thermoplastics can be used for vacuum forming, including high-density polystyrene, ABS, polyethylene, and flexible PVC. Acrylic used in mechanical forming is not recommended for vacuum forming. As can be seen in Figure 11-40, the plastic is clamped over the mold and heated. When the plastic has reached its rubbery elastic stage, the mold is quickly evacuated. As the air is removed, atmospheric pressure (up to 14.7 psi) presses down on the plastic and forces it into the mold. After cooling, the formed plastic can be removed, the flashing trimmed, and any other finishing operations completed.

Modeling Catalogs

Many materials are available for both model making and giving that special finishing touch that will make your work distinctive. Beyond the materials already mentioned, an unbelievable variety of miniature and specialized parts, such as hinges, tubing, electrical components, pumps, switches, batteries, motors, as well as special tools, are available from these sources and many model, art, or hobby stores.

Electronics Modeling

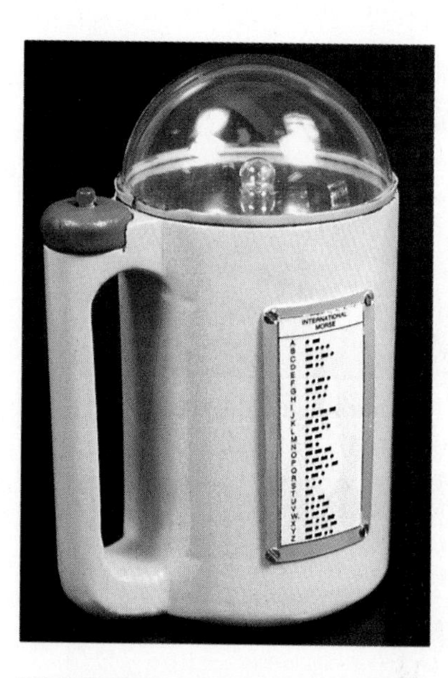

Fig. 11-39

The lamp lens in this model has been formed in acrylic by blow molding. Model courtesy of Todd Engle

Printed circuit boards (PCBs) are part of almost all modern electronic products (Figure 11-41). They are ideal for mass production of complex circuits and are reliable and durable. The thin *traces* of copper on a PCB take the place of individual wires, connecting one component to another. Many electronic components are designed specifically to be installed and soldered to printed circuit boards. Before PCBs were developed, almost all wiring was done by hand.

Printed circuit boards are thin, laminated boards made from many layers of paper and epoxy or from fiberglass cloth and polyester resin with a thin foil of copper on one or both sides. Higher-quality boards are made from fiberglass, but the epoxy paper boards are less expensive and excellent for your project work. Two-sided boards are difficult to develop in the school lab because of the problem of aligning the

VACUUM FORMING PROCESS

a. Examples of vacuum formed polystyrene

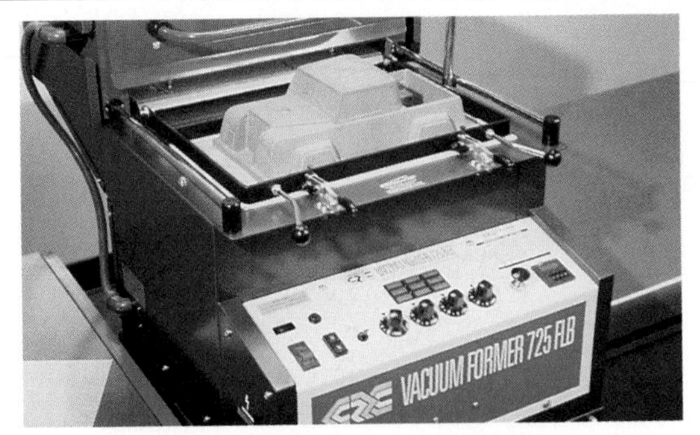

b. Mold of a toy truck on raised platen. The mold must have a number of small holes to allow vacuum to draw the air from corners and deep areas.

c. Polystyrene sheet is clamped into place. The mold has been lowered into the machine.

d. The electric heating element is pulled into position to heat the plastic sheet.

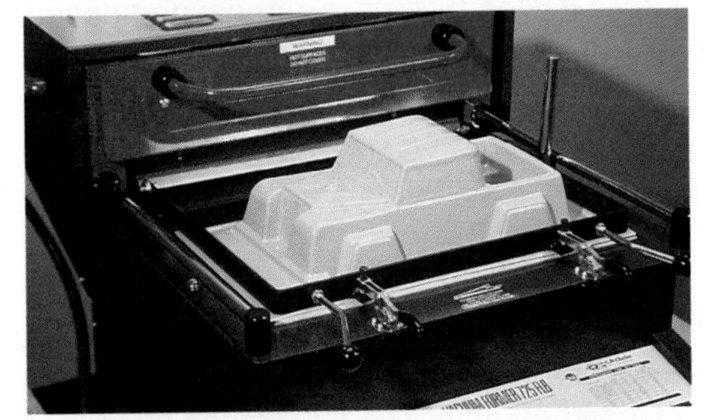

e. The mold is raised and vacuum is applied while the polystyrene sheet is hot and very flexible.

f. Air is blown through the mold to release the polystyrene from the mold.

g. The polystyrene sheet is unclamped and removed from the mold. Flashing needs to be trimmed.

Fig. 11-40 a-g

Making a vacuum-formed part using thermoplastic materials (Courtesy of Paxton/Patterson & C.R. Clarke and Company, Ltd.)

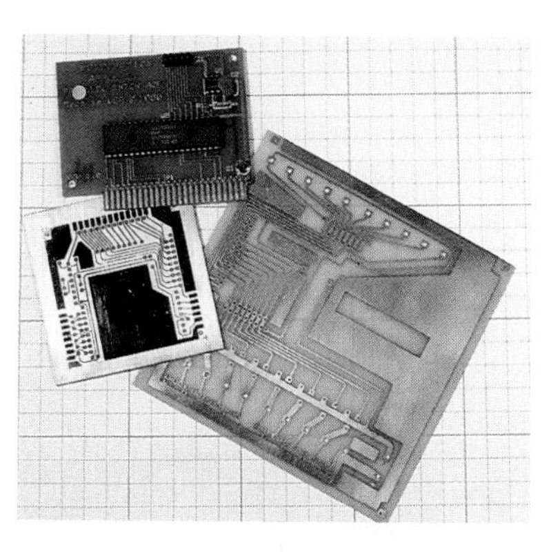

Fig. 11-41

A variety of printed circuit boards

top and bottom traces, but one-sided boards are relatively easy and ideal for most basic circuits.

PCBs are made by masking the board with a pattern that will act as wires between components. The pattern acts as a *resist;* that is, it prevents the chemical from contacting the copper where you want the copper to remain. The board is then dipped in an etching solution, which dissolves the copper foil on the board, except where the mask protects the copper from the solution. When the unwanted copper is etched away, the remaining copper is cleaned with fine steel wool, holes are drilled in the appropriate locations, and components are then mounted and soldered to the copper foil (Figure 11-42).

Making a Pattern

Before making a pattern or a mask, you should make a model of the circuit design. A prototyping board can be used to make certain that the design works and to determine where all connections between components must be made. Making a PCB without first trying out the circuit is almost always a bad experience.

The design of a pattern of traces for a PCB is generally time consuming. This is because, unlike hand wiring, you are working in only two dimensions. You cannot cross two traces on a PCB. You can only run one trace to a point near the other intersecting trace and then run a jumper wire across it to where the first trace continues. This is the advantage of a double-sided board, because you can run traces on both sides so jumper wires are minimized. Look at some commercial PCBs to see how someone else solved the problems of connections, component location, and mounting.

To design a PCB, you will need to sketch your circuit in pencil a number of times. The first time, begin with several major components, such as integrated circuits. Draw lines from pin to pin and from pin to components. Allow room for resistors, capacitors, and other components, and do not crowd. As you are drawing, you will find that you cannot get all connections made without crossing traces, but you will see that some connections work and others do not. Next, redraw on a new page, working from the first sketch and avoiding the problems of the first where possible. On this second drawing, you should get further before you run into problems. You may need to redraw several times, but eventually you will have a usable pattern. Jumpers are fine, but try not to use too many. (Figure 11-43)

Fig. 11-42

A student-built computer interface on a PCB with components mounted

Refine the final drawing by squaring corners and allowing for the size of individual components. If you are in doubt about the distance

Fig. 11-43

Developing a PCB circuit layout

between holes for a capacitor or other component, find a sample and measure the distance between the leads.

Making the Mask

Developing a mask for your circuit can be done in a number of ways. The most straightforward, but definitely not the best way, is to draw the pattern layout directly on the copper foil with a *resist pen.* A resist pen is a permanent marker, such as a Sharpie™, which provides a protective coating on the foil, which resists the etchant solution. PCBs can be made quickly with this technique, but fine pattern lines are difficult to make and the results are not very neat.

A second method of making a mask directly on the board is to use "press-on" tape and dots for lines and component "pads." Pads are small donuts that are later drilled through to accept a component lead. The pad provides a surface for the solder to adhere. Although neater than drawing with a resist pen, this technique is very time consuming and tedious. In addition, you only get one try with both of these methods. If the board does not come out as you would like, then you will have to repeat the whole process and etch another board.

Packages with dot pads and other useful "press-on" resist images are available. They have patterns spaced for integrated circuit pin pads, so the results are very neat and accurate (Figure 11-44).

Fig. 11-44

An assortment of "press-on" figures for developing a PCB mask

Developing a Transfer Pattern

Another technique for creating PCB layouts is to develop the finished pattern on paper and then transfer it to the foil-clad board. This will allow you to carefully draw or use the tape-and-dot method on paper, so you will be able to develop neat patterns with fine-line detail. This is especially important if you want foil trances to go between integrated circuit pins and other close quarters.

CAD and PCB Layout Programs. Using a computer-aided drafting (CAD) or specially designed PCB layout program to develop a pattern is an excellent way to obtain fine detail and neat boards. The development of a pattern design with CAD will be a bit time consuming, but, if you have already done some work with the system, then it should not take you too long. You will need to do some basic sketches for your layout on paper in order to assure that your pattern has all the correct connections and that there are no overlapping traces.

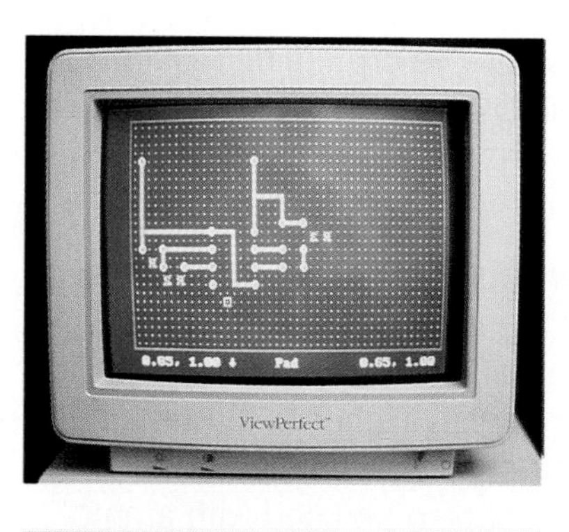

Fig. 11-45

CAD programs to develop PCBs often have autorouting capabilities, which make the development of a circuit layout more efficient and accurate.

PCB layout programs have the advantage of being designed to develop a workable mask quickly. Most programs require that you enter information about each component in the circuit and then enter the connection for each pin. For example, a circuit using a 555 timer chip would require that you specify that pin 7 on the 555 chip goes to pin 1 on a 47K resistor. After you have entered all connections, the program will automatically develop a pattern for a PCB layout. This *autorouting* is especially helpful in complex circuits, for which it would take you much time to develop a pattern that did not have crossed traces. Some programs will develop several alternative patterns and let you choose the best (Figure 11-45).

Transferring the Pattern to the PCB

There are several methods of pattern transfer, some of which are easier than others. One method involves transferring the image by coating the copper-clad board with a photosensitive chemical. The pattern that you have developed on paper is then made into a transparency on a copying machine. The transparency with the image is placed over the photosensitive board and exposed to a bright light for a certain period of time. The exposed board is then developed with another chemical and rinsed in water. This board now has a resist remaining on the board in a pattern matching the original pattern on the transparency. The board is then ready for etching. Boards can be purchased that are already coated with a photosensitive chemical, although they are considerably more expensive than uncoated boards.

Fig. 11-46

Using Press-n-Peel™ to transfer an image to a copper-clad board

Another method of image transfer involves using a material called Press-n-Peel™ (Figure 11-46). A copy of your original pattern is made onto this material in a plain-paper copying machine. The image can also be made directly on the Press-n-Peel™ material with a laser printer if you are generating the pattern on a CAD or PCB autorouting program. The Press-n-Peel™ material is then placed over a clean, copper-clad board, and a standard clothes iron is used to heat the material for about 25 or 30 seconds. Lifting a corner of the material will tell you if it has adhered to the board. Peeling back the material then leaves the pattern on the board. The board is then etched.

Etching the PCB

There are two etchant solutions commonly available for PCB work (Figure 11-47). *Ferric chloride* is available in both liquid and dry powder form (anhydrous), which is mixed with water. The main disadvantage of ferric chloride is that it stains everything—your clothes and your skin. Although it will wear off from your fingers in a few days, it does not come out of cloth. (*Safety Note:* Eye protection must be worn while you are using this chemical, and you must use it in an area with good ventilation, since the fumes are unhealthy!)

Fig. 11-47

Etching a PCB board. Eye protection is a must!

The second solution is *ammonium persulphate,* which usually comes in a dry powder form. You should mix a batch of this chemical at least a day before you need to use it so that it completely dissolves. It should be heated to 100 degrees F or slightly higher to speed up the etching process. This etchant does not stain like ferric chloride and cleans up with water. Eye protection must also be used with ammonium persulphate. Both chemicals are toxic if ingested.

In addition to eye protection, an apron and plastic or rubber gloves should be used when you are etching a PCB because of the staining of ferric chloride and the irritation possible with the ammonium persulphate. Do not use pliers or other tools to remove boards from the etchant, because the solution will attack the metal. Plastic or wooden tongs should be used.

Actual etching should be done in a plastic or glass tray, with at least ½ inch of solution. It is a good idea to drill a ⅛-inch hole in one corner of the board and use it to tie an 8-inch piece of plastic-covered wire. The wire then gives you something with which to agitate the board and to remove the board from the solution. When you can see that the copper is almost gone from the bare areas, watch closely so that the solution will not get under the masked off areas and attack the protected traces.

Carefully remove the board from the solution, and allow the excess to drip back into the tray for several seconds. Next, wash the board thoroughly with water. Clean the remaining resist with very fine steel wool and drill component holes with a small drill, from 0.040 inch to 0.060 inch in diameter. A Dremel™ drill press is ideal, but many standard drill presses will allow you to use this very small drill bit. Place the drill in the chuck so that only about ½ inch is showing in order to avoid bending or breaking the bit. Place a wooden board under the PCB while you are drilling (Figure 11-48).

Soldering PCBs is delicate work, because overheating will cause the foil to lift off the board. Use a 25-watt soldering iron or one that is heat controlled. Place the iron tip so that it rests against both the component lead and the foil pad, and, after about 3 or 4 seconds, touch a piece of small diameter rosin-core solder.

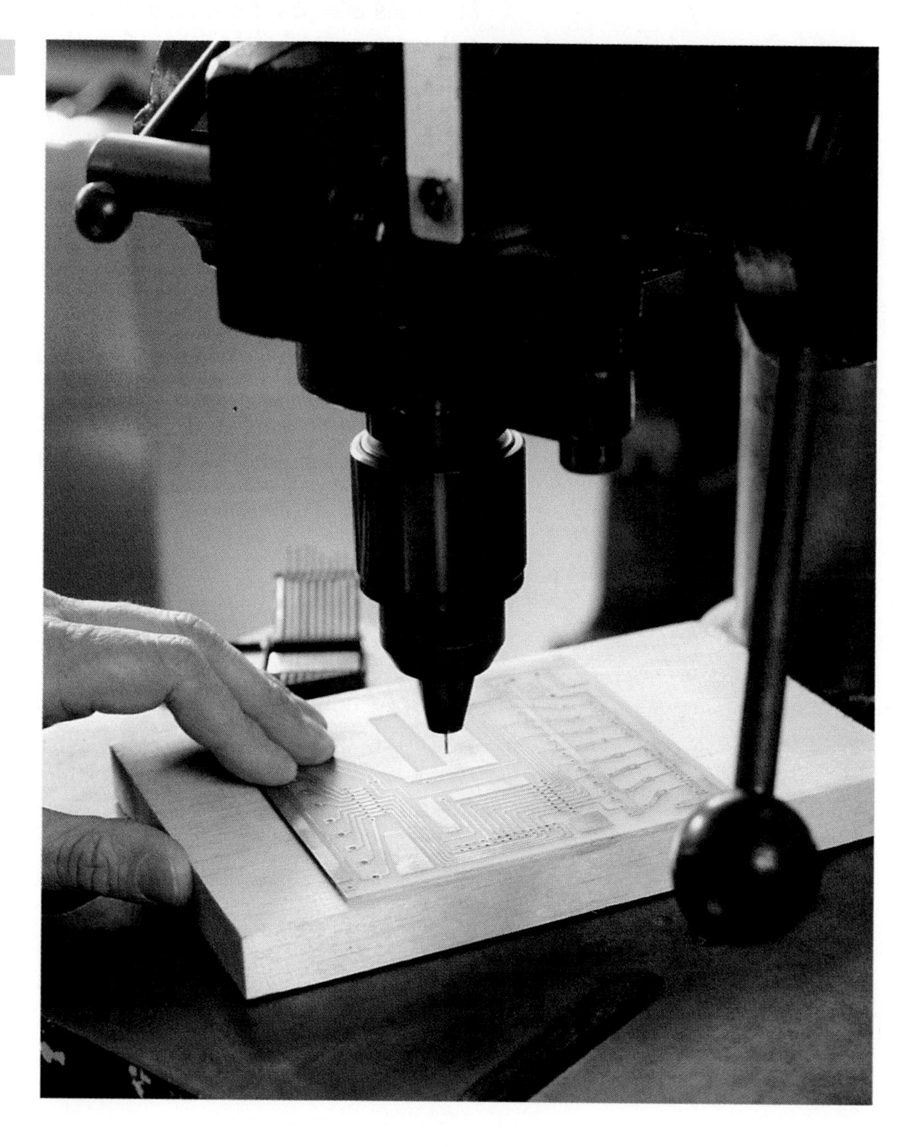

Fig. 11-48

Drilling a PCB to accept component leads after it has been etched

Tips on Making PCBs

The etching solution used to remove unwanted copper will become ineffective after a period of time. The period of time for which it remains usable will depend on how much copper is dissolved. One method of extending the life of the etchant is to design PCB boards with little "open space"; that is, mask off areas that are not part of the circuit and let the copper remain on the board. These areas can be connected to ground in the circuit or not connected to anything. By only removing the minimum amount of copper on a board, the etching solution will remain active longer.

Summary

Materials are so important to human progress that early historical periods were named after them (e.g., the Stone Age, the Bronze Age). The availability of appropriate materials is a critical limiting factor in technological innovation and design. Materials are selected for their physical and aesthetic qualities. Physical factors describe how the material will function under various conditions and include strength, ductility, elasticity, and other measures.

Paper, metals, ceramics, polymers, composites, adhesives, and finishes are commonly used materials in products and models. Graphics add to finishes by providing line, texture, and important messages on packages, products, tools, and machines. Modeling techniques vary greatly. The two broad categories include appearance and prototype models. Techniques are given for modeling in paper, Styrofoam™, and sheet plastic.

Electronics modeling lends itself to the development of printed circuit boards (PCBs), which are used in commercial electronic products. Although techniques vary for their development, all PCB designs require that a circuit be transformed into a pattern, which consists of traces acting as the wires between component connections. The pattern is used directly on the copper-clad board as a resist, or a mask is made from the pattern and transferred to the board. Etching the board in a chemical solution leaves the copper traces intact while removing the rest of the copper. Holes are drilled and components are mounted and soldered for a finished PCB.

Observation/Analysis/Synthesis

1. Identify five consumer products, and make an extensive list of all the materials used to produce those products.
2. Using one product identified in Problem 1, develop a matrix listing the materials used, the material characteristics, and probable justifications for the material being selected.
3. Find at least three examples of a particular product made from three different materials (e.g., metal, plastic, and wood dowels; metal, wood, and plastic lawn rakes). Explain how each example uses the particular properties of the material from which it was made and how it is similar or differs from the other examples.
4. Using each part identified in the matrix for Problem 2, make a list of all production operations needed to produce and assemble the parts into a finished product.
5. For a product you are designing, make a matrix of probable materials, list the material characteristics and uses, and rate the advantages and disadvantages for using each material selected.
6. Develop an appearance model using one of the techniques described in this chapter.
7. Develop a prototype model using techniques described in this chapter or supplemented from another resource book.
8. Make a printed circuit board of an astable timing circuit using a 555 timer chip and an LED output display.

CHAPTER 12 Testing and Evaluating

Introduction

At some point in the process of designing, you must ask yourself how well you have done. It is not easy to answer that question. Seldom does a design fulfill each and every requirement of a problem, and always you must deal with the trade-offs and risks of technology. In addition, you have probably been working within limitations imposed by the amount of time allowed for your work and the resources available to you.

The first question you will want to address deals with how well your design and final product has solved the problem you identified back in the early stages of the process: Does it solve the problem? Although this may seem to be a simple question, it is not.

Fig. 12-1

Developing Appropriate Tests

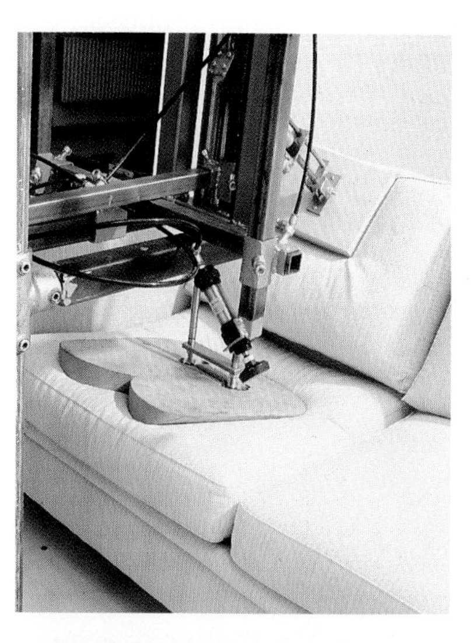

Fig. 12-2

Coming up with schemes to test a solution can be a creative and challenging problem. Courtesy of IKEA U.S., Incorporated

Developing appropriate tests for your design work is another application of the design process. It will require that you define exactly what it is you want to test, investigate testing possibilities, devise a number of possible tests, choose one, and implement it (Figure 12-2). Too often, testing and evaluation are just steps to get through as swiftly and easily as possible, but this phase of the design process is as important as any other step.

Testing is very important. In science, you have a hypothesis or a guess about what you think will happen. Devising and conducting the tests to determine if your hypothesis will hold up is the creative work of the scientist. In technology, testing is used to validate a design.

Testing a Toy

At several points throughout the book, the design of a toy was mentioned. If you undertook such a project, you probably developed a design brief, which provided you with direction, such as the intended age of the child and other considerations. If this activity was designed to meet some other need, such as helping you to develop your knowledge of simple mechanical systems, then the design brief or specifications may have outlined the incorporation of a cam as a requirement of the solution.

Let us assume for a moment that you began with a problem situation you identified in your younger sister's nursery school. You had noticed that the children seemed to quickly break many of the mechanical toys that you saw as more interesting and educational. From this observation, you developed a design brief that stated that you intended to develop a toy which would be educational, hold the interest of the preschooler, and be durable. You have designed and developed a toy prototype, and you are now being asked to test and evaluate it. What do you do (Figure 12-3)?

To begin with, you will need to return to your design brief and specifications to find the standards against which you will measure the success of the toy. Going back to the design brief, you find that you intended the toy to be educational. Great! But how do you measure this? Second, you find that you promised that your toy would be interesting to a preschool child. How will you measure this? Third, you specified that the toy would be durable. How can durability be measured?

All of these characteristics of the toy can be measured, but the testing process is at least as creative as the ideation you went through in the generating and developing ideas stages of the design process. Let's look at some possible testing strategies for your toy and see if these will give you the information you need.

Measuring the educational value of a toy may not be too difficult if you have a clear idea in your mind of what the toy will teach. For example, suppose you have created a toy that will help a child learn to tell time. First, you could let a child play with the toy for a certain length of time and then ask the child to look at a clock and tell you what time it is. Do you think this is a good test? Might it have been better to learn if the child could tell time before he or she used the toy? Why?

Fig. 12-3

What is the true test of a toy?
Courtesy of Fisher-Price

Suppose your toy did not seem to help the child tell time after the child played with it for an hour. Does that mean that it is not educational? Do some concepts take longer to learn than others? Do children of preschool age need longer to learn some concepts than others? Is it possible that telling time is beyond the reach of some preschoolers? These are valid questions that have a bearing on the measure of success for the toy.

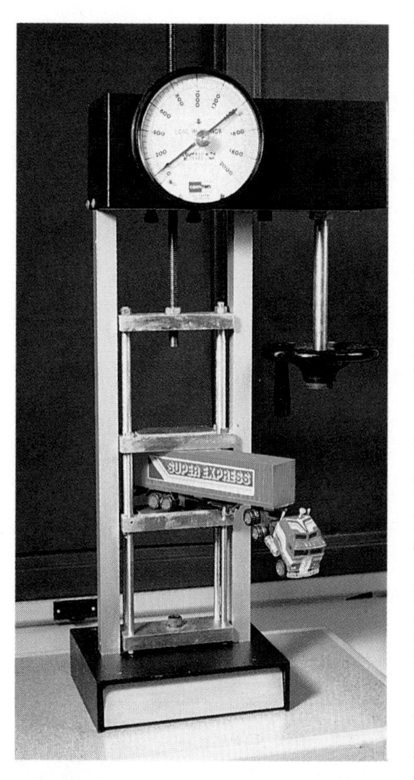

Fig. 12-4

What is the force required to crush a toy truck?

There are alternative schemes for evaluating the educational value of a toy, such as asking the opinions of a person who is an "expert" on both children of this age and educational concepts. This individual may provide you with some insights into the educational value of your toy. Can you think of other testing strategies?

In addition to the educational value of the toy, your design brief stated that it should hold the interest of the child. What does this mean? Does it mean that the child will play with this toy to the exclusion of all others? Does it mean that the child will play with it for only 10 minutes? What is a fair measure of the interest the child has in the toy?

Most children play with one toy for a while and then go on to another toy. If there are many toys from which to choose, they may not pick up the same toy again for quite some time, if at all. The other toys from which the child has to choose will determine whether your toy gets any attention. It would hardly be a fair test to give a child only your toy, or your toy and some beat-up old broken toys to see if he or she played with your toy. Can you think of a fair test?

The last area of testing for your toy is durability. Suppose you gave the toy to your kid sister and let her play with the toy for a day. At the end of the day, the toy was still intact. Could you say with confidence that the toy is durable? Certainly, the conditions under which your sister played with the toy and the kind of treatment the toy received at the nursery school are very different. What are some of the ways in which you might test the durability of a toy? Here are some suggestions:

- Study the kinds of things that happen to toys, and the kind of stresses they are subjected to. Find ways to simulate these events (Figure 12-4).
- Research how toy companies test toys, and what they expect and require.
- What are mandated safety features of toys? Are there government regulations or consumer group guidelines?

The testing of the toy has been used as an example, but the testing of any problem solution should follow similar lines.

Presenting Test Results

Your portfolio is a record of your design work. To communicate your efforts in testing and evaluation, you have a number of possibilities. You can use descriptions, numbers, checklists, and testimonials in presenting your results.

Descriptions

Describing a test and its results can be done through a written description or a combination of graphic and written description. Graphical descriptions may include an illustration of the testing apparatus and a graphical summary of the test results (Figure 12-5).

PROJECT: "SNACKLES" SNACK FOOD
Consumer Interview Response Sheet

Please describe and evaluate the product in your own words.

Consumer # 1:

I thought that the new snack was a little hard to handle. The taste was good and the texture was okay, but it fell apart in my hands. I think small children would have a very hard time and would make a mess.

Consumer # 2:

I liked the two sweet flavors of the snack food — the fruit and the almond. I could even see having a chocolate variety. The salty flavors, like cheddar and onion, were okay, but I guess I like sweet snacks better anyway. I also liked the crunchiness, but they didn't hold up very well. I think you'd have to be very careful with the packaging or they'd never get to the store in one piece.

Consumer # 3:

Skip the sweet ones — the salty varieties are the best. I prefer onion, but how about a beef jerky flavor, and maybe garlic and herb or pizza flavor? Can you make them less crumbly?

Fig. 12-5

Numbers

Test results most often take the form of numbers, such as the number of minutes a child plays with a toy, the amount of force a component or a product can withstand without crushing, or the percentage of people rating the product "very good" and "excellent." Presenting this information in a narrative or description can be tedious. Graphic presentation through charts, graphs, diagrams, and tables is an effective technique. Be certain that figures are well labeled and clear for the reader (Figure 12-6).

Fig. 12-6

Checklists

An effective way of testing a solution against the design brief and specifications is to develop a checklist. The checklist provides a visual summary of the extent to which the solution meets the design requirements (Figure 12-7).

Testimonials

Testimonials are opinions of those who have tried your product and will endorse it. They may like it because of its overall qualities or because of one quality, such as aesthetics or ergonomics. These individuals

PROJECT: "SNACKLES" SNACK FOOD

Please rate these products on a scale of 0 (POOR) to 9 (OUTSTANDING)

Cheddar flavor

taste	0	1	2	3	4	5	6	7	8	9
texture	0	1	2	3	4	5	6	7	8	9
color	0	1	2	3	4	5	6	7	8	9
appearance	0	1	2	3	4	5	6	7	8	9
packaging	0	1	2	3	4	5	6	7	8	9
name	0	1	2	3	4	5	6	7	8	9

Onion flavor

taste	0	1	2	3
texture	0	1	2	3
color	0	1	2	3
appearance	0	1	2	3
packaging	0	1	2	3
name	0	1	2	3

Almond flavor

taste	0	1	2	3
texture	0	1	2	3
color	0	1	2	3
appearance	0	1	2	3
packaging	0	1	2	3
name	0	1	2	3

Fruit flavor

taste	0	1	2	3
texture	0	1	2	3
color	0	1	2	3
appearance	0	1	2	3
packaging	0	1	2	3
name	0	1	2	3

PROJECT: "SNACKLES" SNACK FOOD

Which of the following best describes the samples of snack food you tried?

CRUNCHY ______
CHEWY ______
SMOOTHE ______
CHUNKY ______
TOO SWEET ______
TOO SALTY ______
DELICIOUS ______
BORING ______
MADE ME WANT MORE ______
MESSY ______
EASY TO EAT ______
TOO SMALL ______
GENEROUS ______
SOPHISTICATED FLAVOR ______
APPEALING TO KIDS ______
CATCHY NAME ______
ATTRACTIVELY PRESENTED ______
NEW AND DIFFERENT ______
HEALTHY-SEEMING ______
UNAPPEALING ______
TRUE TO FLAVOR NAME ______
ENVIRONMENTALLY FRIENDLY PACKAGING ______
GOOD VALUE FOR PRICE ______
TOO EXPENSIVE ______
SEEMS FATTENING ______
ARTIFICIAL TASTING ______
TASTES LIKE REAL CHEESE (ONION, NUTS, FRUIT) ______
I'D BUY IT! ______

Fig. 12-7

Fig. 12-8

have given permission to have their names associated with their comments about the product. Testimonials are presented through direct quotes (Figure 12-8).

Evaluating Project Results

Evaluating Design Work

In a world of so many consumer choices, an often asked question is, "What is good design?" Good design can be assessed by carefully evaluating a product or system against a number of standards. In this way, you can analyze a number of possible solutions to a problem (or a final solution design) by rating how each stands up to these standards. These criteria include *aesthetics, ergonomics, performance, durability, cost,* and *impacts.*

Aesthetics. Aesthetics refers to the appearance of an object—with form, scale, color, and so on. An object that has good aesthetics is pleasing to look at.

Because aesthetics is a matter of taste, it would be appropriate to get the opinions of a number of people for this test. A survey or a questionnaire of how well people liked the appearance of the product might give you more usable results if you asked several questions, dealing with such things as color, shape, style, and use of design elements such as texture or line. In this category, the overall quality of the finished product is emphasized, including evidence of careful crafting.

Ergonomics. Ergonomics is often called "human factors engineering," because it has to do with designing products and environments that work well with people.

The test of ergonomics is often if the product "feels" right. Ease of use or comfort while using the product can indicate that the device is appropriately sized, weighted, and that the controls are in an appropriate place. Because of the broad nature of ergonomics, you will need to think about the tests necessary to measure success in this area.

Performance. Performance refers to whether the product does what it is supposed to do. The actual functioning of the product or system determines if it performs as intended. You may need to consider simulations or controlled trials if there is a potential for accident or injury. If you are testing a toy for a child, you may need to first try out the product yourself to make certain that no injury to the child could result. You should also have your teacher look the toy over carefully so that any potential dangers can be identified and corrected, such as sharp edges or corners, small parts that could be swallowed, or unsafe materials.

Durability. Durability refers to the life expectancy of the product and whether it is realistic to the purpose, cost, and other considerations.

A solution may function as intended but only for a very short time. Does this mean that the solution is a good one? When we purchase a product, we expect that it will last for some length of time. To determine if your solution is durable, you may want to test it a number of times or simulate such conditions.

Life-Cycle Cost

Compare two refrigerators: Refrigerator A costs $500; refrigerator B costs $650. Which is the cheaper refrigerator? Not so fast! Refrigerator A has a lower efficiency rating than B, which means that it uses more electricity than B. If A uses $0.70 of electricity per day ($255.50 per year) and B uses $0.45 of electricity per day ($164.25 per year), and the life expectancy of both refrigerators is 10 years, then the cost of owning refrigerator A is

$$\$500 + (\$255.50/\text{year} \times 10\text{ years}) = \$500 + \$2{,}555 = \$3{,}055.00$$

The cost of owning refrigerator B is

$$\$650 + (\$164.25/\text{year} \times 10\text{ years}) = \$650 + \$1{,}642.50 = \$2{,}292.50$$

The life-cycle cost of refrigerator B is actually cheaper than A.

Cost. What do you get for your money? Cost is what you have to pay to buy the product or the *cost of making it*. The product may need regular maintenance or energy in the form of batteries or line current. This is called the **life-cycle cost**.

In the same way appliances are rated for energy efficiency, the cost of your product or system compared to how well it performs is called its **cost-effectiveness**.

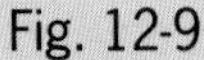
Fig. 12-9

Many products we used in the past—and continue to use today—have necessitated huge "superfund" expenditures in the billions of dollars to clean up toxic wastes from manufacture. These are real costs to the public and are also part of the price of the product. These are called the *hidden costs* of the product and represent a Type IV impact.

Testing in this area means more than adding up the cost of materials. You will want to look at the cost of materials, the time you spent developing and making the product, and the hidden costs of environmental impact. In many commercial products, when these costs are figured in, the cost may be much too high to be cost-effective (Figure 12-9). The disposal of the product should also be figured into the final cost.

Impacts. Impacts describe many positive and negative qualities of an object, including its purpose, function, the materials from which it was made, and the processes used in its manufacture. There may be *environmental impacts, personal impacts, social impacts,* or *legal impacts.* There are a number of questions related to impacts that you will need to consider in evaluating your work:

- How does a product impact on the environment? For example, what happens to the product when its useful life is over? If the object is designed and made with little or no regard to its impact when it is discarded, then the object cannot be considered "good design." Are the materials from which it is made or the chemicals used in its manufacture environmentally hazardous? Many substances have no environmentally sound method of disposal.
- What are the personal impacts of the product? Will the product change your daily life? For example, do you have to travel less now that you own a FAX?
- What are the social impacts of the product? Has the introduction of television resulted in fewer people reading books and newspapers or developing interpersonal communication skills? Are people more aware of current events because of television?
- What are the legal impacts of the product? Did you infringe on a patent or a copyright? Are there potential dangers with the product that could result in legal action?
- What are the ethical responsibilities on the part of the designer? Could you apply the Golden Rule, "Do unto others (including the environment) as you would have them do unto you"?

The criteria of good design are all interrelated, and compromises often are made to assure that a product is worth making and marketing. At the same time, there must be other values at work in the design and development of products and systems besides profit.

Evaluating Your Design Skills

To this point, you have critically looked at the product or system you developed. You had to step back from your project and say that this part was well done and this part was not. Now you must look at your own role in the designing and making of the solution.

You had the responsibility for the decisions made and the management of time and resources that went into the final results. You now need to look critically at how well you did these things. Here are some questions you may use as a guide in the evaluation of your work.

Design Brief and Specifications. Did the design brief provide a solid direction for the project? Were specifications too vague or too restrictive?

Investigation and Research. Did you collect appropriate information about the problem? Did you apply this information to the problem? Did you find out about how others had solved similar problems? Did you return to this step as problems came up in other steps of the process?

Generating Alternative Solutions. Did you use the first idea that you thought of? Were the alternatives presented workable and well thought out? Were creative strategies used to help develop solution ideas? Did you combine attributes of several ideas?

Choosing a Solution. Can you defend the solution chosen in terms of "good design"? Do you have reasons you can list for choosing one solution over the others?

Developmental Work. Did you work out structural, mechanical, electronic, or pneumatic problems before you developed working drawings? Were working drawings developed? Did the working drawings leave unanswered problems in the design?

Modeling and Prototyping. Was the model and/or prototype well crafted? Were appropriate materials, adhesives, fasteners, and finishes used?

Testing and Evaluating. Were the tests that were developed appropriate? Were the results of the tests presented clearly and honestly? Were your evaluations critical and honest? Did you give yourself credit when it was due; self-criticism when you deserved it?

Time Management. Did you use the time you had to work on this project appropriately? Did you attempt a project that was too difficult or too simple for the time allotted (Figure 12-10)?

Summary

Testing and evaluating your ideas and solutions is an important step in the design process. It is not always apparent, however, how to evaluate the success of your work. You need to develop appropriate tests that can assess the effectiveness of a solution in terms of the design brief and specifications you developed in the early stages of your design work. It is often necessary to apply indirect tests for certain aspects of a solution, such as the specification that a toy be educational.

In presenting test results, you may choose to provide descriptive, numerical, checklist, and testimonial results. The presentation of these results then become an important part of a major design problem portfolio.

The evaluation of your work should take two forms: evaluation of your design work, usually the end product, and evaluation of your design skills, including the planning, thought, and effort you put into the project.

Fig. 12-10

Companies test their products and consumers use the results.

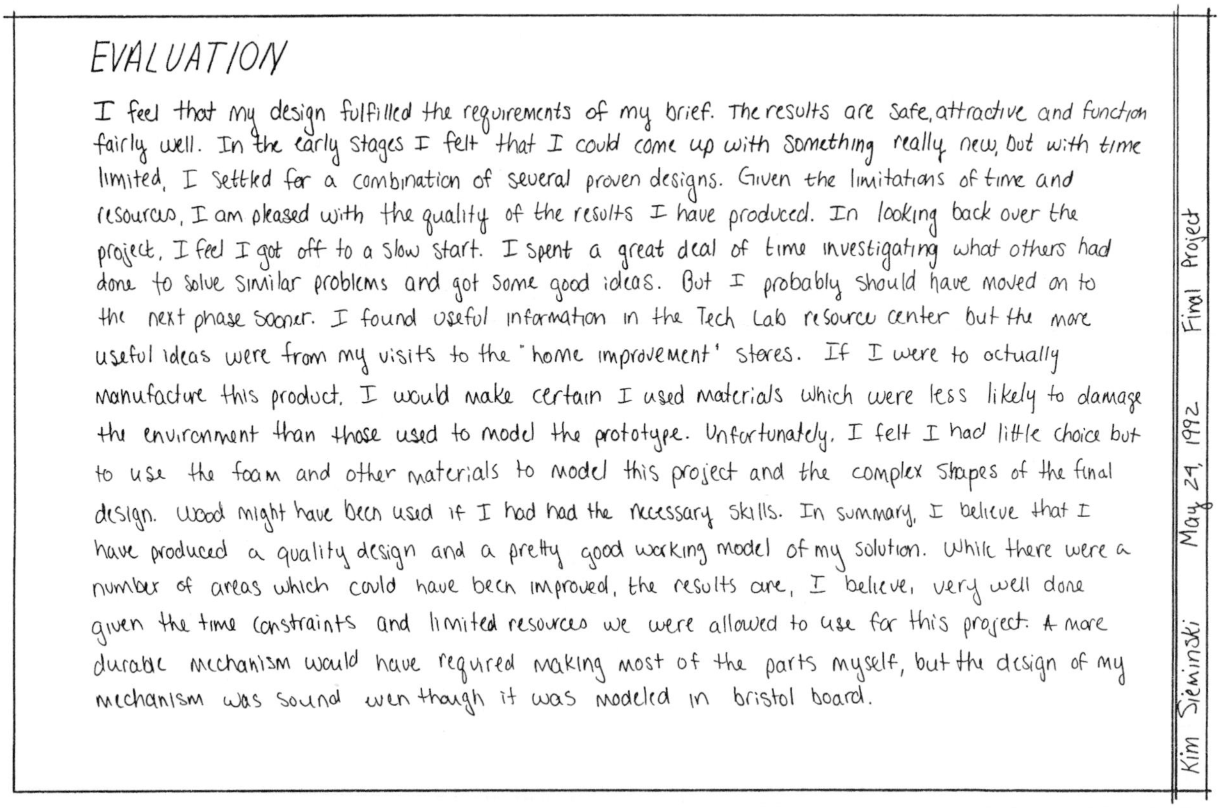

EVALUATION

I feel that my design fulfilled the requirements of my brief. The results are safe, attractive and function fairly well. In the early stages I felt that I could come up with something really new, but with time limited, I settled for a combination of several proven designs. Given the limitations of time and resources, I am pleased with the quality of the results I have produced. In looking back over the project, I feel I got off to a slow start. I spent a great deal of time investigating what others had done to solve similar problems and got some good ideas. But I probably should have moved on to the next phase sooner. I found useful information in the Tech Lab resource center but the more useful ideas were from my visits to the "home improvement" stores. If I were to actually manufacture this product, I would make certain I used materials which were less likely to damage the environment than those used to model the prototype. Unfortunately, I felt I had little choice but to use the foam and other materials to model this project and the complex shapes of the final design. Wood might have been used if I had had the necessary skills. In summary, I believe that I have produced a quality design and a pretty good working model of my solution. While there were a number of areas which could have been improved, the results are, I believe, very well done given the time constraints and limited resources we were allowed to use for this project. A more durable mechanism would have required making most of the parts myself, but the design of my mechanism was sound even though it was modeled in bristol board.

Kim Sieminski | May 29, 1992 | Final Project

Observation/Analysis/Synthesis

1. Develop a way to test the solution to a problem based on these design briefs:
   a. Design and make a game for 7–8-year-olds. The children should learn some important historical facts while playing the game.
   b. Design and make a product that helps a person with arthritis to remove the lids on jars.
   c. Design and make an emergency light for a person who has been thrown overboard at sea at night.
2. Make a list of the kinds of stresses a toy for a 2-year-old has to withstand. Devise a test for each one.
3. Find a commercial product, and analyze it for "good design," using the criteria outlined in this chapter.
4. Discuss in detail how a chosen product rates on the cost and ethics criteria for "good design."
5. Conduct a product analysis on three similar products from different manufacturers, detailing how each rates on the "good design" criteria. Present your results in a graphic form.
6. Find two examples of the same product from two different manufacturers. Come up with at least five criteria for evaluating their use. Develop a test for each criteria and describe it. Evaluate these products.

Production and Beyond

Introduction

The design and problem-solving process presented in the preceding chapters of this book only partially describe the Technological Age in which we live. How many problems would be solved if only one newspaper or magazine were printed? Who would be fed if only one loaf of bread were baked, and how many people could get to school or work if only one car or bus were produced? The genius of human invention and innovation is only partly seen through the design of new products. The other half of technological genius can be seen through the *organizations* we create to bring design ideas to fruition.

Fig. 13-1

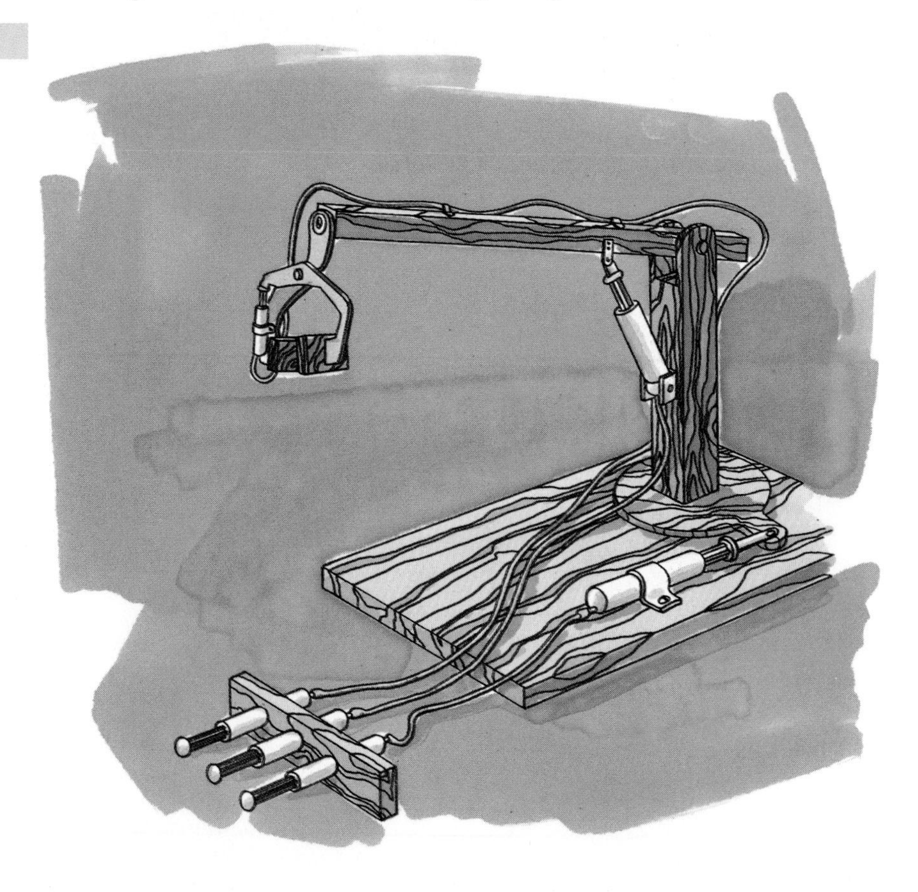

Organizations are necessary, in the form of *managed production systems*, to bring together resources for the purpose of producing products. Resources include

- Capital
- People
- Knowledge
- Materials
- Energy
- Tools and machines
- Processes

Recently, we have heard that we live in a *global community* and that we are in an Information Age. Although this is true, the greater economic imperative is the dominant force of production of goods necessary to fuel the economic engine.

The purpose of this chapter is to introduce concepts concerning the social organization broadly defined as *industry* and the impacts of production activity on the individual, the world community, and the environment. To make this analysis, it is helpful to consider the life cycle of a product (Figure 13-2).

Product Life Cycle

Products begin as a design solution to a specific need or as something that a designer believes people will want. We have stated that technology meets our wants and needs and extends our human capabilities. As the information in the preceding chapters attests, technological design and problem solving is a formal process that involves asking many question and weighing the benefits of many options. There is never one right answer to a technological problem, and all possible solutions involve trade-offs and risks. If the process is properly executed, the design solution chosen is the best one possible, given the situation, and is the one that satisfies the design brief. What remains at the conclusion of the creative design process are decisions relative to producing the product, as well as the impact of the product on the user, society, and the environment.

Selecting the Manufacturing Method

Prior to the Industrial Revolution (1750–1850), products were produced by craftsmen one at a time. This form of producing goods is known as *custom production.* Custom production is not very efficient, but it is good when the demand for a product is low. This form of production would be the method of choice for model making, as well as for producing a bicycle for Greg LaMond, but not much else. Today, new materials and production processes are bringing exciting new products to the marketplace (Figure 13-3).

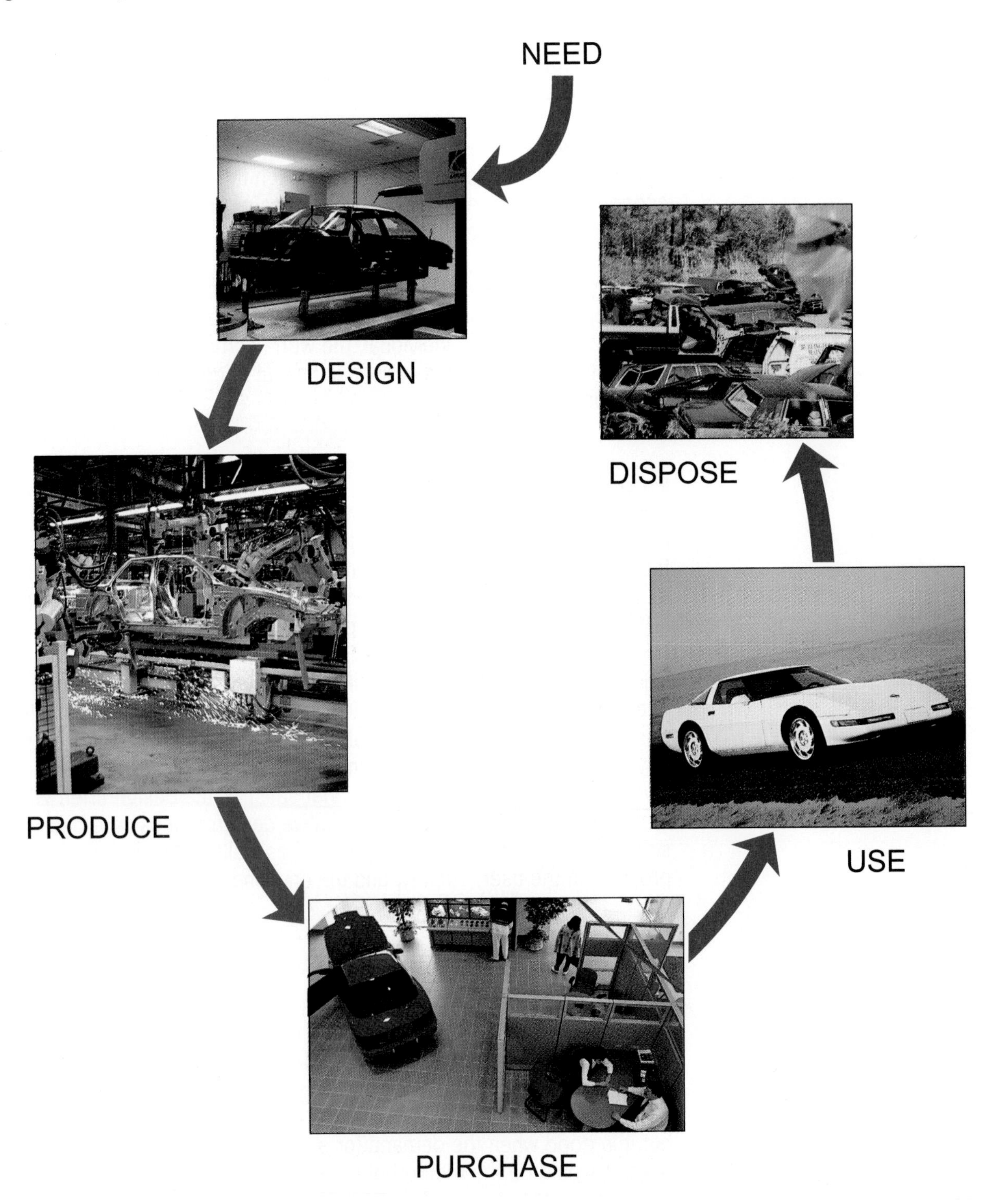

Figs. 13-2

Product life cycle. Figures a, b, c courtesy of Saturn Corporation. Figure d courtesy of General Motors

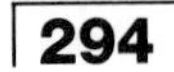

Fig. 13-3

Prince™ tennis racquet. Different materials are used to produce racquets for players of varying skills. Courtesy of Prince Manufacturing, Incorporated

Economic System

Manufacturing methods function within an economic system. In North America and many countries in other parts of the world, the system is called *free enterprise*. The cost of a product is reflected in four broad measures, namely

- Direct material
- Direct labor
- Overhead or burden
- Profit

Material costs can reflect 40 percent of the cost of the product. The free enterprise system allows producers and consumers freedom to choose what to produce and what to buy. The driving force behind this system is *supply and demand.* Supply reflects those things available to be purchased by the consumer. Demand reflects those things that the consumer wants to buy. Products that are expected to last at least three years are called *durable goods,* and those that are expected to last less than three years are called *nondurable goods.* The system functions to make a profit. In a market-driven economy, designs are selected for production based on their potential to generate a *profit.*

Identifying Consumer Demands. Before a design is selected by management to be produced, a market survey is usually completed. A series of questions are asked of a sample group within an identified market population. Questions are selected to gather specific information related to the product's potential demand, as well as to determine consumer preferences. The following are a series of questions that the decision maker would want answered:

- Who will buy this product?
- What are the characteristics of the population that want to buy this product?
- What is the long-term potential for this product?
- Are there other similar products?
- What is the product's potential market share?
- What is the potential for other new competition?
- How many products can I expect to sell?

This last question is extremely important in planning for production, because it tells the decision maker how many products must be manufactured. Once a product is selected for production, it is necessary to select or design the best method of production to ensure the success of the product.

Planning Production

Production is a system that takes as its input all available resources, including materials, and changes those inputs into a desired output or product. These changes that occur during manufacturing create a new product that is more valuable than the sum of all the resources put into the system. This concept of changing resources to make them more valuable is known as *value added*.

Production systems function best by standardizing all processes and components needed for the final product. Since the late eighteenth century, when Eli Whitney first attempted to manufacture rifles with interchangeable parts, industry has searched to achieve the elusive concept of consistent, repeatable, and measurable output. As with all technological endeavors, there are trade-offs and risks associated with the manufacture of a product. Concurrently, each product or system output has the potential to impact the individual, society, or the environment in both desirable and undesirable ways, as well as in expected and unexpected ways.

The size of the potential market is used to determine the manufacturing resources that must be made available, as well as the social and environmental impacts. If the market is larger, then new facilities, equipment, and processes can be considered. In addition, the organization may choose to invest in more expensive and more productive tooling. These decisions are based on cost estimates. New plants may need to be built, and additional workers may need to be hired and trained.

Break-even Point

The expression "Time is money" is never truer than when planning production. In addition to time, materials and all other inputs into the production system must be carefully planned. Management must plan, organize, and control production. *Planning* is the management function associated with setting short- and long-term goals. Companies deciding to invest in new equipment reflect this planning function.

Companies satisfy their *organizing* role by establishing a hierarchical structure, such as president, department heads, production personnel, and worker. This structure means that people, materials, and machines will be in the right place at the right time for efficient production. How well companies organize will determine to a large extent how successful they become. New production structures, loosely grouped under the category computer-integrated manufacturing (CIM), will be presented later in this chapter.

Fig. 13-4

Consumer Reports

The final role of management is to control production. This *controlling* function is best observed through quality control concepts. Not only is it important that the product be made efficiently, but also it must be made well. As the publication *Consumer Reports* attests, some products are better made, last longer, and require less service while in operation than other products (Figure 13-4). Consumers are usually willing to pay more for a better made product.

When only a few products are to be made, the break-even point is relatively high, because all the costs of production are divided by a small number of products. When the quantity of the production run is high, specialized tooling can be made that will reduce the cost of each item. These production decisions are based on an analysis of fixed and variable costs.

Fixed and Variable Costs. Some costs associated with producing a product occur only once and are called *fixed costs.* Fixed costs are mostly associated with tooling up for production. *Tooling up* for production includes acquiring special equipment, jigs, and fixtures, or designing an automated operation. High fixed costs can only be justified when the production run is sufficiently long enough to distribute costs over many products. *Variable costs* are those expenses that are incurred when each product is produced, regardless of production quantity. Materials, labor, transportation, and storage are examples of variable costs. A break-even point is a measure of the relationship between fixed costs, variable costs, and production quantity. Calculating the break-even point will help decision makers decide on special tooling or new machine purchases and ultimately, on the best method of production (Figure 13-5).

Material Processing

Most design solutions call for materials to be changed so that they will function in some desired manner. For example, a block of steel and a piece of wood are shaped and assembled into a hammer. The goal of the student in design and problem solving should be to develop an understanding of basic industrial material-processing techniques so that they can be applied to design decisions. As an example of this principle, let us consider the cameras modeled in Chapter 11 (Figure 13-6). In order to make a decision about materials, the designer would need to consider the material characteristics as well as the potential for efficient production if that material were selected. Understanding basic industrial material-processing techniques will give the designer the confidence to know that most plastics could be easily formed into the desired camera shapes. From this knowledge, the decision to make the camera body in plastic with a forming process

Fig. 13-5

Calculating the break-even point

	(M_1) Production Method#1 (Without Automation)	(M_2) Production Method#2 (With Automation)
Fixed Costs (F) Tooling-Up	$5,000	$10,000
Variable Costs(V) "Resources"	$15/Unit	$10/Unit

The Cost of Each Method of Production:

$M_1 = F_1 + QV_1$

$M_2 = F_2 + QV_2$

At The Break-Even Point (Q_b) $M_1 = M_2$;
Therefore:

$F_1 + Q_bV_1 = F_2 + Q_bV_2$

$$Q_b = \frac{F_2 - F_1}{V_1 - V_2}$$

$$Q_b = \frac{\$10{,}000 - \$5{,}000}{\$15 - \$10} = 1000 \text{ Units}$$

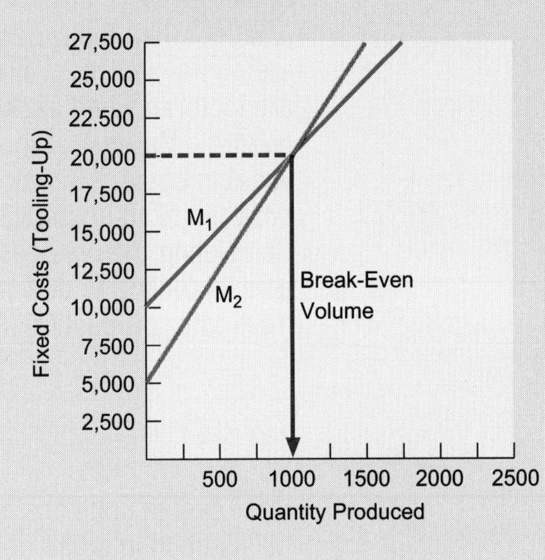

would have to be decided upon. This decision, however, cannot be finalized until production quantities have been established as a result of market planning.

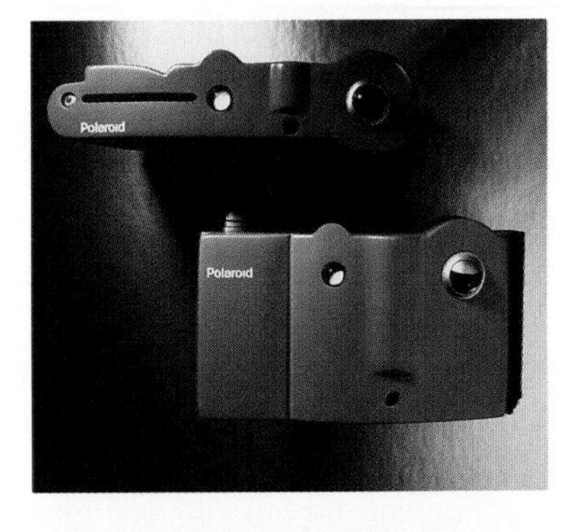

Fig. 13-6

Product models. Courtesy of the University of the Arts, Philadelphia

Material Production Cycle

All materials begin a production cycle in their basic form. *Raw materials*, the most basic form of materials, can be extracted from the earth, water, or air, or they can be grown as genetic products. Silica, iron ore, timber, petroleum, and cotton are examples of raw materials. Most raw materials are refined or processed into *standard industrial materials* by primary industries. *Primary industries* in textiles, metals, paper, petroleum, chemical, lumber, and energy produce products that are, in turn, used by other industries as resources to produce most consumer goods. The lumber industry, for example, processes trees to make dimensional stock for construction, as well as plywood and a number of other composite materials. Trees, as a renewable resource, are also processed by paper-making industries to produce newsprint, book papers, and cardboard. These industries also produce paper towels and facial tissues. Some of these products must be processed by other industries, such as newsprint by the local newspaper, before they become useful products. Other products, such as facial tissue, can be used immediately by the consumer.

Most products made by primary industries are used as standard stock and become one of the inputs into the manufacturing process. Manufacturing industries produce components that are to be assembled into a final product. Components such as automobile tires, gears, belts, ball bearings, glass bottles, and cardboard boxes may be used to make many different types of products. A ceramic coffee mug, on the other hand, is an example of a manufactured component that is also the end product. Most complex products require many components (Figure 13-7).

Fig. 13-7

Many different components are needed to make a complex product

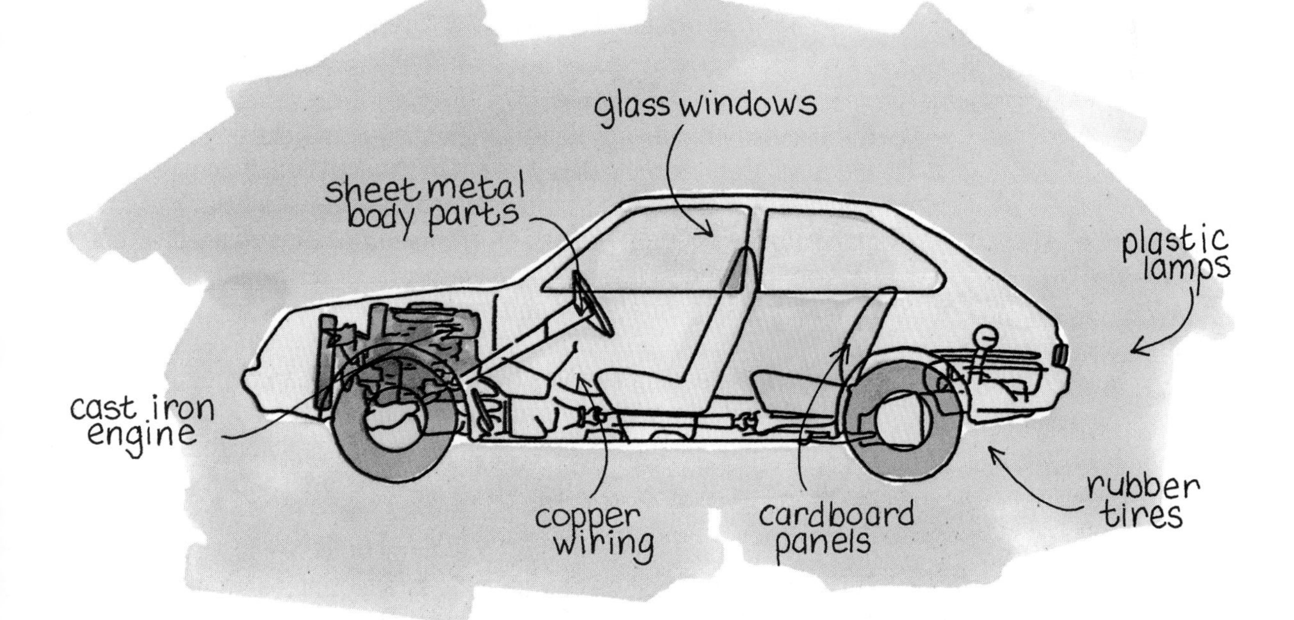

The combining of many components is called product *assembling.* Recently, consumers in the United States have learned that it is difficult to determine where a complex product, such as an automobile, is actually made, because the components in most cars are produced in many countries. Usually, product-assembling industries require workers with lower skills than industries in which primary and component parts are manufactured.

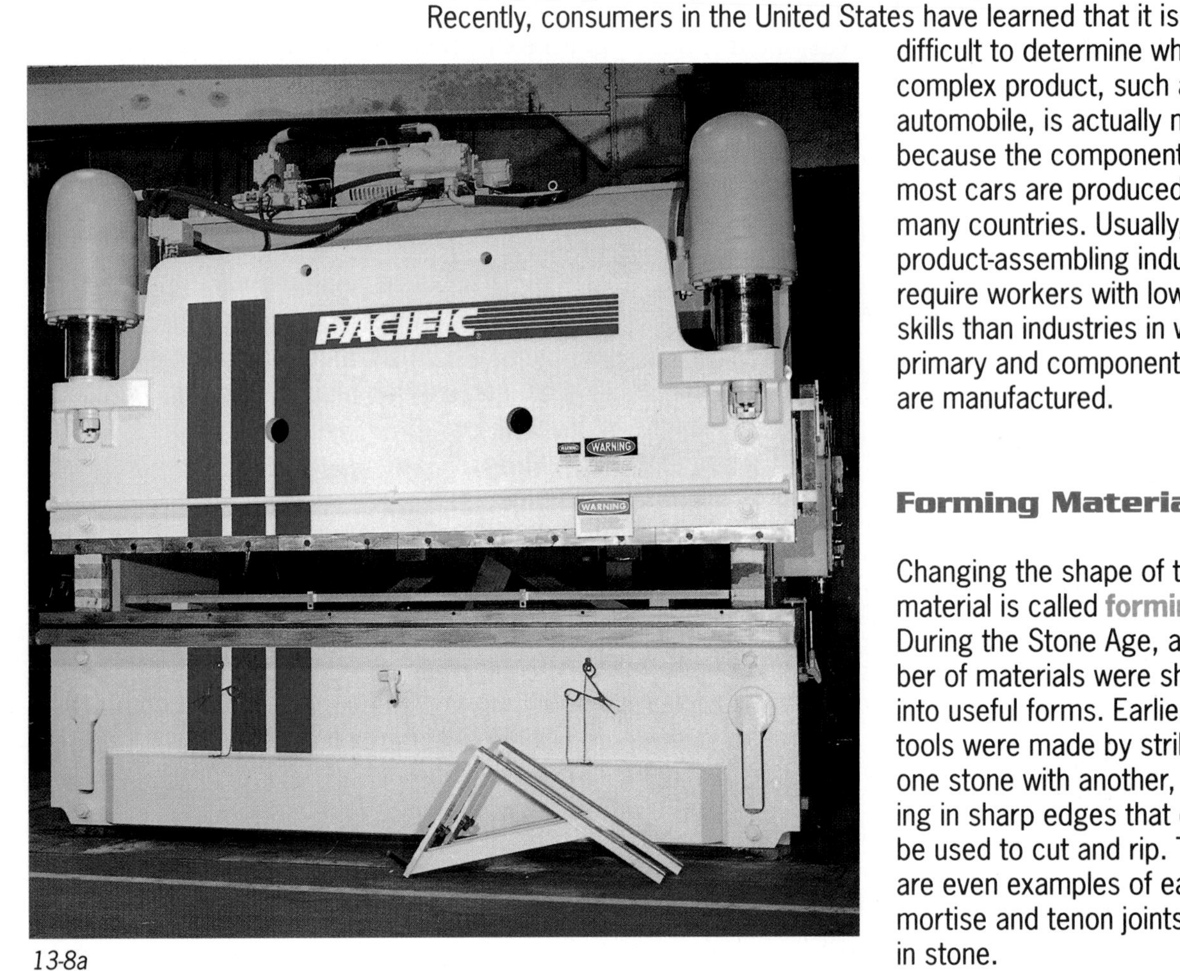

13-8a

Forming Materials

Changing the shape of the material is called **forming.** During the Stone Age, a number of materials were shaped into useful forms. Earliest tools were made by striking one stone with another, resulting in sharp edges that could be used to cut and rip. There are even examples of early mortise and tenon joints made in stone.

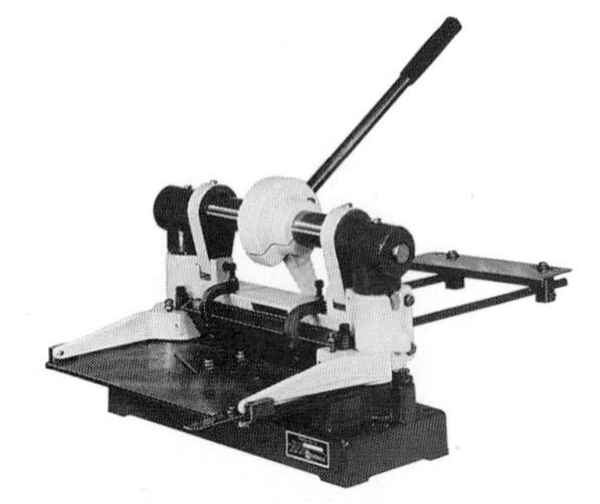

13-8b

Fig. 13-8

Metal brake.
13-8a Hydraulic-press brake. Courtesy of Pacific Press and Shear Incorporated
13-8b Box-and-pan brake. Courtesy of Strippit Corporation

The blacksmith during Colonial times shaped metal using a hammer and anvil. Considerable skill was necessary, and each item was given a unique and distinctive shape. Products also took a long time to make. Today, modern processes are used to shape many different types of materials. The two primary means of forming materials are by compressing–stretching and casting.

Bending. Bending involves both compressing and stretching the material, and is a relatively simple forming operation. Most materials are bent cold, whereas some, such as thermoplastics and glass, are heated prior to bending. Bending is changing the shape of the material in one direction. Cardboard is bent at right angles to make a package. Metal can be bent using a metal brake (Figure 13-8). Thermoplastics must be heated and softened on a strip heater before they can be bent. Wood and glass require more effort to bend, whereas textiles bend very easily. Bending usually distorts the material, and this must be accounted for in the design.

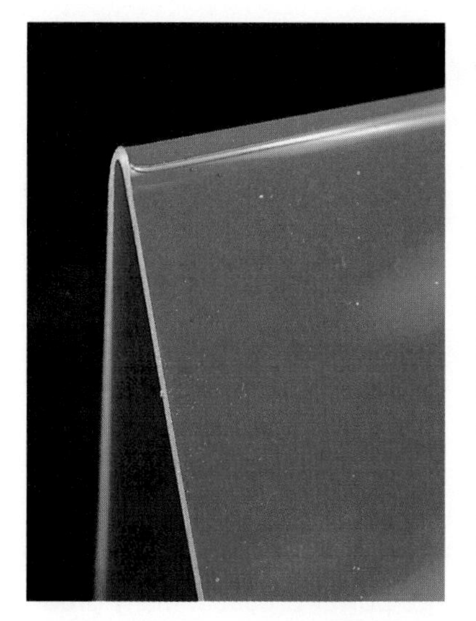

Fig. 13-9

Distortion caused by bending a thermoplastic material

If you will recall from the section on structures, when a material is put under a bending force, part of the material is under compression and part of the material is under tension. Because of these compression and tension forces, material on the outside of the bend is stretched, while the material on the inside of the bend is compressed (Figure 13-9). These distortions are usually visible, although not overly objectionable, unless the material begins to tear. Pipe or other tubular materials must be bent with a special tool that keeps the material from collapsing at the bend. Heavy paper and cardboard are usually scored to help make the bend smoother. Wood and paper are best bent with the grain, while metals are bent across the grain (Figure 13-10).

Pressing. Pressing is another way in which materials are compressed and stretched, and it usually involves processes that are more complex than bending. In pressing, male and female *molds,* sometimes called *dies,* are used. Making the molds is expensive and contributes to the fixed costs of the product. Design specifications must state the mold dimensional tolerance and surface finish. Most molds are custom produced using special metal alloys and are hardened to withstand the wear of production.

Most metals are pressed cold, whereas materials such as thermoplastics and glass must be pressed hot. In production, pressing cycles must be adjusted to the material being processed. Because metals are pressed cold, the production cycle is more rapid. When pressing thermoplastics and glass, a cool-down cycle is required. Most composites require a slower cycle to give the resin time to cure. Ceramics and paper need cycling time for water used in the process to be removed.

A simple punch and die illustrate the making of a simple V-bend in metal (Figure 13-11). The extensive use of formed sheet metal throughout this

Fig. 13-10

Sample products made with materials formed by bending. Courtesy of Steelcase, Incorporated

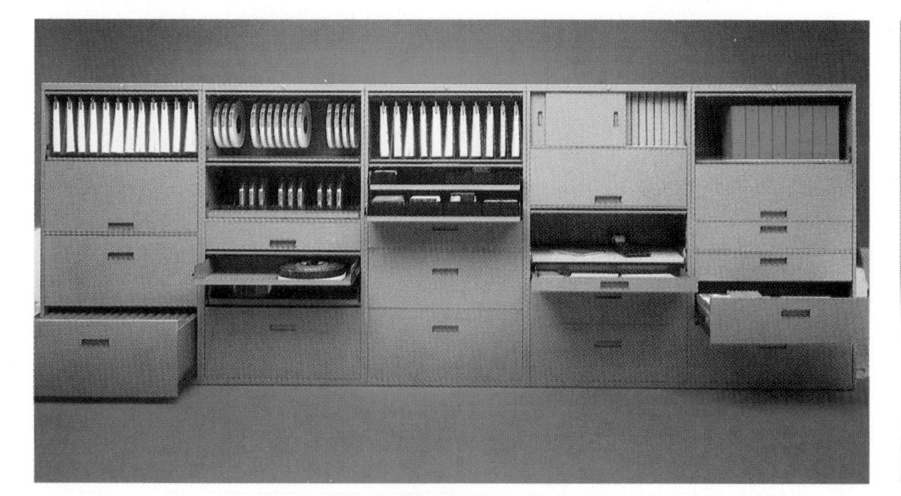

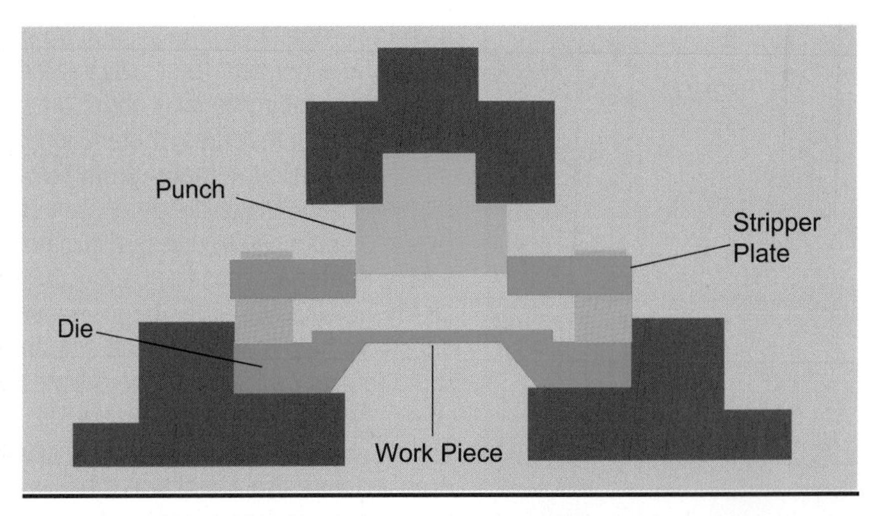

Fig. 13-11

Illustration of punch and die

century has led to the development of large power presses. Many products, such as automobile body parts and appliance parts, utilize complex three-dimensional shapes to give relatively lightweight sheet metal parts good stiffness qualities (Figure 13-12). When the part to be formed has considerable depth, multiple forming stages will be used, each of which stretches the material to a practical limit until the desired shape is achieved.

Fig. 13-12

An automotive body part formed by pressing. Courtesy of Honda of America

A variety of materials are formed by many different forming techniques. Vacuum, drape, and blow forming are used to form plastic, glass, and some special metal alloys (Figure 13-13). Some forming processes, such as the deep-draw technique used to make aluminum cans, use both male and female dies, whereas other processes use only one die or forming mold. Explosive hydroforming uses only a female die, whereas rotational spinning, shear, and flow forming use only a male die (Figure 13-14).

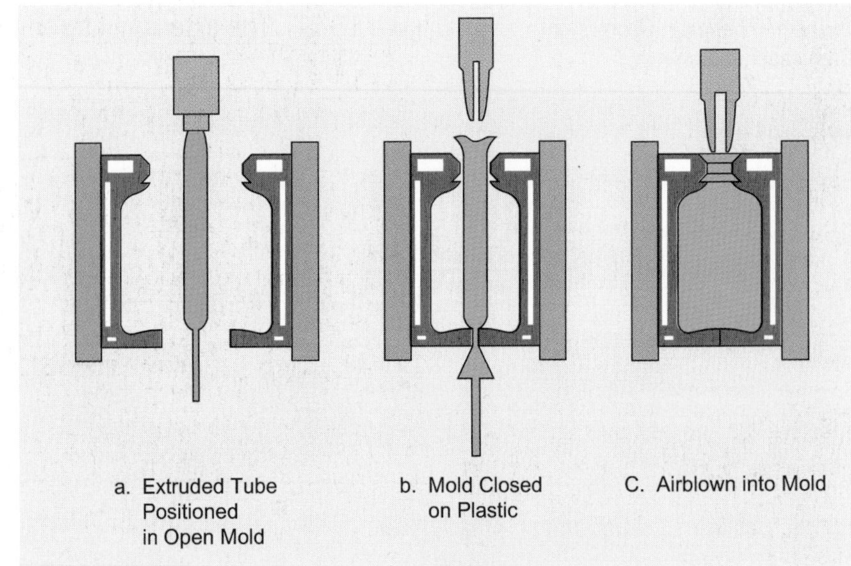

Fig. 13-13

Forming plastic bottles

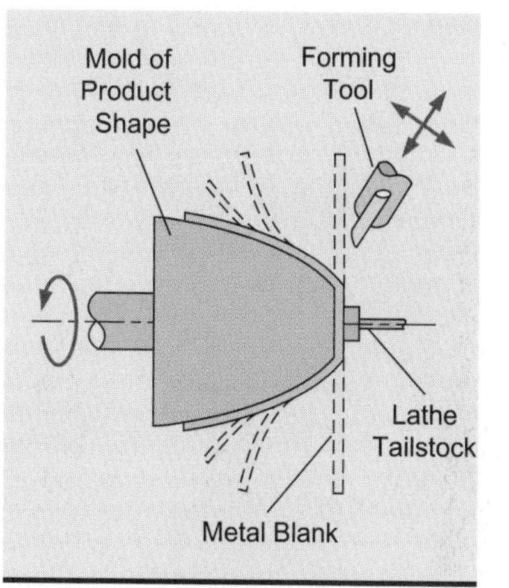

Fig. 13-14

Forming by spinning

Forging. While pressing usually involves sheet material, forging takes a solid blank and processes it into a desired shape using very high pressure (Figure 13-15). Drop forging is the most commonly used process. Generally speaking, products formed by this process have low-dimensional tolerance and a poor finish. Forged products are strong because of the unique grain structure created as the material is formed into the desired shape (Figure 13-16). Copper alloys are usually hot forged, whereas ferrous metals are cold forged.

Extruding and Drawing. The shape of toothpaste as it comes out of the tube has been caused by an extruding process. *Extruding* is the process of forcing a material through a die that imparts a predetermined shape (Figure 13-17). In extruding, the material is pushed through the die. Many shapes can be created in aluminum, copper, cast iron, thermoplastics, and clay by using this process. Window and door frames, as well as tubing and plastic-coated wire, are examples of extruded products (Figure 13-18).

Drawing is the opposite of extruding in that the material is pulled instead of pushed through the die. Products such as metal or plastic rod and wire are made by this process. In drawing, the product is usually formed in a series of steps. For some applications, such as making pipes, the material is initially formed by extrusion and then finished by drawing.

Casting. When the material to be formed is in a liquid state, casting operations are used. In *casting*, a liquid material is poured into a mold that accounts for the processing sometimes being referred to as *molding*. A pattern must be produced that represents the shape to be taken by the cast material. Patterns are made to exacting specifications, including the molding technique to be used and shrinkage factors. Usually, when a material is formed as a liquid, it shrinks as it solidifies. The pattern is used to make a cavity in the molding material and usually contains two or more parts (Figure 13-19). With few exceptions, the pattern is removed before the liquid material is poured into the mold. While most people know that iron and steel are cast, any material that has a liquid state can be formed by casting.

Fig. 13-15

Forming by forging. Figure at left courtesy of Simpson Industries. Figure at right courtesy of Pressure System Industries

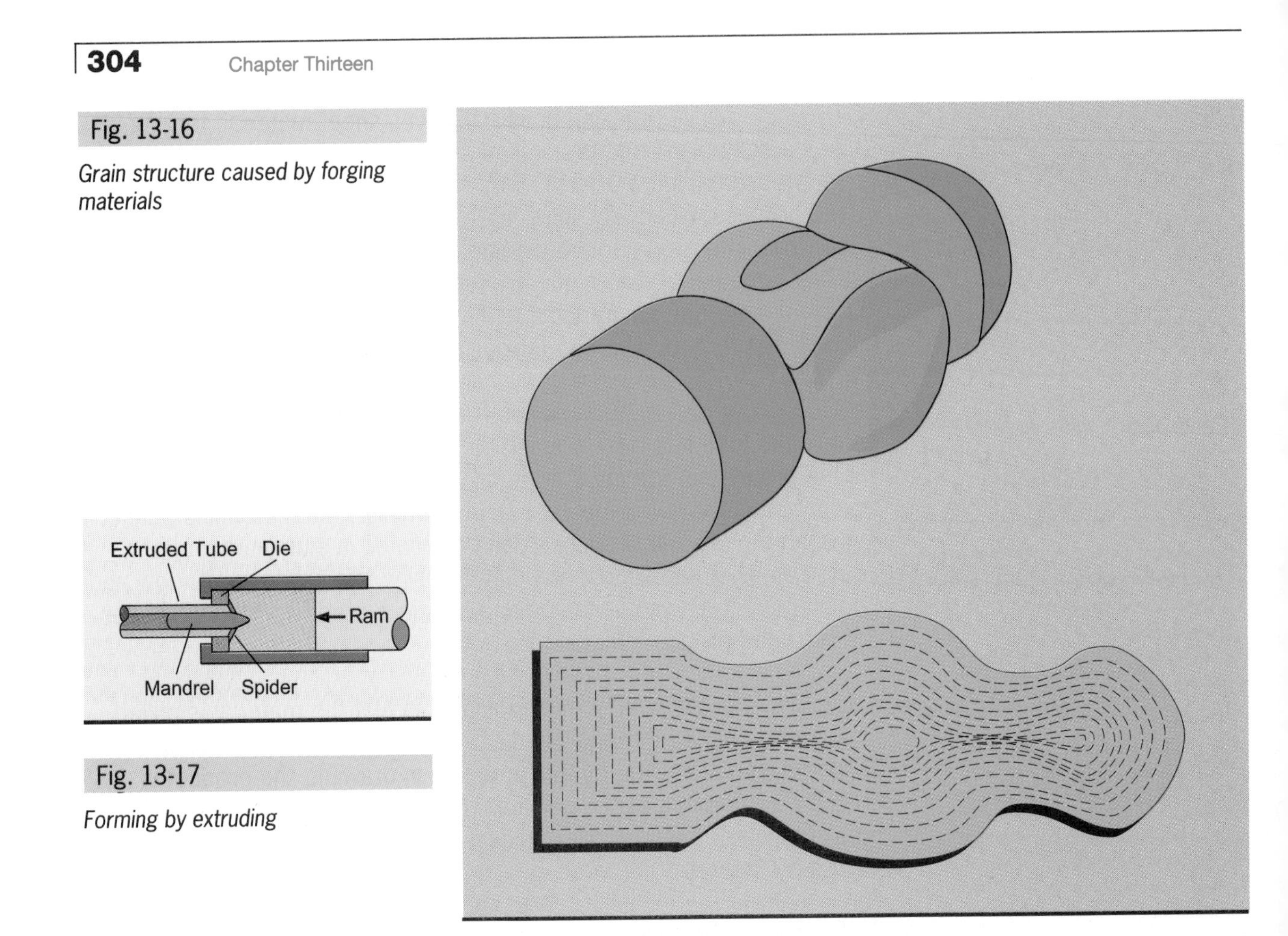

Fig. 13-16

Grain structure caused by forging materials

Fig. 13-17

Forming by extruding

Types of Molds. The cheapest types of molds are called *one-shot molds*. Sand casting, shell mold casting, and investment casting are commonly used one-shot molds (Figure 13-20). These molds are made from a pattern and filled with the molten metal or other liquid materials to be formed. In removing the product from the mold, the mold is destroyed. When the product is to be hollow, such as in the case of a brass spigot or pipe fitting, a *core* must be placed in the mold.

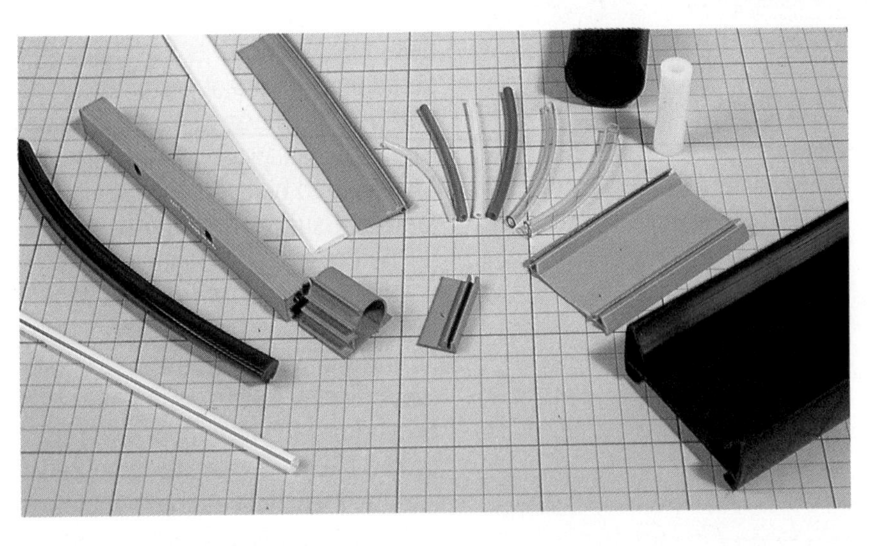

Fig. 13-18

Sample extruded products

Unlike one-shot molds, *permanent molds* can be reused. Die casting, transfer molding, centrifugal casting, injection molding, and lay-up forming represent types of permanent molds. Permanent molds are typically used to form thermoplastics and ceramics (Figure 13-21).

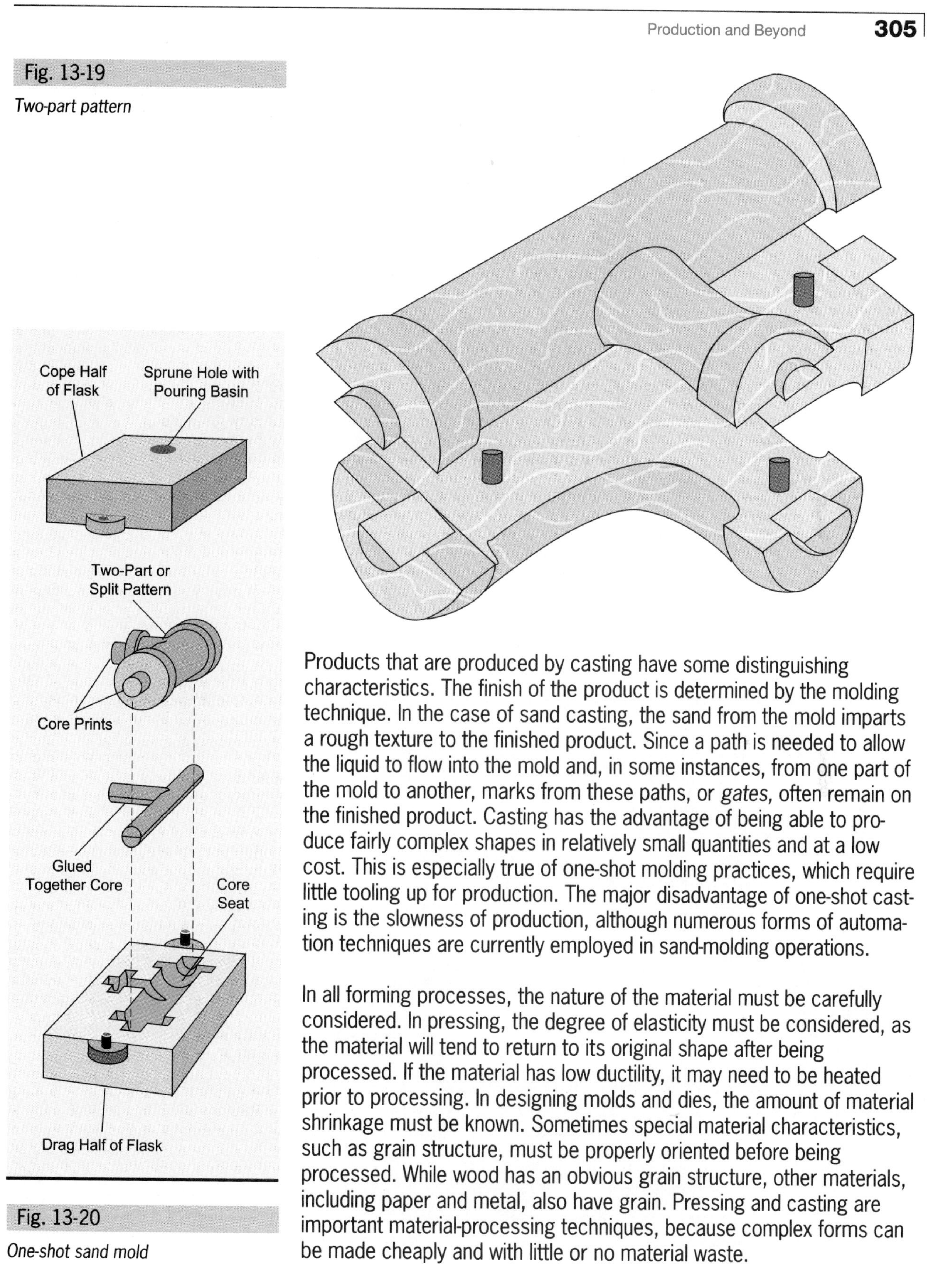

Fig. 13-19

Two-part pattern

Fig. 13-20

One-shot sand mold

Products that are produced by casting have some distinguishing characteristics. The finish of the product is determined by the molding technique. In the case of sand casting, the sand from the mold imparts a rough texture to the finished product. Since a path is needed to allow the liquid to flow into the mold and, in some instances, from one part of the mold to another, marks from these paths, or *gates*, often remain on the finished product. Casting has the advantage of being able to produce fairly complex shapes in relatively small quantities and at a low cost. This is especially true of one-shot molding practices, which require little tooling up for production. The major disadvantage of one-shot casting is the slowness of production, although numerous forms of automation techniques are currently employed in sand-molding operations.

In all forming processes, the nature of the material must be carefully considered. In pressing, the degree of elasticity must be considered, as the material will tend to return to its original shape after being processed. If the material has low ductility, it may need to be heated prior to processing. In designing molds and dies, the amount of material shrinkage must be known. Sometimes special material characteristics, such as grain structure, must be properly oriented before being processed. While wood has an obvious grain structure, other materials, including paper and metal, also have grain. Pressing and casting are important material-processing techniques, because complex forms can be made cheaply and with little or no material waste.

Fig. 13-21

Permanent injection mold

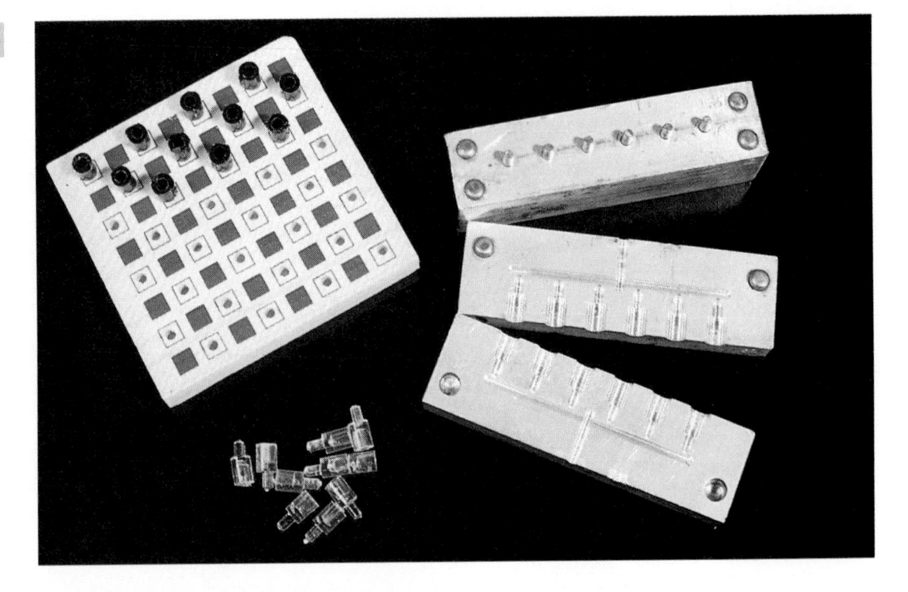

Separating Materials

Separating processes are used to take something away from the material being processed. Unlike forming processes, in which little waste is created, **separating** has the potential of creating considerable waste. Although some of this material can be recycled as *preconsumer waste*, it is important to minimize the amount of material to be wasted or removed. The designer can help to minimize waste by planning products around standard stock sizes. Removing materials can be expensive, because it typically requires extensive time and machine processing. Nearly any shape can be created by separating techniques; however, these processes are usually not as efficient or productive as forming techniques. For this reason, separating processes are often used in combination with forming processes to produce many finished products. For example, the automobile engine block is formed by casting to produce its basic shape, and then it is processed using material separation techniques to its final specifications (Figure 13-22). The casting process was employed to

Fig. 13-22

Cast engine block being machined. Courtesy of Cincinnati Milacron

reduce the amount of material that needed to be removed to produce the final product. Separating techniques vary from a simple operation, like cutting a 2" x 4" to length, or drilling a hole in a part for future assembly, to sophisticated laser, chemical, and ultrasonic processing. Separating techniques are used on all materials.

In planning for separating techniques, three problems must be solved:

- How is the material going to be removed?
- How is the material going to be held during material removal?
- How is the cutting tool going to be controlled?

Material separating can be achieved through mechanical, electrical, chemical, and thermal processes. Mechanical separating is the most commonly used of these processes.

Mechanical-Separating Techniques. In our discussion of pressing, it was shown how thin metal sheet and other materials can be shaped to yield a product with a relatively high degree of stiffness. Because of a large demand for products of this type, many special-

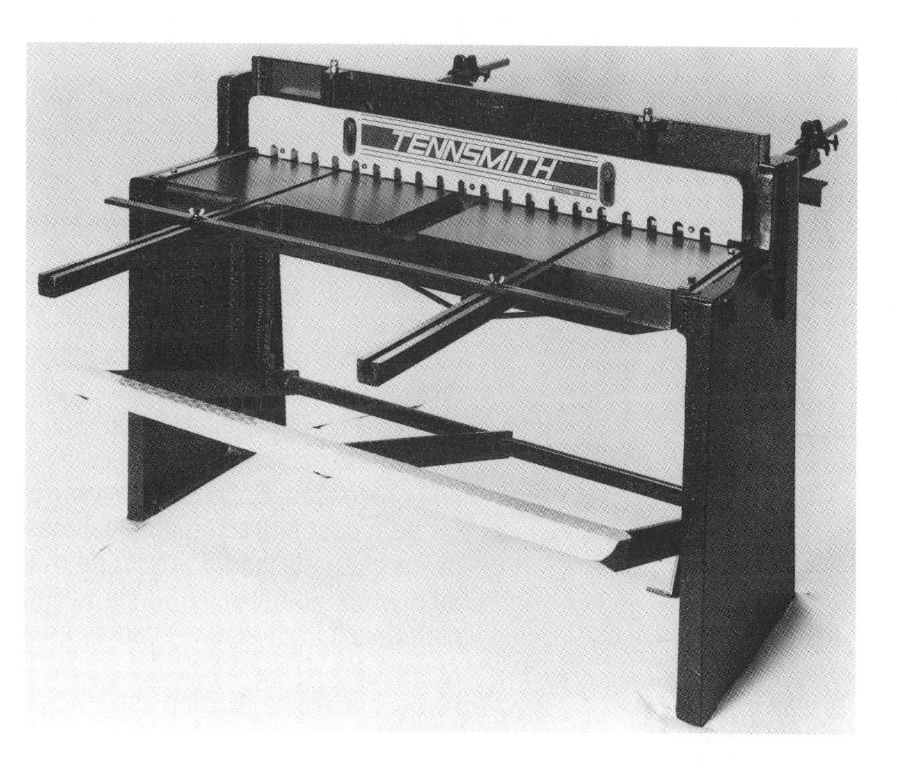

Fig. 13-23

Separating by shearing. Courtesy of Tennsmith, Incorporated

ized mechanical-separating techniques have been developed for sheet materials. Most of these processes combine a cutting action that causes plastic deformation at the tool edge and eventual material failure by shearing (Figure 13-23).

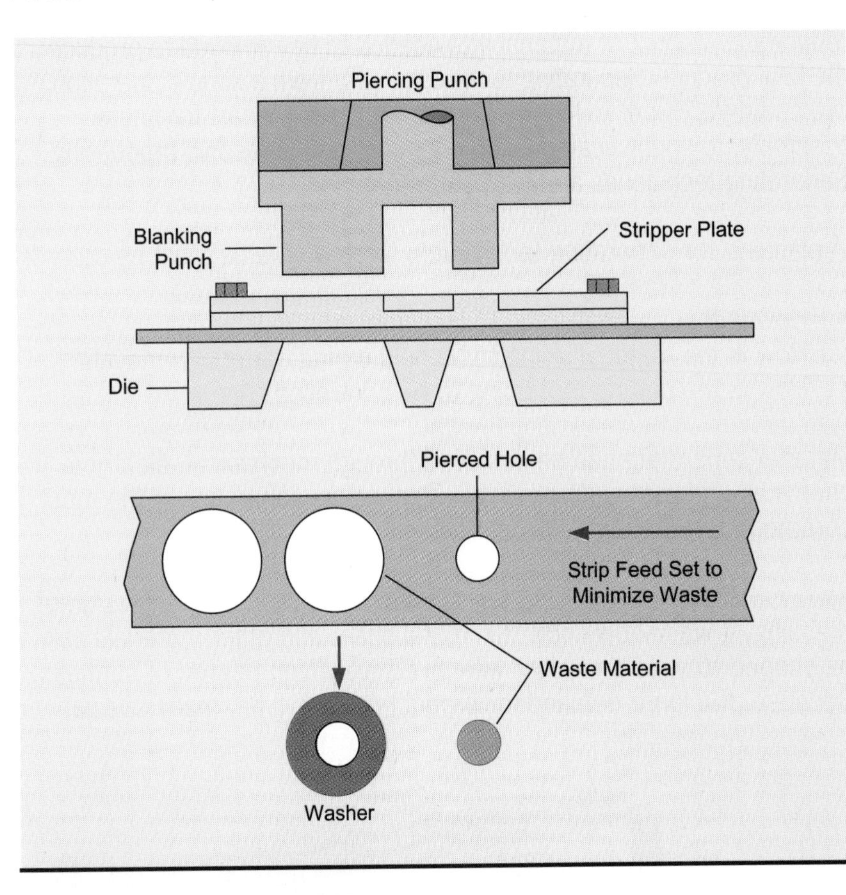

Fig. 13-24

Producing a common washer using shearing processes

Anyone who has used scissors has separated materials by shearing. In Chapter 6, shear was shown as one of the basic forces acting on materials. In product design, materials must be selected to withstand an expected shearing force. In material processing, shearing machines must be designed to exert enough shearing force to separate the material. Properly sheared edges will show a slightly burnished surface opposite the two outside surfaces, with a somewhat duller center area. Shearing operations are commonly used to trim excess material from finished products, such as flashing resulting from pressing and casting operations.

Piercing operations, which also incorporate shearing principles, are used to make small holes in thin materials. Common drilling operations, which will be discussed in the section on chip removal, are not appropriate for making holes in thin stock. *Blanking* operations are similar to piercing except that the material being removed is the desired product. The common washer is made by combining both operations (Figure 13-24). In addition to metal, paper and cardboard products are commonly sheared by several operations, including die cutting, slitting, and perforating. All special shapes, such as envelopes, packaging materials, and cardboard displays, use die-cutting processes. Because economy of materials is always important for both cost and environmental reasons, shearing operations must be carefully planned. Parts to be made by blanking operations should be oriented to minimize waste while still taking into consideration material characteristics such as grain.

Heavy machining is often associated with separating by *chip removal.* Unlike shearing, by which the product was removed from a stock sheet, machining removes the unwanted materials and leaves the product. Common hand tools, such as the chisel, drill, and saw, use chip-removing techniques as a way of separating materials. Most chip removal, however, is done by heavy machines, including the drill press, the machine lathe, the power saw, and the milling machine

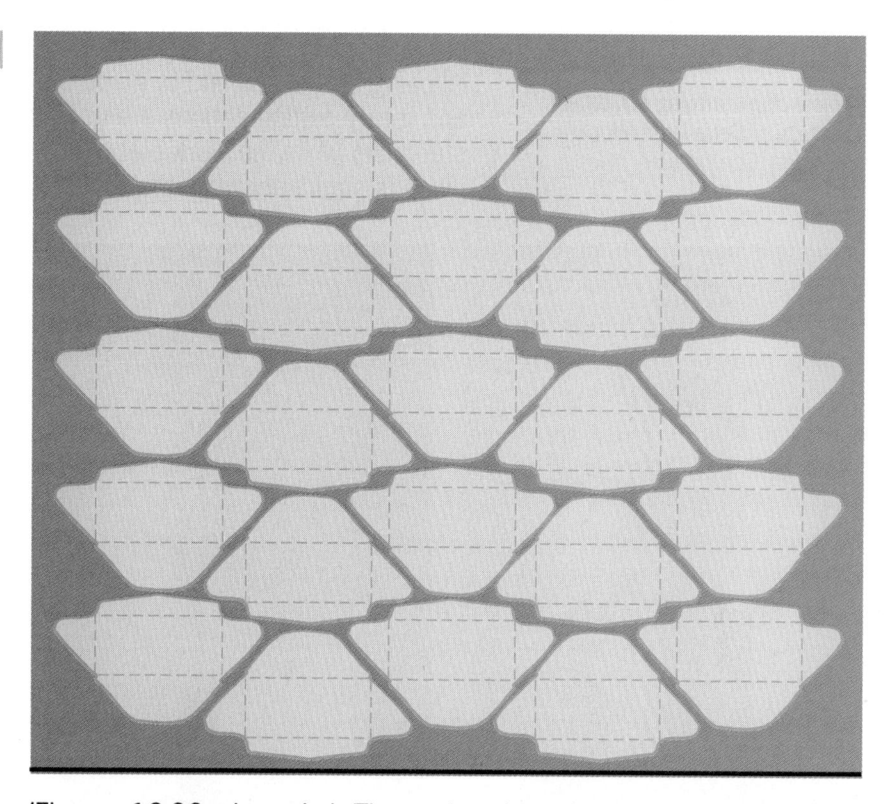

Fig. 13-25

Layout for efficiently cutting multiple #10 business envelopes from sheet material

(Figures 13-26a, b, and c). The wedge-shaped cutting tool is the basis for most chip-removing processes.

The essential features of the wedge-shaped cutting tool are the clearance angle and rake angles (Figure 13-27). The *clearance angle* is necessary to prevent the cutting tool from rubbing against the material being cut. The *rake angle,* which actually controls the amount of cutting action, must be carefully designed. The efficiency and quality of the cut will be determined by using the proper rake angle. This angle must be selected based on the material being separated and the process being used. The rake angle will affect the amount of energy being used and the length of the cutting tool's life. When a particular shape is planned, a *form-cutting tool* can be used. Router bits are good examples of specialized form-cutting tools (Figure 13-28). Metal, plastic, wood, and other composites are best suited for chip-removing processes.

Other material-separating processes are used in the production of commonly used products. The most important of these processes use abrasives as the basis of the chip-removing process. Abrasives were introduced in Chapter 11. *Grinding* and *ultrasonic machines* use abrasives to separate and smooth materials. *Electrical* and *chemical* methods are used to erode material from the work piece. *Thermal* methods are used to remove materials by melting. A hot-wire plastic cutter, oxyacetylene setup, and laser are examples of thermal material-separating processes.

Figs. 13-26a, b and c

Typical chip-removing equipment. Courtesy of Delta International Machinery Corporation

In order to separate materials, it is necessary to hold both the material being processed and to guide the tool doing the cutting. Both linear and rotary cutting actions are used. In a lathe operation, the tool is stationary while the work piece moves. In a milling operation, the work piece is stationary while the tool moves. In other operations, both the tool and the material move. To aid in holding the work piece, a variety of clamping devices have been invented. Many of these devices use the toggle mechanism described in Chapter 7.

13-26a. bandsaw

13-26b. drill press

13-26c. machine lathe

Fixtures consist of a broad range of devices that hold the work to be processed (Figure 13-29). *Jigs* are devices designed to guide the tool in production. In modern production, the tool is often controlled by a computer. In production settings, material-separating processes have been highly automated, with both transfer and holding of the work piece, and tool operation automatic. Modern industrial organization methods will be covered later in this chapter. When a new production line is needed, industrial engineers are often asked to design new jigs and fixtures.

Combining Materials

In order to make products, it is sometimes necessary to combine components that have been made by forming or separating processes. **Combining** techniques are often associated with product assembly or *fabrication.* Joining components can be done by mechanical or chemical techniques. In addition

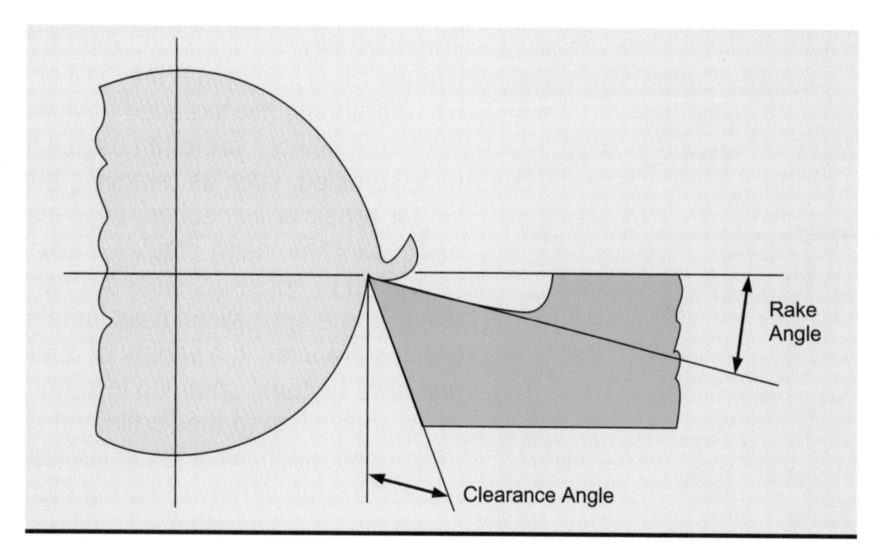

Fig. 13-27

Principles of the wedge-shaped cutting tool

Fig. 13-28

Set of router form-cutting tools

to assembling components, combining techniques are used to coat products with paint or to put a printed image on paper or some other substrate. Broadly viewed, combining materials are necessary for assembling and finishing products.

Mechanical Fastening. One of the oldest methods of fastening materials together was by making various and ingenious joint configurations. Mortise and tenon joints have been discovered from prehistoric times. These and other wood joints form a basis for many modern material joints (Figure 13-30). Some joints, such as the dovetail, are self-locking but are very difficult to cut. In order to simplify the locking of joints between two components, other types of mechanical fasteners have been developed.

Fig. 13-29

A simple fixture used to hold paper for printing on a memo pad

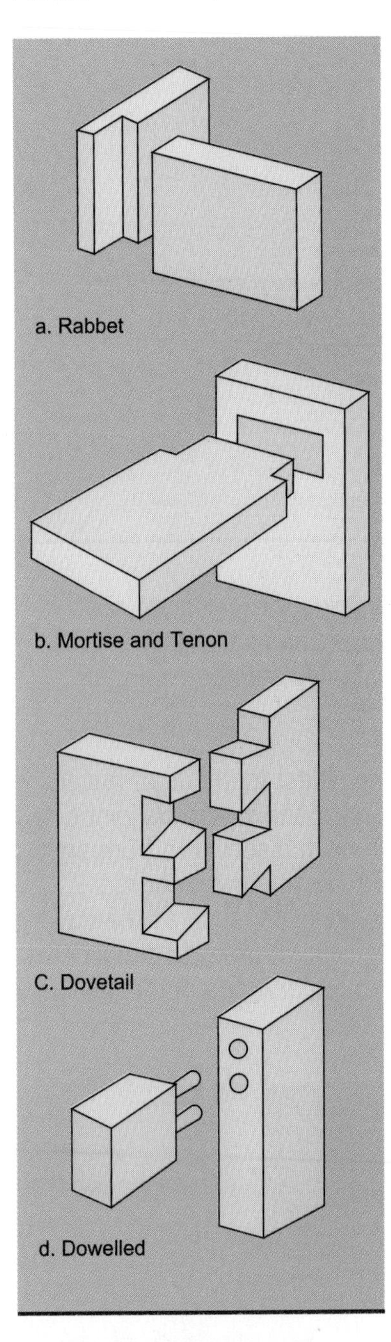

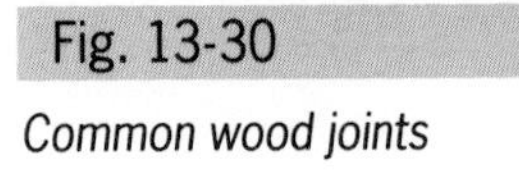
Fig. 13-30

Common wood joints

Threaded devices are a common type of mechanical fastener (Figure 13-31). Although they are convenient to use in many assembly operations, the threaded device itself is an expensive component. Threaded devices are often used when the product may need to be disassembled, such as removing the tire from an automobile or bicycle for repair or inspection. Less expensive mechanical fastening devices include nails, staples, sewing, and lacing. Nails are mostly restricted to construction and other applications where wood is used. Some specialized nails can be used with masonry products. Staples are used in a variety of applications in the printing industry and have become common in surgical procedures. Sewing, because of its application to the textile and shoe-making industries, is the most used mechanical fastening technique. In addition to these techniques, a large variety of spring clips, rings, pins, rivets, spiral and comb binders, and other special mechanical devices have been invented (Figures 13-32a, b, and c). Modern material developments have provided the designer with additional fastening techniques.

Chemical Fastening. Soldering, brazing, and welding use thermal processes to fasten components together. In Chapter 11, these and other bonding techniques were discussed. Thermal bonding joins two materials together by melting adjacent areas, thus allowing a covalent bond to form when the material cools. Thermal bonding provides a very strong bond but is relatively expensive and has the potential to distort the material due to uneven heating. Thermal fastening is restricted to metal and plastic materials (Figure 13-33).

Adhesives also employ chemical-bonding principles. Although natural adhesives have been used for centuries, modern developments have led to the invention of very strong synthetic adhesives. These adhesives are being used to glue ceramic tile on the space shuttle, the rearview mirror on automobile windshields, and assembled furniture and utility pipes. Some materials, such as thermoset plastics, can only be fastened by adhesives. Modern adhesives are used to fasten metal, plastics, ceramics, leather, paper and cardboard, wood, and nearly any other material. (See the list of adhesives in Figure 11-18.) Adhesives may prove to be an excellent material-fastening technique for many design applications. Adhesives are often stronger than the material itself and, if applied carefully, will provide a very clean joint.

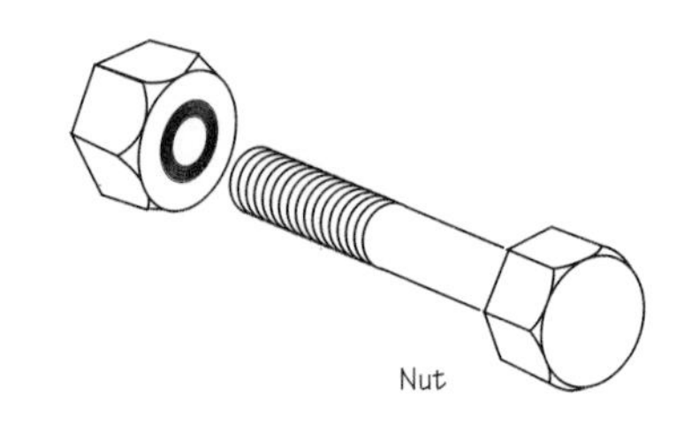

Fig. 13-31

Threaded fastener

The selection of the type of combining technique is determined by analyzing questions of application, including the type of material, anticipated stress, cost, and aesthetics.

- How permanent is the joint to be?
- What are the characteristics of the materials to be joined? (Often dissimilar materials are difficult to join with adhesives.)
- How strong does the joint need to be?
- Is the joint to be stiff or flexible?
- What forces will be present at the joint (vibration, heat, corrosives, etc.)?
- What are the aesthetic considerations? (Can the joint be seen, etc.?)
- What are the environmental impacts of the joining material on the worker, consumer, and environment?

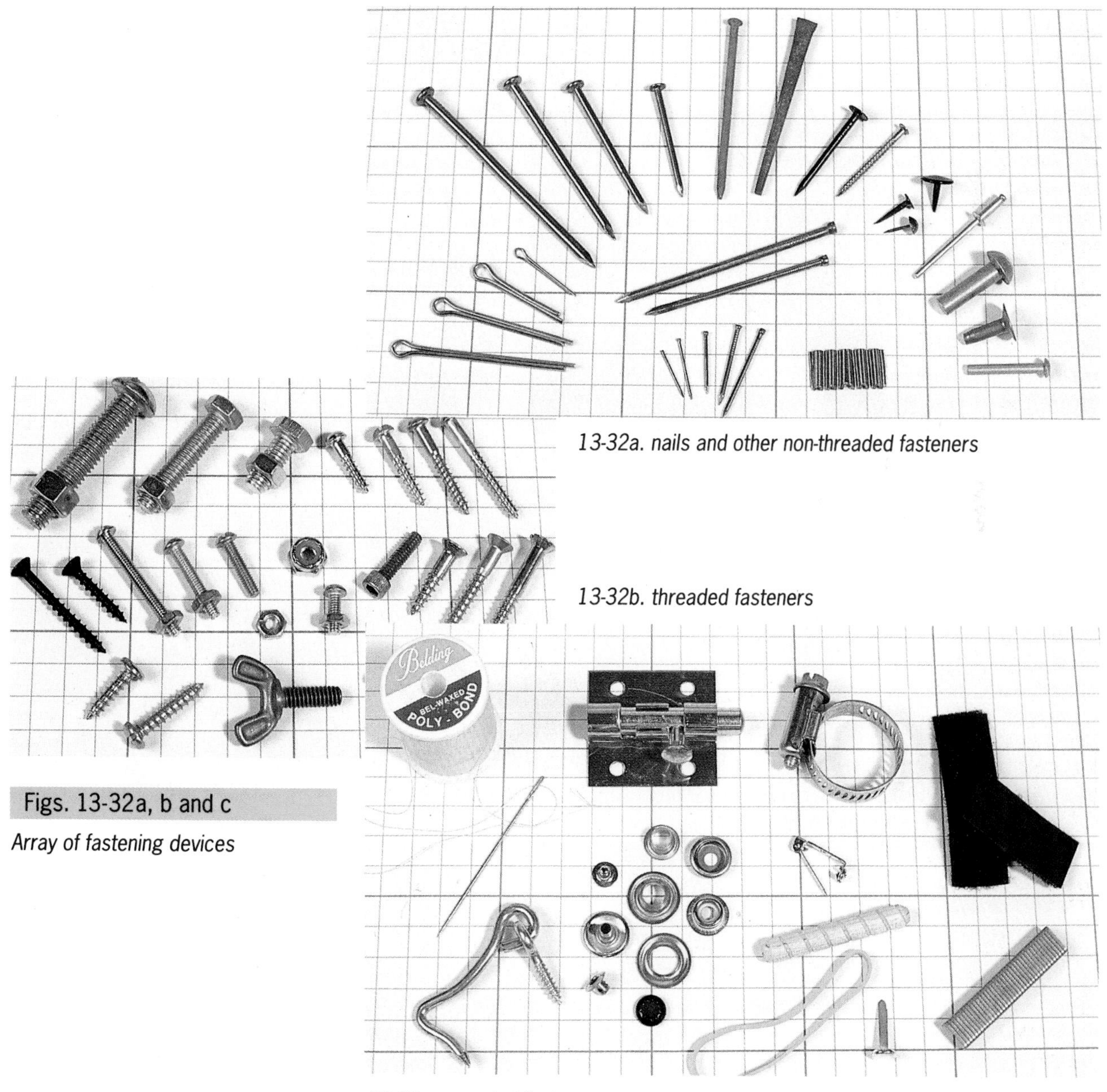

13-32a. nails and other non-threaded fasteners

13-32b. threaded fasteners

13-32c. assorted fasteners

Figs. 13-32a, b and c

Array of fastening devices

Organizing for Production

The movement during the Industrial Revolution from craft or custom production to the *factory system* of the twentieth century caused a fundamental shift in society. As people moved into cities, they became better informed and more interested in politics. *Labor unions* were formed, working conditions were improved, and, at the same time, leisure time increased. Most industrial countries became more democratic. With more efficient production systems, productivity increased and the *standard of living* improved for the middle-class workers. To continue increasing efficiency, more complex machines were developed to replace older, less productive machines. Usually, these new machines did more of the work that was previously done by the worker. *Automated machines*, being more complex, control the tool and work piece to a greater degree. Automation usually has the effect of replacing the need for skilled machine operators.The nature of the work place and production in a technological society is constantly changing.

"Buy American" is becoming an increasingly difficult goal. We live in a global economy,

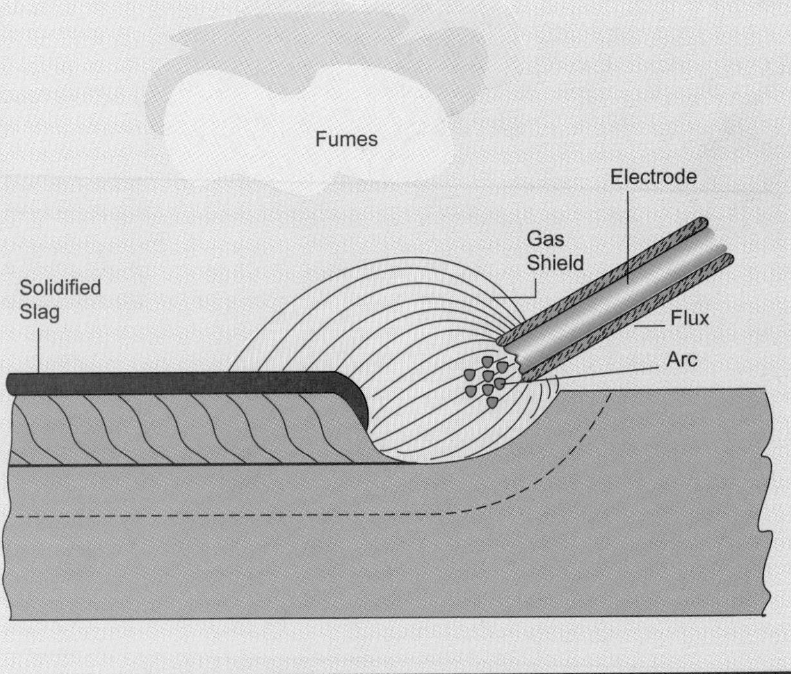

Fig. 13-33

Thermal fastening by welding. Photo courtesy of Southern Illinois University at Carbondale and American industry must be able to compete in that arena. The "Steel Belt" of the post-World War II period became the "Rust Belt" of the 1970s and is now becoming the high-technology "Automation Alley" of the twenty-first century. These are all terms used for the northern, eastern and midwestern regions of the United States. For a number of years, most notably in Japan, a number of new organizational concepts emerged and are now being accepted in the West. Interestingly, the successes of many foreign competitors can be traced back to the management principles developed by W. Edwards Deming, an American statistician.

Computer-Integrated Manufacturing

Computer-integrated manufacturing (CIM) is an umbrella term used to describe a full range of contemporary manufacturing techniques. CIM (pronounced *SIM*) encompasses a broad range of modern processes, including computer-aided design (CAD), computer-aided manufacturing (CAM), automated material handling (AMH), total quality control (TQC), robotics, and others. In actuality, it is management and the involvement of everyone in the organization that makes the system successful (Figure 13-34). The computer is the technological tool that drives CIM. CIM has the potential to provide all the necessary information in real time (current) to maximize all production operations. Areas of concern include

Fig. 13-34

Critical CIM links prepared by the Society of Manufacturing Engineers

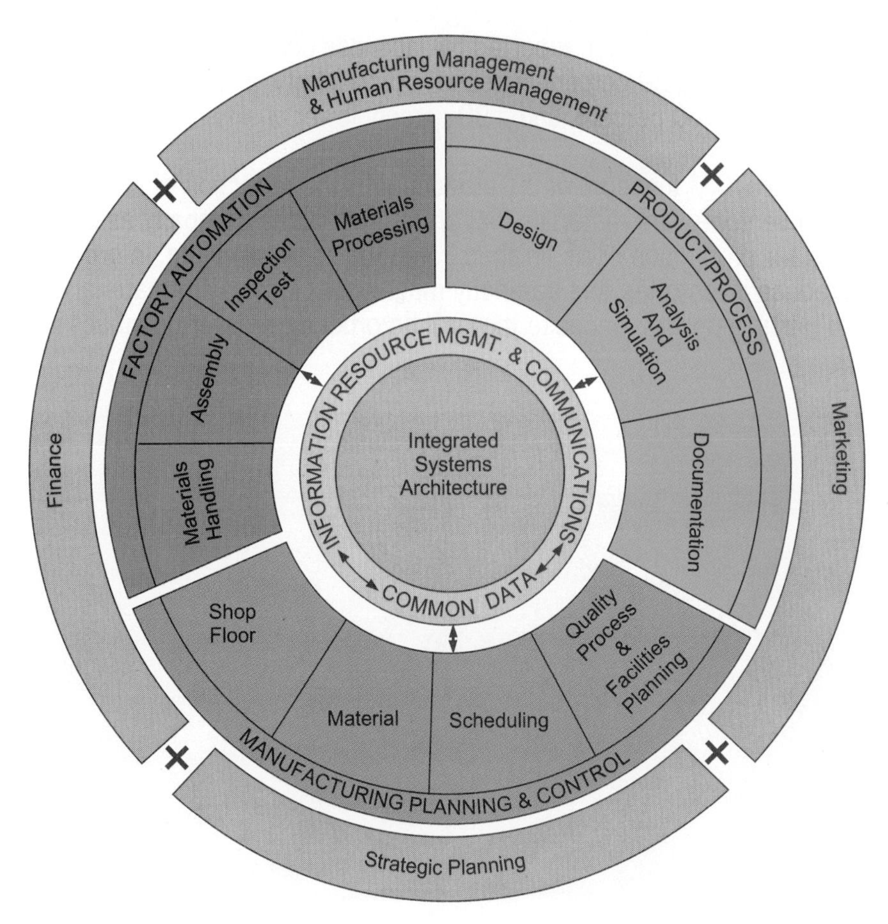

- The impact of product design decisions on production
- Converting design ideas into production drawings
- Translating specifications into parts lists and production routing
- Preparing schedules
- Controlling machine operations
- Providing current cost of materials
- Providing real-time information to managers
- Performing routine clerical tasks

CIM techniques are intended to increase productivity and quality. By carefully planning, organizing, and controlling production, wasted time and materials can be reduced. These savings reduce raw material and labor costs. They also reduce the time a product is in the production cycle.

A CIM technique called *group technology (GT),* is being incorporated into traditional batch production to increase productivity. As much as 75 percent of production is done by batch processing. Under GT principles, management groups together parts or components that can share common machine operations and setups. Through careful planning, savings can be achieved by sharing expensive jigs and fixtures, by reducing setup time, by needing fewer machines, and by improving the flow of parts. One of the newest developments in GT is the introduction of the manufacturing cell. The cell is a collection of tools and material-handling equipment grouped together to produce a family of parts (Figure 13-35).

Just-in-Time (JIT). Just-in-time is a CIM technique largely associated with Japanese manufacturing. With this technique, material handling, inspections, and storage are considered wasteful, leaving only production processes as a means of adding value to the product. Over a five-year period, JIT has been reported by Japanese companies as being responsible for a 30 percent increase in productivity, a reduction in inventories by 60 percent, and a 90 percent lower quality rejection rate.

The critical CIM function for JIT production is the reduction of inventories. *Inventories* are materials that are stored by the company as either stock for production or as finished components or products. In order to reduce inventories, the company must work closely with material and parts suppliers, because they will be ordering smaller quantities of

Fig. 13-35

CIM cell

materials more frequently. One of the disadvantages of JIT is the inability of the system to react quickly to demand shifts. Companies that use JIT techniques have also developed quality circles and have forged new relationships between management and the labor force.

CAD/CAM. One area of development that has affected nearly every industry is the introduction of computer-aided design and manufacturing (Figure 13-36). CAD/CAM has been most successful in reducing the amount of time needed to bring new products to the marketplace. In some instances, these CIM techniques have been used to reduce duplication of effort in preparing prototype models and product specifications. With CAD/CAM, information from the design stage can be directly input into prototype and production equipment, saving considerable time and money.

Flexible Manufacturing System (FMS). Similar to the cell concept in group technology, this new CIM technology may be the factory of the future. The flexible manufacturing system (FMS) incorporates between four and thirty-two work stations into an integrated manufacturing environment. In FMS, all scheduling, material handling, process and tool management, quality control, and data keeping are integrated into the system. Robotics and other forms of automation are key elements of flexible manufacturing.

Fig. 13-36
CAD/CAM

Robots were first used, and continue to be used, to do many monotonous or dangerous jobs in industry. With increased use of microprocessors, robotics are moving into more sophisticated operations, including material processing, assembly operations, finishing, material handling, and inspection operations with such techniques as machine vision. Currently, 90 percent of all robots are used in the automobile industry. With robots costing up to $100,000 each, companies must look at both the short- and long-term implications. Some advantages of using robotics include consistency, reliability, and, most important, cost-effectiveness. The safety factor of using robots in dangerous and unsafe environments is important to consider. While robots are incorporated into flexible manufacturing systems, the major advantages of FMS are realized in the reduction of material-handling time and inventory reductions.

Future Impacts

It is extremely important for the health of the economy in a free enterprise system that companies remain viable and competitive in producing goods for the global marketplace. Competition begins with appropriate design solutions. Product development through efficient production systems is the key element in maintaining a viable economy.

It is estimated that over 70,000 different natural and synthetic materials are available for production. As a society, we must be concerned about the depletion of needed natural resources. Critical resources can be saved by finding or designing alternative materials or, more important, by conserving those that already exist. The current trend to design new products using fewer materials (source reduction) is one way to save valuable resources. Recycling and waste reduction are others.

In the quest to remain competitive, our society will need to consider important environmental questions. In 1970, the EPA established standards for air and water quality. These regulations have been subsequently revised. The current concern over the depeletion of the ozone layer points to the need of all citizens to be involved in the decision-making process in balancing the trade-offs and risks of technological progress. Good design always considers the impacts of production and consumption on the individual, society, and the environment.

Summary

Manufacturing products involves a managed production system that brings together as system inputs capital, people, knowledge, materials, energy, tools and machines, and processes. Products begin their life cycle as a design to a perceived need and end the cycle as a discarded piece of waste. In between, the product impacts the individual, society, and the environment. Manufacturing is part of our economic system. This system, known as free enterprise, is based on the concepts of supply and demand. Planning production is important so that the best method of production can be found. Materials are processed in the system by forming, separating, and combining. Business in the United States must compete in a global economy. New techniques for managing production involve computer-integrated manufacturing (CIM) technologies. The way in which our society identifies and solves technological problems will continue to have a major impact on individuals, societies, and the environment.

Observation/Analysis/Synthesis

1. Select one of the products found in the scene in Figure 13-37, and list all materials and processes needed to produce the product.
2. Given the product you designed, list all materials and production processes that will be necessary to produce the product.
3. Take your design, including the list of materials and processes, to a local manufacturing company, and ask a manager about the viability of producing the product.
4. Select one of the least efficient processes necessary to produce your product, and suggest alternative, more-efficient processes.
5. Suggest ways in which the design could be changed to better conform to standard materials and industrial processes.
6. Evaluate your product for each of the following criteria:
   a. Manufacturing viability
   b. Durability
   c. Maintenance
   d. Aesthetics
   e. Material source reduction
   f. Environmental impact
   g. Social impact

Fig. 13-37

The artifacts of technology, made from different materials and formed using different processes, are all around us, serving many purposes

Glossary

actuator—in electronic systems, a output component, such as a motor, solenoid, or electromagnet, that converts an electrical signal into physical phenomenon.

adhesion—the process of joining two materials by mechanical or physical bonding; gluing is an adhesion process; using adhesives to interlock molecules of materials being joined.

amorphous—Pertaining to a solid that has no specific form or structure; a noncrystalline material.

annotated sketch—a drawing with numerous notes describing various features of an object intended for analysis and as a careful record.

anode—the positive terminal of an electrical component or power supply.

anthropometrics—data concerning physical characteristics of humans, such as height, weight, foot size, and a comprehensive list of other dimensions, including distances from joint to joint, while sitting, standing, and so on.

appearance model—a model developed to work out or display the visual characteristics of a product.

armature—the rotating part of a motor; the main current-carrying winding of a generator or alternator that may be rotating or stationary; the movable part of a solenoid or relay.

astable timer—a circuit that has no stable state and therefore alternates continuously between two states (on and off); a circuit that produces a pulse signal.

bar chart—a graph that uses horizontal or vertical lines or bars to represent a quantity; a bar chart that compares the number of television sets in the United States in 1960 with the number today would have a short bar for 1960 and a long bar for today; provides a visual display of a quantity.

beam—a horizontal structural element designed to resist compression and bending forces.

bell crank—a mechanical link that changes the direction of force and motion 90 degrees.

bending—when external forces tend to distort a structural element, such as a beam.

binary number—a base two number; binary numbers have only 0's and 1's; the number 7 is represented as 0111 in the binary number system.

bipolar transistor—a transistor that has an emitter, a base, and a collector in either an NPN or PNP format; N and P material is produced by "doping" silicon with an impurity with either a valence 5 or valence 3 element.

brainstorming—a specific technique that uses a small group of people for generating a lot of ideas and is controlled by specific rules.

brittleness—the lack of ductility; a material that fractures without any noticeable deformation, such as glass.

call number—a number on a library card catalog card that directs you to a specific location in the library where the book is stored.

cam—usually, a rotating part that is not circular and that causes a cam follower (a part that rides against the cam) to move in some predetermined way; part of a system with a cam follower to convert rotary motion to reciprocating motion.

capacitor (c)—an electrical component that contains two conducting plates (usually metal) separated by a thin insulator (dielectric); capacitors are able to maintain an electrical charge after a voltage source is removed; standard unit of capacitance is a farad, but practical component values are measured in microfarads, picofarads, and nanofarads.

cathode—the negative terminal of an electrical component or power supply.

Chaos theory—a theory that states that natural systems have so may variables that it is impossible to track each one and that is therefore not possible accurately to predict an outcome; in the best-selling book *Jurassic Park* by Michael Crichton, chaos theory was explained by describing that a butterfly flapping its wings in Japan affects the weather in Washington.

chart—a diagram that shows the relationship between two or more numerical quantities, such as the age and average height of females.

chroma—a description of the intensity or brightness of a particular hue (color).

clutch—a device that allows engagement and disengagement of a rotating shaft from input power.

coating—surface finish using paint, varnishes, penetrating compounds, and many others; a protecting and/or appearance finish.

cohesion—a chemical bonding resulting in the joining of two similar materials; atomic or molecular bonding is the strongest form of joining two materials together. Welding is a type of cohesive bonding.

combining—joining materials by mechanical fastening (nails, screws, etc.), or by chemical fastening (gluing, soldering, brazing, etc.).

commutator—used with brushes in a DC motor or generator to transfer current from the rotating windings and the stationary terminals.

complementary color—a color that has the greatest visual contrast to another color and falls on the opposite side of the color wheel.

complemenary metal oxide semiconductor—CMOS; integrated circuit chips which are sensitive to static electricity and may operate on voltages between 4.5 and 16.

composites—materials made from two or more different materials, such as plywood, fiberglass, carbon fiber, and others.

compression—when forces are exerted on a material that tend to push a material together.

computer-integrated manufacturing (CIM)—the integration of computer-aided design (CAD), computer-aided manufacturing (CAM), robotics, and many other techniques.

computer model—a computer program, usually consisting of certain mathematical algorithms (precise instructions for doing something) that simulate a process or an environment; a computer model to study automobile body shape drag would need to account for body shape, friction from air at various speeds, interaction between body shape and road surface, and so forth.

concept tree—a mind map; a thinking tool that involves putting an idea on paper and developing related ideas by graphically connecting the ideas in some logical way.

conductor—in electronic circuits, materials that carry electricity with little resistance; typically, copper or aluminum, but silver and gold are often used for contacts where corrosion is possible.

Cornucopian—a term used to describe someone who believes that our natural resources are plentiful and that we are in little or no danger of running out of them.

cost-effectiveness—the performance of a product relative to its cost.

cottage industry—a term used to describe a decentralized, often home-based industry; popular before the Industrial Revolution, during which time factories eliminated most of these.

crating—a technique of drawing used in perspective and isometric drawing that begins with a box or crate in which the object or some feature of the object is drawn; for example, a camera may be drawn by drawing a crate for the body and another for the lens.

current—the flow of electrical current that refers to the rate at which electrons are flowing in a circuit; the higher the rate of electron flow, the higher is the quantity of electrons passing a point; current is measured in amperes.

custom production—making one of something, such as a device, product, or home; hand-made as opposed to mass produced.

cutaway view—a sketch or drawing that represents an object as if it had been sawn in half or as if part of the object had been removed; intended to show subparts or assemblies that are hidden.

Darlington pair—two transistors connected together in such a way that a signal on the first switches the second; usually a combination of a sensitive, low-power transistor and a power transistor; can be made from two separate transistors or purchased as one component.

descriptive statistics—numbers that summarize or represent some characteristic about a population, such as the income of college graduates on their first job after graduation who live in the United States; deals with des-cribing populations (complete groups).

design brief—a statement that describes both what a designer must do and what a successful design must accomplish.

design loop—a process employed to solve a problem with technology; a process that brings you back to the starting point so that you can evaluate your results against your identified problem.

developmental sketch—a drawing that is intended to help work out problems in material, size, fastening, aesthetics, and other developmental issues.

dimensional stability—the quality of a material to resist changing shape due to changes in temperature, humidity, and other factors.

display—in electronic systems, an output component or subsystem intended to provide human users with information, such as lights, meters, or digital number displays.

double-acting cylinder—pneumatic or hydraulic cylinders that allow fluid pressure to be applied to either side of the piston.

ductility—the ability of a material to be shaped without it fracturing.

durable goods—the name given by the U. S. Department of Commerce to describe products intended to last more than three years.

dwell—in cam design, the number of degrees of the cam lobe where the cam follower is pushed out to its maximum travel.

dynamic load—a force or load in motion, such as the load on a roof as the wind is gusting.

eccentric cam—a very smooth-acting cam that is simply a circular disk with the axle placed off-center.

elasticity—the ability of a material to return to its original shape after being stretched, compressed, or otherwise deformed.

elastomers—polymers that can stretch over 200 percent without permanent deformation.

electromagnet—a wire coil, often wrapped around an iron core, which becomes a magnet when a current is flowing through the coil.

equilibrium—the condition of a structure that exists when the external loads or forces and the internal forces are equal and the structure is stable, or not moving.

ergonomics—the study of the relationships between humans and the equipment and environments they use; a term used for human factors engineering.

exploded view—a drawing that is intended to show the relationship between a number of parts that must be assembled or disassembled.

external forces—loads or forces that are exerted on a structure from outside the structure, such as the weight of gravity or a load that the structure is intended to support.

faultfinding—a process used to determine a malfunction in a system; a process employed to discover the reason for an engine not starting or an electronic system not functioning.

five-port valve—a control valve used to operate a double-acting pneumatic cylinder.

fluid power—having to do with the use of gases and liquids in systems to do work.

forming—changing the shape of a material by bending, pressing, forging, extruding, drawing, or casting.

frequency—in electronics, the number of complete cycles of an alternating current sine wave occurring in 1 second; inverse to wavelength.

fulcrum—the pivot point of a lever or linkage.

functional model—a model that has been developed to work out the functional or operational problems or ideas without regard to appearance.

Gaussian distribution—known also as a "normal" distribution curve; the distribution of some characteristic, such as height, over an entire population ranges from a few extremely short to a few extremely tall individuals, with most people falling in the middle or "average" height category; a chart of the frequency of occurrence of some characteristic that takes on a "bell-shaped" appearance because of the few occurances at the extremes and the large number of occurrences toward the middle.

graph—a diagram that shows how some numerical quantity changes over time, such as the increase of world population from 1793 to 1993.

Green Label—a term used to describe a product which has passed stiff tests of environmental acceptability; products which have passed such tests in Europe are often given a "Green Label" by environmental groups.

heavy metal—a term given to a number of metals that are difficult to dispose of in an environmentally acceptable manner, especially mercury, chromium, lead, arsenic, zinc, gold, silver, cadmium, cobalt, manganese, and molybdenum, which are quite toxic.

hertz—a unit of frequency that indicates the number of cycles per second; AM broadcast signals fall between 500 kilohertz and 1,600 kilohertz; FM broadcast signals fall between 88 megahertz and 108 megahertz.

highlight—on a drawing of an object, a light or white area, line, or dot that represents the reflection of light from a bright source.

hue—a "pure color," such as blue, green, or red.

human factors engineering—the study of the relationships between humans and the equipment and environments they use; also called "ergonomics."

hydraulic—having to do with a system that uses liquid under pressure to do work.

ideation—the process of developing ideas.

incineration—the process of reducing the volume and weight of solid waste by burning it at very high temperatures.

Industrial Age—the name given for the period of time when industry, in the form of factories, became the dominant employer and generator of wealth. In the United States, the period of time was from about 1907 to 1957; in Great Britain, where the Industrial Revolution is said to have begun, this period of time began somewhat earlier.

Industrial Revolution—the time period when rapid technological development caused large numbers of people to move from agricultural occupations to factory jobs; generally believed to have been fueled by the development of a practical steam engine.

Information Age—a name for the time period from the late 1950's to the present; named for the fact that "white collar" jobs outnumbered "blue collar" jobs at that during that time; also named because of the tremendous growth of information systems (computers).

innovation—an improvement of an existing idea or product.

insulator—a material that has high electrical or thermal resistance.

integrated circuit (IC)—an electronic system capable of performing at least one function made from a single piece of silicon that has been etched to produce an array of transistors, resistors, diodes, and other components, or a substrate that has had these components deposited on its surface.

internal forces—those forces generated by the molecular structure of a material to counter external forces.

International System (SI for System International)—commonly referred to as the metric system, SI is the standard for most industrialized nations.

invention—a new idea or product.

ion—an atom or molecule which has lost or gained one or more electrons causing it to have a positive or negative charge.

kinematics—the study of motion on bodies without regard to force or mass; part of the study of mechanisms.

landfill—a government-controlled disposal area for solid waste.

lateral thinking—creative thinking; the ability to look at a situation in a nontraditional way; also known as divergent thinking.

life-cycle cost—the cost of a product over its expected lifetime, including the purchase price, maintenance costs, and costs of disposal.

light-dependent resistor (LDR)—see photoresistor

light-emitting diode (LED)—a diode that creates red, green, amber, or infrared light using very little electrical current; discrete components are clustered into display devices such as seven-segment displays.

linear motion—straight-line motion in one direction.

logo—a symbol for a company or a product; an easily identified figure or symbol that represents a team, group, company, or product.

Luddite—the nickname for someone who believes that technology is evil.

Malthusian—after the English writer Thomas R. Malthus (1766–1834) who believed that population would increase faster than the food supply, resulting in war and poverty; the name given to an individual who believes that there are finite natural resources and, due to increasing population and consumption, that these resources are in danger of extinction or very high demand–supply shortages.

Maslow's Hierarchy—a hierarchy of human needs that begins with the basic physiological needs of water, food, clothing, shelter, and so on. When these needs have been met, then humans have other needs at the next level of the hierarchy that should be met.

mass production—production of a product through the use of assembly lines and automated machinery.

mean—a numerical average found by adding up all the values and dividing by the number of cases.

mechanical advantage (MA)—the gain in force by a basic machine, which results from giving up motion, such as in a lever.

mechanism—an assemblage of components that takes a given input motion or force and creates a desired output motion or force.

microphone—a device that changes sound vibrations into an electrical signal, most often a voltage.

mind map—a thinking tool that involves putting an idea on paper and developing related ideas by graphically connecting the ideas in some logical way; a concept tree.

model—a representation of the real thing; a proof-of-concept exercise.

monostable timer—a circuit that has one stable state; a monostable circuit with a stable off state will produce an on condition for a duration of time and then return to the stable off state.

nondurable goods—the name given by the U. S. Department of Commerce to describe products intended to last less than three years.

ohm (R)—a unit of electrical resistance in a circuit or a wire; 1 ohm is a relatively small amount of electrical resistance, so it is common to refer to higher resistances in kilohm (thousands of ohms) and megaohm (millions of ohms) values.

Ohm's Law—a law that refers to the relationship between voltage, current, and resistance in electrical circuits; current in a circuit is directly proportional to the voltage and inversely proportional to the resistance; the formulae

$$V = I \times R$$
$$I = V/R$$
$$R = V/I$$

orthographic projection—drawings that include a front view, a top view, and at least one side view of an object; may include other views as needed; intended for development of construction or production of an object.

oscillating motion—back and forth motion in an arc, such as the movement of the agitator in a top-loading washing machine.

percentile—a point in a distribution of characteristics, test scores, or frequency of events, below which a certain percentage of the population falls; if your score on a certain test falls in the 80th percentile, then 80 percent of the population scored below your score.

perspective drawing—a drawing that provides realism by incorporating converging lines to represent the fact that things appear smaller the further away they are.

photoresistor—a sensor that changes its electrical resistance in relationship to the amount of light striking it.

photovoltaic cell—a device that converts light energy into electrical energy.

physiological needs—the most basic human needs for survival, such as water, food, clothing, shelter, and so on.

pictograph—similar to a bar chart but uses small pictures or symbols to represent a certain quantity.

pie chart—a chart that is useful for representing percentages; a pie represents 100 percent so the size of a particular slice can represent a portion of that whole.

planned obsolescence—the concept of planning the failure of a product after a certain amount of use.

pneumatic—having to do with a system that uses the force of a gas, usually air, to do work.

point load—a load or a force exerted at a specific location on a structure or structural element.

polymer—organic or inorganic materials made from giant molecules; shellac is a natural polymer, and substances such as nylon and polypropylene are synthetic polymers.

popliteal height—the measure from the floor to the underside of the upper leg of a person while that person is seated.

portfolio—a collection of ideas, drawings, sketches, notes, and evidence of research, testing and evaluation for a particular project or a collection of projects; used by designers, artists, engineers and students of technology.

posture—in anthropometrics, the various measurements of a person or percentile group when sitting, bending, standing, reaching, gripping, and kneeling.

potential difference—in electronics, the voltage between two points in a circuit.

potentiometer (POT)—an electrical component that changes resistance in relationship to the position of a knob, slider, or screwdriver slot; often used for a position sensor.

power supply—a piece of electrical equipment that provides a DC or an AC voltage source.

preliminary sketch—a sketch or a drawing of first ideas about a problem; very rough and tentative sketches aimed at putting ideas down on paper as quickly as possible before they are lost.

primary color—a hue; one of the three basic colors from which all other colors are created; in additive (light) theory, red + green + blue = white (RGB = W); in subtractive (pigment) theory, yellow + magenta + cyan = black (YMC = K); artist's primaries are yellow, red and blue.

prime mover—a machine that is intended to convert an energy source into mechanical motion, such as a steam or diesel engine.

probability—the likelihood of some event happening; a ratio of the number of times an event occurs to the number of trials, such as the number of times heads comes up on a coin toss to the number of coin tosses.

prototype—a one-of-a-kind model or product that faithfully represents a product in both appearance and function.

ratchet mechanism—a mechanism that will allow rotation or movement in only one direction.

reciprocal motion—back and forth linear motion, such as a piston moving inside a cylinder.

resistor—an electrical component that is intended to limit or reduce current flow.

resources—the necessary ingredients for a technological system, including people, knowledge, materials, energy, tools and machines, and capital; the inputs to a system; raw materials, such as energy resources in the form of coal, petroleum, solar energy, and so on.

risk—danger that arises from the use of a technological system; part of the trade-off of using such a system.

rotary motion—rotating or revolving motion, such as a wheel on a car.

safety factor—describes how much a product, or an element within a product, is overbuilt, that is, intended to withstand forces greater than the calculated rating.

schematic diagram—a diagram containing specific symbols that represent components for electrical, pneumatic, hydraulic, or mechanical systems.

scientific method—a strategy for confirming or refuting a scientific hypothesis; used to validate an observation.

sensor—in electronic systems, a component that converts a physical property, such as heat, light, humidity, sound, and others, into an electrical signal; also used in pneumatic and hydraulic systems to convert physical properties into a fluid signal.

separating—removing material by cutting, piercing, blanking, or chipping.

sequence diagram—a captioned or noncaptioned series of drawings whose purpose is to describe a multistep process, such as how to replace a spark plug in a lawn mower.

shade—a color value resulting from adding black to a hue.

shear—a condition in which a material is being pushed in two, opposite directions; forces that tend to cause a material to deform by one part sliding past an adjacent part of the same material.

signal—in electronics, an electrical current that carries information; information can be in the form of on/off or varying values of current or voltage.

single-acting cylinder—pneumatic cylinders that allow air pressure to be applied to only one side of the piston; the return of the piston to the "rest" position is usually accomplished by an internal or an external return spring.

solenoid—a device that has a wire coil and a movable iron armature; when a voltage is applied to the coil, the armature moves in or out; solenoids are used with fluid valves so that an electrical current can control a fluid system, such as a washing machine solenoid valve.

source reduction—a practice of reducing the packaging and associated product materials so that these materials do not enter the waste stream.

standard deviation—the square root of the variance of a population; for a normal or Gaussian distribution, 1 standard deviation (SD) above and below the mean takes in 68 percent of the population, and $\pm$ 2 SD takes in 95 percent of the population.

static load—a load at rest, such as the weight of a person sitting in a chair and not moving.

stepper motor—a DC motor that rotates a small increment, usually 1 or 2 degrees, when it receives a pulse of current; special control circuits are required to operate stepper motors.

strain—the change in shape of a material caused by compression or tension forces.

stress—the force being exerted on the molecules of a material caused by compression or tension forces.

structure—a body that will resist external forces without changing its shape, except for that due to the elasticity of the material or failure of a joint.

strut—a structural element designed to resist compression forces.

techno-fix—a mentality that believes that technology can solve all of our problems; usually ignores the reality that solving one problem usually results in creating a new problem.

Technological Age—a name given to the latter part of the twentieth century, usually after 1957, during which white-collar workers outnumber blue-collar workers.

tension—when forces are exerted that tend to pull a material apart.

thermistor—a sensor that changes electrical resistance in relationship to heat.

thermocouple—a device made from two dissimilar pieces of metal that produces a small electrical charge when exposed to heat.

thermoplastic—polymer-based materials that can be heated, formed, and then heated and reformed many times.

thermosetting—polymer-based materials that are heat formed but cannot be reheated and reformed.

three-port valve—a control valve used to operate a single-acting cylinder, which allows trapped air pressure to escape into the atmosphere when returned to the "off" position.

tie—structural elements designed to resist tension forces.

tint—a color value resulting from adding white to a hue.

torsion—a force that tries to twist material apart.

trade-off—the term given to the fact that solving one problem results in another problem; if you are willing to accept the trade-offs associated with a technology, it means that you are also willing to accept the problems arising from that solution.

transducer—in electronic systems, a component that converts an electrical signal to a physical property, such as a loudspeaker.

transistor—a semiconductor device made from a base material of silicon or germanium that has at least three contacts; used as a switch, amplifier, or detector.

transistor-transistor logic (TTL)—integrated circuit chips that usually operate on 5 volts.

underlay grid—a graphed paper, often laminated, which is slipped under a sheet of drawing paper as a guide for sketching; comes in various perspective formats, isometric and as a standard grid for orthographic projection drawings.

vector—a line that describes both the direction and magnitude of a force being applied to a system; represented by a line of a length proportional to the magnitude of the force and with an arrow to indicate the direction that the force is being applied.

velocity ratio (VR)—the input distance moved by an effort or force divided by the distance the load is moved; emphasis is on the amplification of output distance as opposed to mechanical advantage, which emphasizes the amplification of the output force.

vertical thinking—logical reasoning, also called "high-probability thinking"; thinking that allows us to make assumptions based on past experience, such as the sun will probably come up tomorrow or a book will fall if we let it go; also called convergent thinking.

voltage—electrical pressure resulting from a potential difference between two points in an electrical circuit; measured in units of volts; 1 volt is required to push 1 ampere of current through 1 ohm of resistance.

vulcanization—a chemical reaction under heat in which sulphur and rubber develop a cross-link polymer chain. The resulting material is stronger and more resilient.

wavelength—the distance that electromagnetic radiation travels in one complete sine wave cycle at a certain frequency; an FM station's signal wavelength is about 9 feet (about 3 meters), whereas a higher frequency radar signal has a wavelength of less than 1 centimeter; light travels between 400 and 700 nanometers (1 nanometer is one-billionth of a meter).

working drawings—drawings developed for producing a product and/or assembling a product.

Young's modulus of elasticity—a measure developed by Thomas Young to describe the stiffness of a material; found by dividing stress by strain.

Appendix A

Related Mathematical, Scientific, and Technological Information

List of Formulas

Area of a Circle

area = πr^2
where $\pi = 3.1416$
r^2 = The radius squared

Circumference of a Circle

circumference = $2\pi r$
where $2\pi = 2 \times 3.1416$
r = The radius

Electrical Resistance

(example for three resistors in series)
$R_T = R1 + R2 + R3$
where T = Total resistance

(example of three resistors in parallel)

$$R_T = \frac{1}{\frac{1}{R1} + \frac{1}{R2} + \frac{1}{R3}}$$

where T = Total resistance

Gear Ratio

$$\text{G.R.} = \frac{\text{Driven gear}}{\text{Drive gear}}$$

Mean

$$\text{M} = \frac{\sum n \text{ (Sum of all cases)}}{n \text{ (Number of cases)}}$$

Mechanical Advantage

$$\text{MA} = \frac{L \text{ (Load)}}{E \text{ (Effort)}}$$

Ohm's Law

$$R = \frac{V}{I}$$

$$I = \frac{V}{R}$$

$E = I \times R$
where V = Voltage
I = Current
R = Resistance

Electrical Power

P (Watt) = $V \times I$

Equilibrium

L (Load) $\times$ D (Distance of load) = E (Effort) $\times$ D (Distance of effort)

Moment

$$M = F\ (\text{Force}) \times D\ (\text{Distance})$$

Stress

$$\sigma \text{ or } s = \frac{F\ (\text{Force}) \text{ or } P\ (\text{load})}{A\ (\text{Cross-sectional area})}$$

Safety Factor

$$SF = \frac{\text{Ultimate stress (breaking point)}}{\text{Working stress (expected load)}}$$

Velocity Ratio

$$VR = \frac{D_e\ (\text{Distance moved by effort})}{D_l\ (\text{Distance moved by load})}$$

Strain

$$\varepsilon = \frac{\Delta L\ (\text{Change in length})}{L\ (\text{Original length})}$$

Scientific Notation

Very large numbers such as 21,000,000,000 and small numbers such as .0000000000143 are cumbersome to use in mathematics. *Scientific notation* allows these numbers to be represented using fewer digits and to be added, subtracted, multiplied, and divided more easily.

Here is an example of a number represented by means of scientific notation:

$$960 = 96 \times 10 \text{ or } 96 \times 10^1$$

$$\begin{aligned} 9{,}600 &= 960 \times 10 \text{ or } 960 \times 10^1 \\ &= 96 \times 100 \text{ or } 96 \times 10^2 \\ &= 9.6 \times 1000 \text{ or } 9.6 \times 10^3 \end{aligned}$$

Scientific notation is generally in the form of a *number between 1 and 10, multiplied by a power of 10.* In the example above, 9,600 would generally be represented by 9.6×10^3. Of course, 9,600 is easy to write and to use, so you do not often see a number this small represented in scientific notation. Larger numbers with many zeros are appropriate for scientific notation. Here is an example:

$$7{,}430{,}000{,}000 = 7.43 \times 10^9$$

The number 7,430,000,000 has 9 more digits before the decimal point than does the number 7.43, so it is 10^9 times as large as 7.43.

Here is an example of a number that is less than 1 represented by means of scientific notation:

$$.5 = 5.0 \times 10^{-1}$$
$$.05 = 5.0 \times 10^{-2}$$
$$.005 = 5.0 \times 10^{-3}$$

In this example, the decimal place would be moved one place to the left in order to make 5.0 become .5; therefore, 5.0 is multiplied by 10^{-1}. Similarly, the decimal place would be moved three places to the left in order to make 5.0 become .005; therefore 5.0 is multiplied by 10^{-3}. Here is an example:

$$.00000078 = 7.8 \times 10^{-7}$$

Metric System (SI)

Length

1 angstrom	=	0.0000001	millimeters (mm)
		0.0000000039	inch
1 meter (m)	=	10	decimeters (dm)
		100	centimeters (cm)
		1,000	millimeters (mm)
1 kilometer (km)	=	1,000	meters (m)

Weight

1 gram (g)	=	10	decigrams (dg)
		100	centigrams (cg)
		1,000	milligrams (mg)
1 kilogram (kg)	=	1,000	grams (g)
1 metric ton	=	1,000	kilograms (kg)
		1,000,000	grams (g)

Volume and Capacity

1 liter (l)	=	1	cubic decimeter (dm^3)
		10	deciliters (dl)
		100	centiliters (cl)
		1,000	milliliters (ml)
		1,000	cubic centimeters (cc)
1 kiloliter (kl)	=	1	cubic meter (m^3)
		1,000	liters (l)

Area

1 centare (ca)	=	1	square meter (m^2)
		100	square decimeters (dm^2)
		10,000	square centimeters (cm^2)
		1,000,000	square millimeters (mm^2)
1 are (a)	=	1	square dekameter (dkm^2)
		100	square meters (m^2)
1 hectare (ha)	=	100	ares (a)
		10,000	square meters (m^2)
1 square kilometer (km^2)	=	1,000,000	square meters (m^2)

Other Measures

Time

1 picosecond (ps)	=			=	0.000000000001	s
1 nanosecond (ns)	=			=	0.000000001	s
1 microsecond (μs)	=	1,000	ns	=	0.000001	s
1 millisecond (ms)	=	1,000	μs	=	0.001	s
1 second (s)	=	1,000	ms	=	1/3,600	hr
1 minute (min)	=	60	s	=	1/60	hr
1 hour (hr)	=	60	min			
1 day (da)	=	24	hr			
1 week (wk)	=	7	da			
1 common solar year	=	365	da			
1 leap year	=	366	da			
1 decade	=	10	yr			
1 century	=	100	yr			
1 millennium	=	1,000	yr			

SI Multiples and Prefixes

Prefix and Symbol			Factor by Which the Unit Is Multiplied
exa, E	10^{18}	=	1 000 000 000 000 000 000.
peta, P	10^{15}	=	1 000 000 000 000 000.
tera, T	10^{12}	=	1 000 000 000 000.
giga, G	10^{9}	=	1 000 000 000.
mega, M	10^{6}	=	1 000 000.
kilo, k	10^{3}	=	1 000.
hecto, h	10^{2}	=	100.
deka, da	10^{1}	=	10.
	10^{0}	=	Base unit
deci, d	10^{-1}	=	0.1
centi, c	10^{-2}	=	0.01
milli, m	10^{-3}	=	0.001
micro, μ	10^{-6}	=	0.000 001
nano, n	10^{-9}	=	0.000 000 001
pico, p	10^{-12}	=	0.000 000 000 001
femto, f	10^{-15}	=	0.000 000 000 000 001
atto, a	10^{-18}	=	0.000 000 000 000 000 001

Equivalents

Length

1 in	=	2.54 cm
1 m	=	39.37 in
1 m	=	3.281 ft
1 ft	=	0.3048 m
12 in	=	1 ft
3 ft	=	1 yd
1 km	=	0.621 mi
1 mi	=	1.609 km
1 mi	=	5,280 ft
1 Å	=	10^{-6} m
1 light year	=	9.461×10^{15} m

Area

1 m^2	=	104 cm^2
1 m^2	=	10.76 ft^2
1 in^2	=	6.452 cm^2

Volume

1 m^3	=	10^6 cm^3
1 m^3	=	6.102×10 in^3
1 ft^3	=	1,728 in^3
1 ft^3	=	2.83×10^4 m^3
1 liter	=	1,000 cm^3
1 liter	=	1.0576 qt
1 liter	=	0.0353 ft^3
1 gallon	=	3.786 l

Force

1 N	=	0.2248 lb
1 lb	=	4.448 N

Velocity

1 mi/h	=	1.47 ft/s
1 mi/h	=	0.447 m/s
1 mi/h	=	1.61 km/h

Pressure

1 bar	=	10^5 N/m^2
1 bar	=	14.50 lb/in^2
1 atm	=	760 mm Hg
1 atm	=	14.7 lb/in^2
1 Pa	=	1 N/m^2
1 Pa	=	1.45×10^{-4} lb/in^2

Energy

1 J	=	0.738 ft lb
1 cal	=	4.186 J
1 BTU	=	252 cal
1 BTU	=	1.054×10^3 J
1 kWh	=	3.60×10^6 J

Power

1 hp	=	550 ft lb/s
1 hp	=	0.746 kW
1 BTU/h	=	0.293 W

Liquid Measure Equivalents

1 fluid ounce	=	1.805 cubic inch	1 quart, imperial	=	40 ounces
1 fluid ounce	=	29.6 cubic centimeters	1 gallon	=	8 pints
1 pint	=	16 fluid ounces	1 gallon	=	4 quarts
1 pint, imperial	=	20 fluid ounces	1 gallon, imperial	=	80 ounces
1 quart	=	32 fluid ounces	1 firkin	=	9 gallons
1 quart	=	2 pints	1 hogshead	=	63 gallons
1 quart	=	946 cubic centimeters			

Note: Imperial pints and imperial gallons are standard British measures.

Equivalents

		Square Inches	Square Feet	Square Yards	Square Meters
1 square foot	=	144	1	.1111	0.09290
1 square yard	=	1,296	9	1	0.83613
1 square meter	=	1,550	10.7639	1.19599	1
1 square rod	=	39,204	272.25	30.25	25.293
1 acre	=	6,272,640	43,560	4840	4,046.86
1 square mile	=	——	27,878,400	3,097,600	2,589,999
1 square kilometer	=	——	10,763,867	1,195,985	1,000,000

Temperature

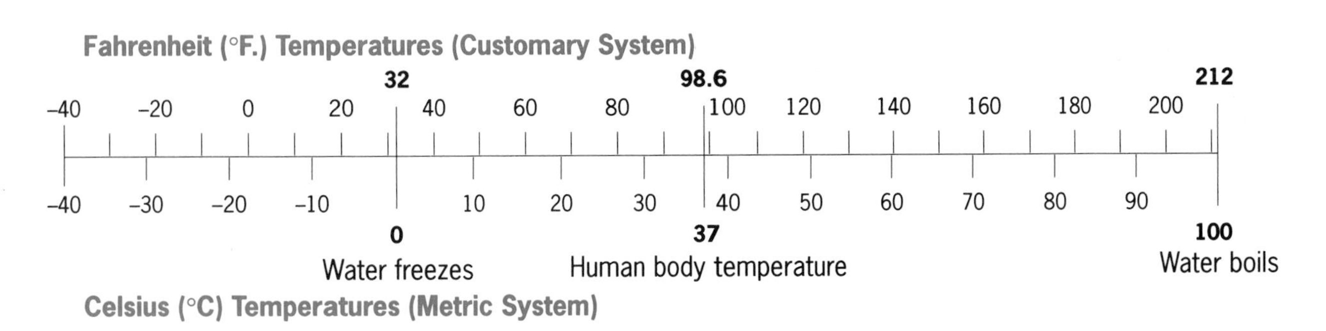

Name	Symbol	Atomic Number	Atomic Weight[a]	Name	Symbol	Atomic Number	Atomic Weight[a]
Actinium	Ac	89	(227)	Lawrencium	Lr	103	(257)
Aluminum	Al	13	26.93	Lead	Pb	82	207.19
Americium	Am	95	(243)	Lithium	Li	3	6.94
Antimony	Sb	51	121.75	Lutetium	Lu	71	174.97
Argon	Ar	18	39.95	Magnesium	Mg	12	24.31
Arsenic	As	33	74.92	Manganese	Mn	25	54.94
Astatine	At	85	(210)	Mendelevium	Md	101	(256)
Barium	Ba	56	137.34	Mercury	Hg	80	200.59
Berkelium	Bk	97	(247)	Molybdenum	Mo	42	95.94
Beryllium	Be	4	9.01	Neodymium	Nd	60	144.24
Bismuth	Bi	83	208.98	Neon	Ne	10	20.18
Boron	B	5	10.81	Neptunium	Np	93	(237)
Bromine	Br	35	79.90	Nickel	Ni	28	58.71
Cadmium	Cd	48	112.40	Niobium	Nb	41	92.91
Calcium	Ca	20	40.08	Nitrogen	N	7	14.01
Californium	Cf	98	(251)	Nobelium	No	102	(254)
Carbon	C	6	12.01	Osmium	Os	76	190.20
Cerium	Ce	58	140.12	Oxygen	O	8	16.00
Cesium	Cs	55	132.91	Palladium	Pd	46	106.40
Chlorine	Cl	17	35.45	Phosphorus	P	15	30.97
Chromium	Cr	24	51.99	Platinum	Pt	78	195.09
Cobalt	Co	27	58.93	Plutonium	Pu	94	(242)
Copper	Cu	29	63.54	Polonium	Po	84	(210)
Curium	Cm	96	(247)	Potassium	K	19	39.10
Dysprosium	Dy	66	162.50	Praseodymium	Pr	59	140.91
Einsteinium	Es	99	(254)	Promethium	Pm	61	(147)
Erbium	Er	68	167.26	Protactinium	Pa	91	(231)
Europium	Eu	63	151.96	Radium	Ra	88	(226)
Fermium	Fm	100	(253)	Radon	Rn	86	(222)
Fluorine	F	9	18.99	Rhenium	Re	75	186.20
Francium	Fr	87	(223)	Rhodium	Rh	45	102.91
Gadolinium	Gd	64	157.25	Rubidium	Rb	37	85.47
Gallium	Ga	31	69.72	Ruthenium	Ru	44	101.07
Germanium	Ge	32	72.59	Samarium	Sm	62	150.35
Gold	Au	79	196.97	Scandium	Sc	21	44.96
Hafnium	Hf	72	178.49	Selenium	Se	34	78.96
Helium	He	2	4.00	Silicon	Si	14	28.09
Holmium	Ho	67	164.93	Silver	Ag	47	107.78
Hydrogen	H	1	1.008	Sodium	Na	11	22.99
Indium	In	49	114.82	Strontium	Sr	38	87.62
Iodine	I	53	126.90	Sulfur	S	16	32.06
Iridium	Ir	77	192.20	Tantalum	Ta	73	180.95
Iron	Fe	26	55.85	Technetium	Tc	43	(99)
Krypton	Kr	36	83.80	Tellurium	Te	52	127.60
Lanthanum	La	57	138.91	Terbium	Tb	65	158.92

[a] The numbers in parentheses are mass numbers of the principal isotopes; all other values are rounded to the second decimal.

Approximate Conversions

If You Know:	Multiply by:	To Find:
Length and Distance		
inches	25	millimeters
feet	30	centimeters
yards	0.9	meters
miles	1.6	kilometers
millimeters	0.04	inches
centimeters	0.4	inches
meters	1.1	yards
kilometers	0.6	miles
Area		
square inches	6.5	square centimeters
square feet	0.09	square meters
square yards	0.8	square meters
square miles	2.6	square kilometers
acres	0.4	hectares
square centimeters	0.16	square inches
square meters	1.2	square yards
square kilometers	0.4	square miles
hectares	2.5	acres
Volume		
fluid ounces	30	milliliters
pints	0.47	liters
pints, imperial	0.568	liters
quarts	0.95	liters
quarts, imperial	1.137	liters
gallons	3.8	liters
gallons, imperial	4.546	liters
milliliters	0.034	fluid ounces
liters	2.1	pints
liters	1.76	pints, imperial
liters	1.06	quarts
liters	0.88	quarts, imperial
liters	0.26	gallons
liters	0.22	gallons, imperial
Weight and Mass		
ounces	28	grams
pounds	0.45	kilograms
short tons	0.9	metric tons
grams	0.035	ounces
kilograms	2.2	pounds
metric tons	1.1	short tons

Temperature

degrees Fahrenheit	5/9 after subtracting 32	degrees Celsius
degrees Celsius	9/5 then add 32	degrees Fahrenheit

Energy and Power

British thermal unit (BTU)	1055	joule (J)
BTU/second	1.05	kilowatt (kW)
foot pound (ft lb)	1.31	joule (J)
foot pound/second	1.35	watt (W)
horsepower (hp)	0.746	kilowatt

Other

mile/hour (MPH)	1.16	kilometers/hr (km/hr)
mile/hour (MPH)	0.447	meter/second (m/s)
pound/square inch (psi)	6.89	kilopascal (kPa)

Appendix B

SAMPLE DOCUMENTATION: "The Can Crusher"

In the design/problem-solving process, documentation represents what the designer has done (and learned) on the way to the solution. Since the process may be different from anything you have done before in school, it is important to be conscientious and complete in recording all related information. This series of slides illustrates what one student did to solve the problem of accumulating bulky aluminum cans for recycling.

You have learned that the design/problem-solving process involves nine steps and is a logical or rational human process. You will find, further, as you begin to solve a problem, that the process is not linear. For example, in the can crusher solution, the student made a number of "cycles" through the design/problem-solving process. The first cycle involved an investigation into mechanisms that culminated in a Mecanno™ model (slide "t"). After evaluating the effectiveness of the mechanism's design, the student began to investigate what "form" the product should take. This cycle included:

- concerns for the human user,
- an analysis of existing kitchen appliances, and
- a study of possible materials.

Using the information gathered, the student rendered a final design solution (slide "y"). While not shown, this cycle would typically continue with a prototype model and testing/evaluation. After all design criteria have been satisfied, another cycle would begin including marketing and production planning.

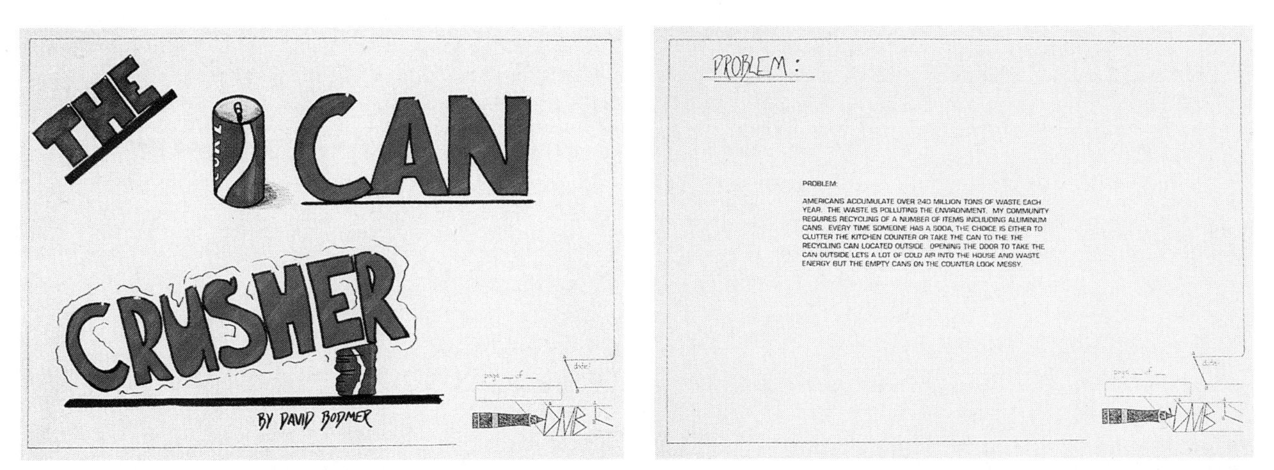

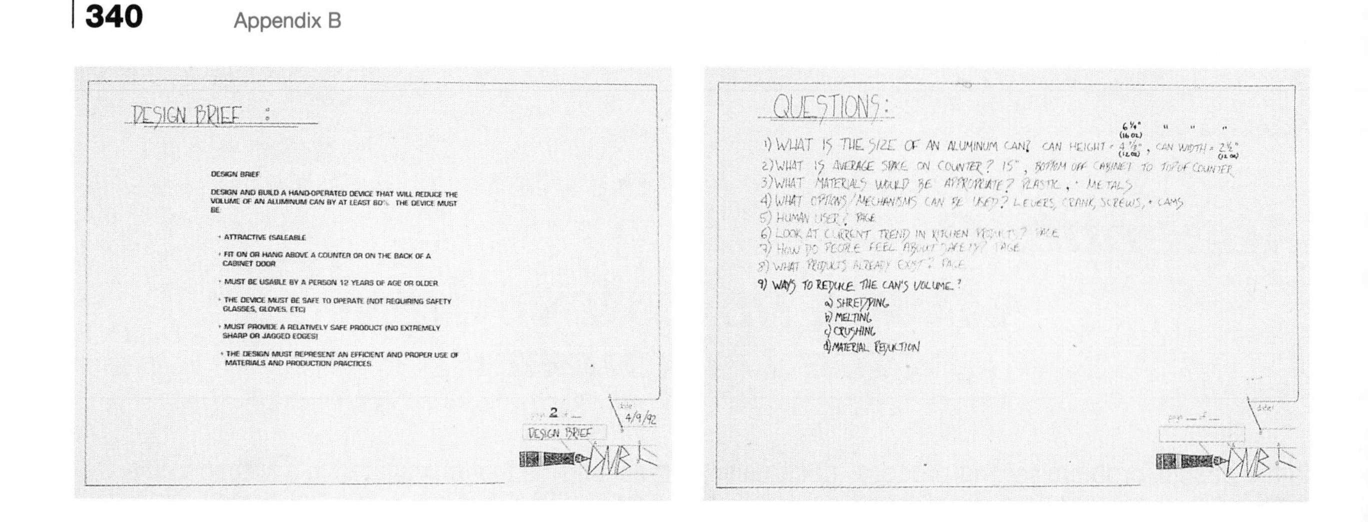
DESIGN BRIEF :
DESIGN BRIEF
DESIGN AND BUILD A HAND-OPERATED DEVICE THAT WILL REDUCE THE VOLUME OF AN ALUMINUM CAN BY AT LEAST 80%. THE DEVICE MUST BE:
• ATTRACTIVE (SALEABLE
• FIT ON OR HANG ABOVE A COUNTER OR ON THE BACK OF A CABINET DOOR
• MUST BE USABLE BY A PERSON 12 YEARS OF AGE OR OLDER
• THE DEVICE MUST BE SAFE TO OPERATE (NOT REQUIRING SAFETY GLASSES, GLOVES, ETC)
• MUST PROVIDE A RELATIVELY SAFE PRODUCT (NO EXTREMELY SHARP OR JAGGED EDGES)
• THE DESIGN MUST REPRESENT AN EFFICIENT AND PROPER USE OF MATERIALS AND PRODUCTION PRACTICES
2
4/9/92
DESIGN BRIEF
QUESTIONS:
1) WHAT IS THE SIZE OF AN ALUMINUM CAN? CAN HEIGHT = 4 7/8" (12 oz), 6 1/4" (16 oz), CAN WIDTH = 2 1/2" (12 oz)
2) WHAT IS AVERAGE SPACE ON COUNTER? 15", BOTTOM OFF CABINET TO TOP OF COUNTER
3) WHAT MATERIALS WOULD BE APPROPRIATE? PLASTIC, + METALS
4) WHAT OPTIONS/MECHANISMS CAN BE USED? LEVERS, CRANK, SCREWS, + CAMS
5) HUMAN USER? PAGE
6) LOOK AT CURRENT TREND IN KITCHEN PRODUCTS? PAGE
7) HOW DO PEOPLE FEEL ABOUT SAFETY? PAGE
8) WHAT PRODUCTS ALREADY EXIST? PAGE
9) WAYS TO REDUCE THE CAN'S VOLUME?
a) SHREDDING
b) MELTING
c) CRUSHING
d) MATERIAL REDUCTION

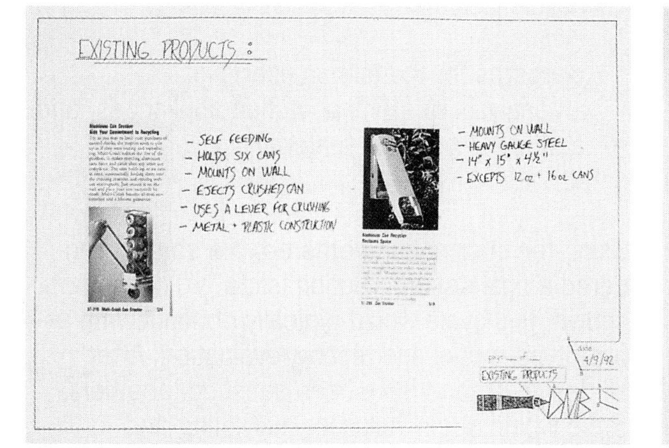
EXISTING PRODUCTS :
- SELF FEEDING
- HOLDS SIX CANS
- MOUNTS ON WALL
- EJECTS CRUSHED CAN
- USES A LEVER FOR CRUSHING
- METAL + PLASTIC CONSTRUCTION
- MOUNTS ON WALL
- HEAVY GAUGE STEEL
- 14" x 15" x 4 1/2"
- EXCEPTS 12 oz + 16 oz CANS
4/9/92
EXISTING PRODUCTS

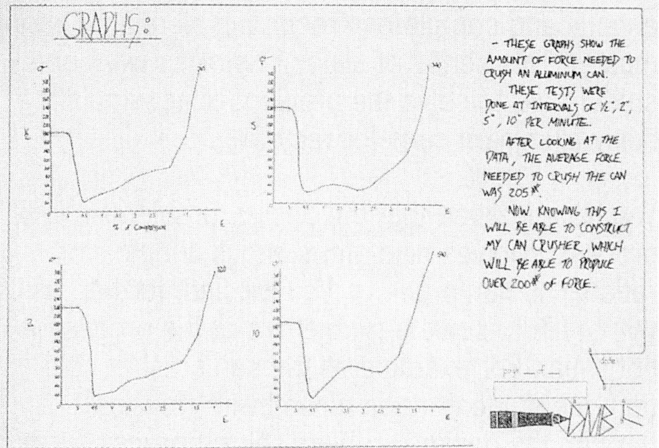
GRAPHS:
- THESE GRAPHS SHOW THE AMOUNT OF FORCE NEEDED TO CRUSH AN ALUMINUM CAN.
THESE TESTS WERE DONE AT INTERVALS OF 1/2", 2", 5", 10" PER MINUTE.
AFTER LOOKING AT THE DATA, THE AVERAGE FORCE NEEDED TO CRUSH THE CAN WAS 205#.
NOW KNOWING THIS I WILL BE ABLE TO CONSTRUCT MY CAN CRUSHER, WHICH WILL BE ABLE TO PRODUCE OVER 200# OF FORCE.

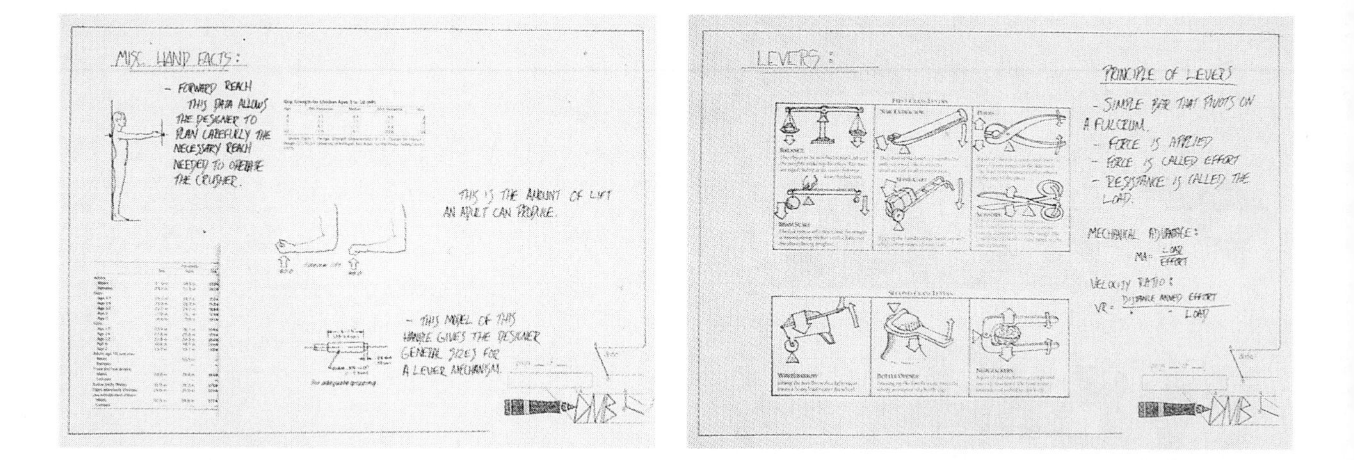
MISC. HAND FACTS:
- FORWARD REACH
THIS DATA ALLOWS THE DESIGNER TO PLAN CAREFULLY THE NECESSARY REACH NEEDED TO OPERATE THE CRUSHER.
THIS IS THE AMOUNT OF LIFT AN ADULT CAN PRODUCE.
- THIS MODEL OF THIS HANDLE GIVES THE DESIGNER GENERAL SIZES FOR A LEVER MECHANISM.
LEVERS:
PRINCIPLE OF LEVERS
- SIMPLE BAR THAT PIVOTS ON A FULCRUM.
- FORCE IS APPLIED
- FORCE IS CALLED EFFORT
- RESISTANCE IS CALLED THE LOAD.
MECHANICAL ADVANTAGE:
MA = LOAD / EFFORT
VELOCITY RATIO:
VR = DISTANCE MOVED EFFORT / - LOAD

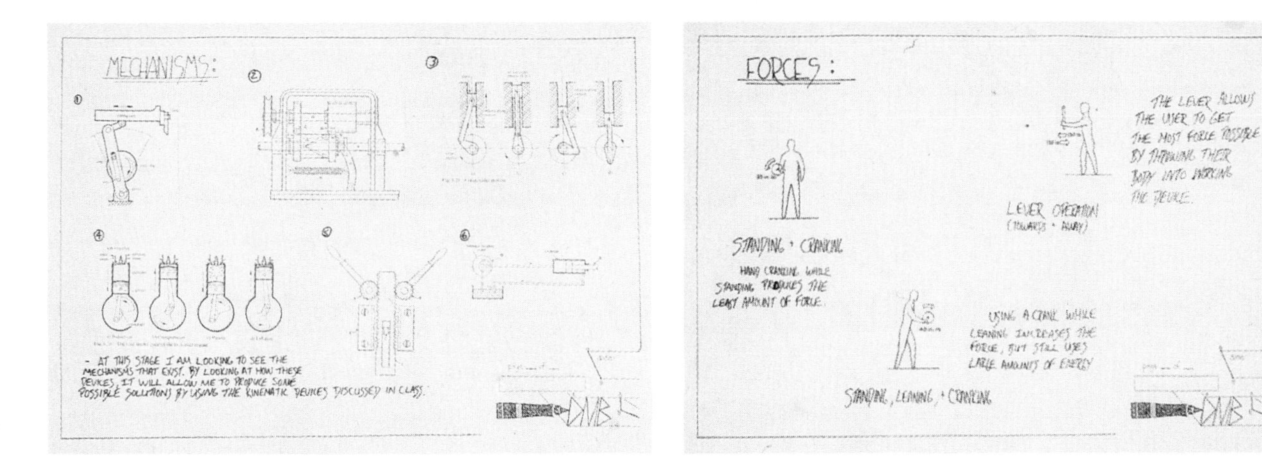
MECHANISMS:
- AT THIS STAGE I AM LOOKING TO SEE THE MECHANISMS THAT EXIST. BY LOOKING AT HOW THESE DEVICES, IT WILL ALLOW ME TO PRODUCE SOME POSSIBLE SOLUTIONS BY USING THE KINEMATIC DEVICES DISCUSSED IN CLASS.
FORCES:
STANDING + CRANKING
HAND CRANKING WHILE STANDING PRODUCES THE LEAST AMOUNT OF FORCE.
LEVER OPERATION (TOWARDS + AWAY)
THE LEVER ALLOWS THE USER TO GET THE MOST FORCE POSSIBLE BY THROWING THEIR BODY INTO WORKING THE DEVICE.
USING A CRANK WHILE LEANING INCREASES THE FORCE, BUT STILL USES LARGE AMOUNTS OF ENERGY
STANDING, LEANING, + CRANKING

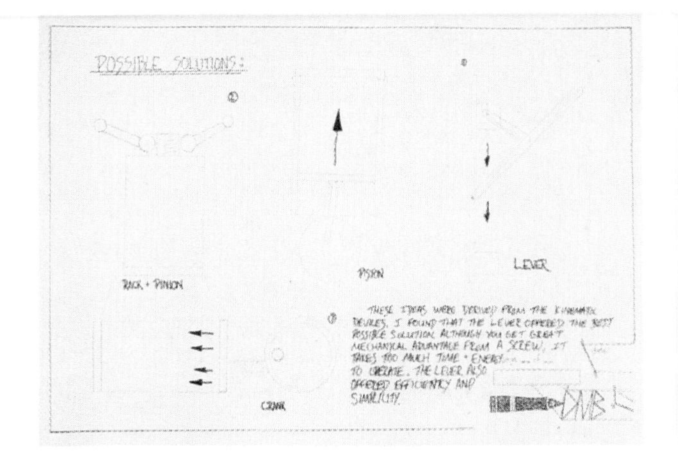
POSSIBLE SOLUTIONS:
RACK + PINION
PISTON
LEVER
CRANK
THESE IDEAS WERE DERIVED FROM THE KINEMATIC DEVICES. I FOUND THAT THE LEVER OFFERED THE BEST POSSIBLE SOLUTION. ALTHOUGH YOU GET GREAT MECHANICAL ADVANTAGE FROM A SCREW, IT TAKES TOO MUCH TIME + ENERGY TO OPERATE. THE LEVER ALSO OFFERED EFFICIENCY AND SIMPLICITY.

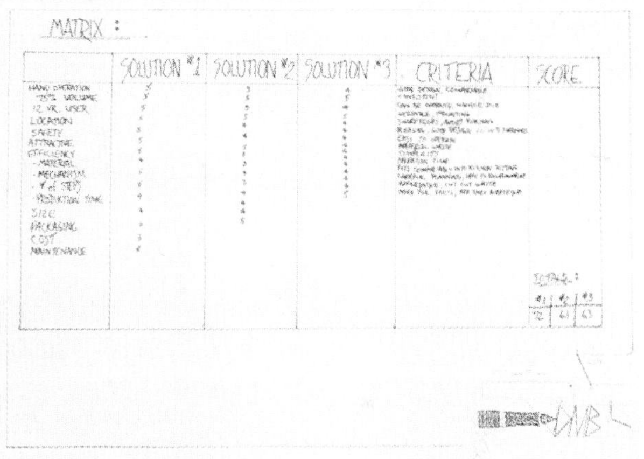
MATRIX:
SOLUTION #1
SOLUTION #2
SOLUTION #3
CRITERIA
SCORE
HAND OPERATION
75% VOLUME
12 YR. USER
LOCATION
SAFETY
ATTRACTIVE
EFFICIENCY
-MATERIAL
-MECHANISM
-# of STEPS
-PRODUCTION TIME
SIZE
PACKAGING
COST
MAINTENANCE
TOTAL:
#1 #2 #3

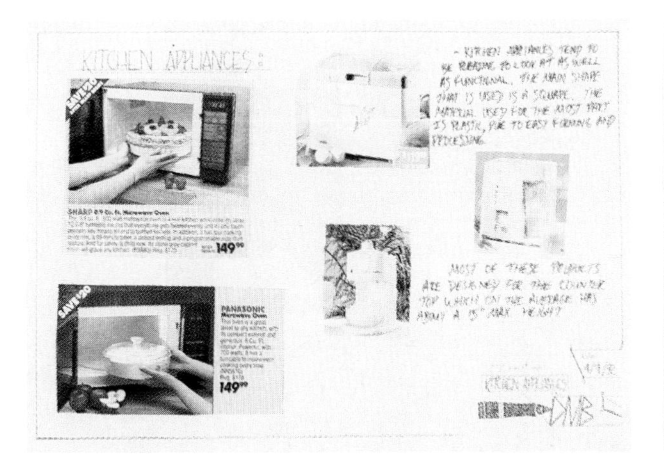
KITCHEN APPLIANCES:
- KITCHEN APPLIANCES TEND TO BE PLEASING TO LOOK AT AS WELL AS FUNCTIONAL. THE MAIN SHAPE THAT IS USED IS A SQUARE. THE MATERIAL USED FOR THE MOST PART IS PLASTIC, DUE TO EASY FORMING AND PROCESSING.
MOST OF THESE PRODUCTS ARE DESIGNED FOR THE COUNTER TOP WHICH ON THE AVERAGE HAS ABOUT A 15" MAX. HEIGHT.
SHARP 0.9 Cu. Ft. Microwave Oven
149.99
PANASONIC
Microwave Oven
149.99

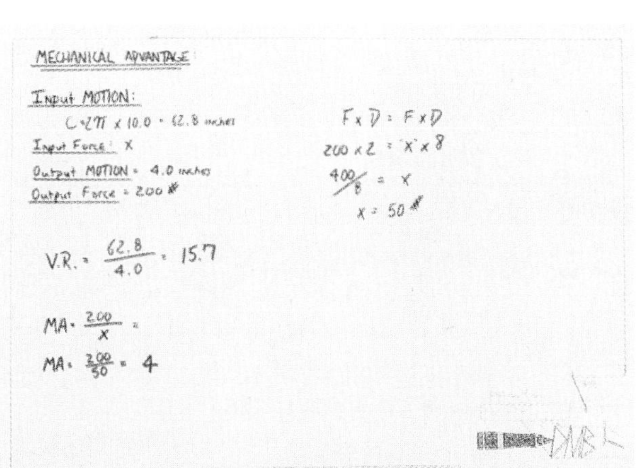
MECHANICAL ADVANTAGE
Input MOTION:
C = 2π x 10.0 = 62.8 inches
Input Force: X
Output MOTION = 4.0 inches
Output Force = 200 #
F x D = F x D
200 x 2 = X x 8
400/8 = X
X = 50 #
V.R. = 62.8/4.0 = 15.7
MA = 200/X =
MA = 200/50 = 4

SAFETY:

- SAFETY IS ONE MAIN DESIGNING FACTOR THAT MUST BE DEALT WITH. THIS CHART LETS THE DESIGNER UNDERSTAND HOW DIFFERENT AGES THINK ABOUT SAFETY.

TEENS: RESTLESS, RESISTANCE TO AUTHORITY, ERRATIC BEHAVIOR, AND DARING BEHAVIOR.

YOUNG ADULTS: MAKE USE OF BODY STRENGTH, TAKE MORE CHANCES, AND TAKE ON MORE DEMANDING FEATS.

ADULTS: TAKES LIMITED RISKS, ADJUSTING TO NEW LIFE PATTERNS, AND ORIENTED TOWARDS EFFICIENCY.

STRENGTH TEST RESULTS:

PLASTICS TO BE USED:
A) POLYALLOMER - MILKY WHITE
B) EPOXY - CLEAR w/ BLACK DIE PELLETS

Compression Strength
A) $1/4^2$ CAN WITH STAND 437.5 lbs per square 1/4"
B) 35,000 P.S.I., DEFINITLY ENOUGH STRENGTH

Tensile Strength
A) $1/4^2$ CAN WITH STAND 219 lbs of Force
B) 17,000 P.S.I., DEFINITLY STRONG ENOUGH

* THESE RESULTS ARE FOR THE SMALLEST AMOUNT (AREA) OF PLASTIC ON MY PRODUCT. I FIGURE THAT IF THE SMALLEST AREA CAN WITH STAND THE FORCES, THAN OBVIOUSLY A GREATER AREA WOULD BE THAT MUCH STRONGER.

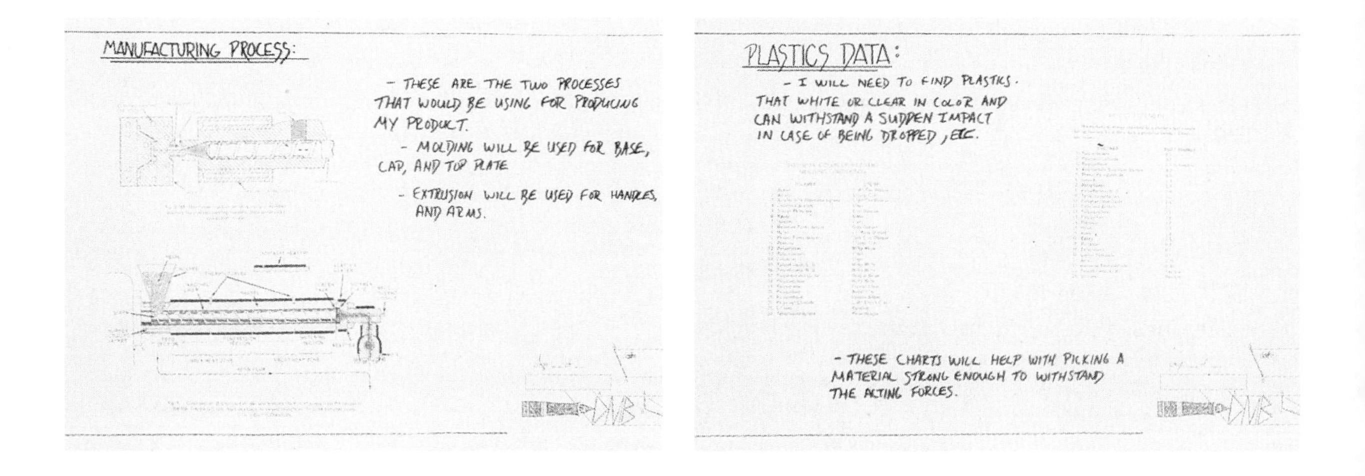

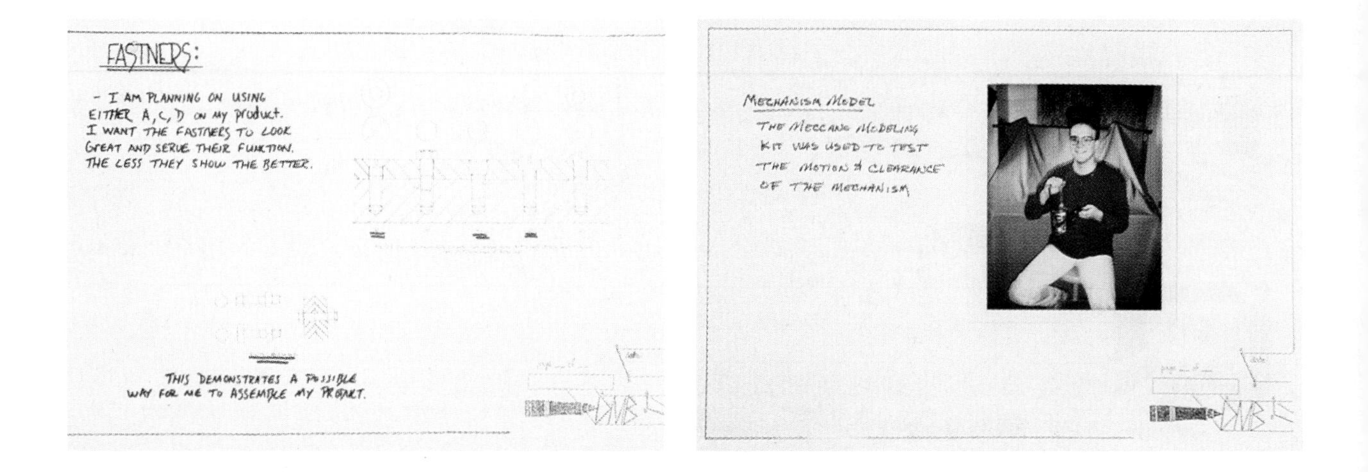

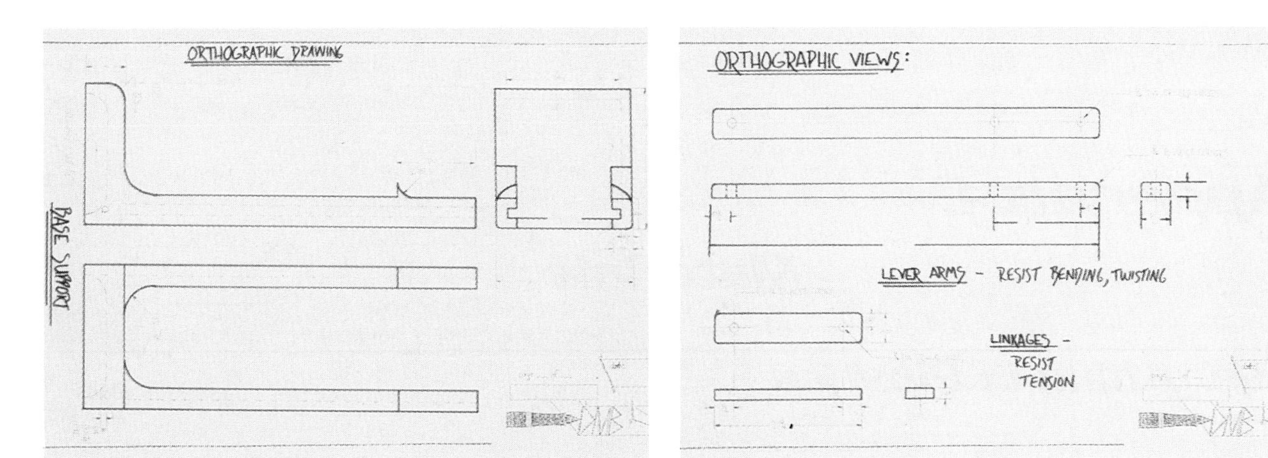

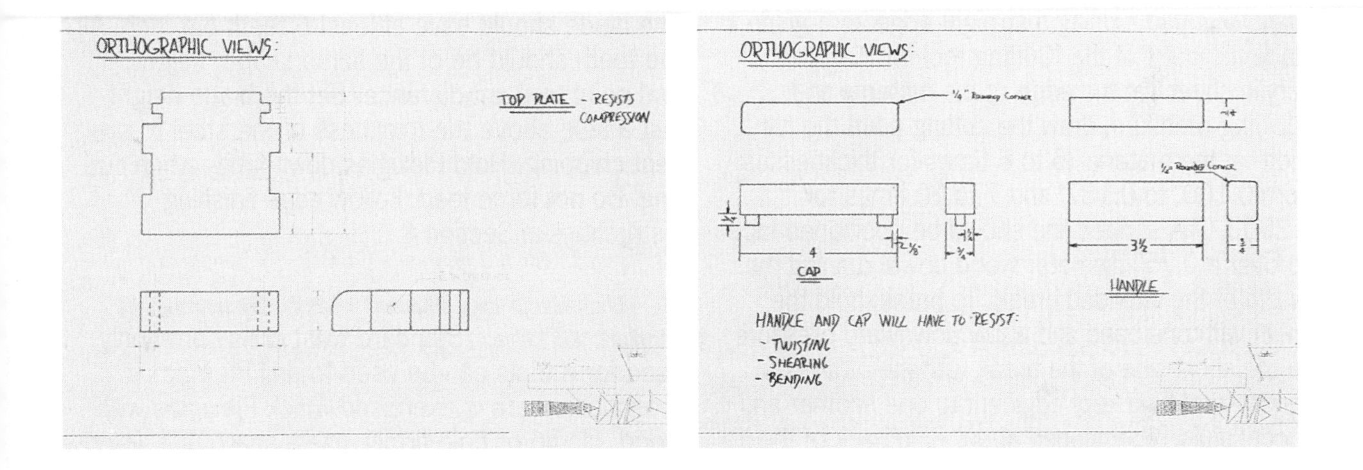

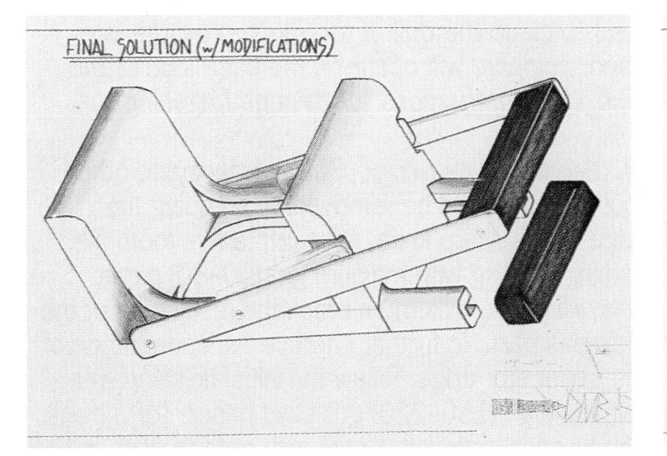

CONCLUSIONS:

AFTER ALL THE RESEARCH AND DEVELOPMENT, I FEEL THAT I HAVE A DURABLE, VISUALLY PLEASING PRODUCT. THE FINAL SOLUTION FULFILLS ALL OF THE SPECIFICATIONS THAT WERE ASSIGNED. THIS WAS MY FIRST PROJECT THAT DEALT HEAVILY WITH HUMAN FACTOR ENGINEERING, I WAS NOT AWARE OF ALL THE DIFFERENT AREAS THAT MUST BE INVESTIGATED. BY DOING THIS PROJECT, IT OPENED MY EYES TO ALL THE DIFFERENT POSSIBLE SOLUTIONS THAT EACH PROBLEM POSSESSES. OVERALL THIS WAS AN EXCELLENT LEARNING EXPERIENCE, AND HOPEFULLY BY DOING THIS WORK IT HAS ALLOWED ME TO THINK OPENLY AND TO LOOK AT BROAD IDEAS RATHER TUNNEL VISION.

DAVID BOMER

Appendix C

Materials

Everything in Plastics

Working with Plexiglas

1. Scribing and Breaking (up to 1/4" thickness). Using a straight edge as a guide, place the point of the "Cutting tool for Plexiglas acrylic sheet" at the edge of the material and, applying pressure, draw the cutting point the full width of the material (5 to 6 times for thicknesses from 0.100" to 0.187" and 7 to 10 times for 0.250"). The scribed line should be positioned face up over a 0.75" diameter wood dowel running the length of the intended break. To break, hold the sheet with one hand and apply downward pressure on the short side of the break with the other. The hands should be kept adjacent to one another and successfully repositioned about 2" in back of the break as it progresses along the scribed line. The minimum cut-off width is about 1.5". Patterned Plexiglas cannot be scored or broken. Follow edge finishing instructions in Section 4.

2. Cutting with Saws. Do not remove protective masking paper before cutting. If cutting unmasked sheet is unavoidable, apply masking tape on both sides of intended cut to reduce friction and gumming behind blade.

Curved shapes are easily cut with sabre, band, and reciprocating jig saws. Sabre and reciprocating jig saw blades should have at least 14 teeth per inch. Straight cuts can be made with a sabre or hand jig saw by guiding the tool along a straight edge. Band saws should have at least 10 teeth per inch. Hold Plexiglas down firmly when cutting. Do not force feed. Follow edge finishing instructions in Section 4.

Circular saws are ideal for straight cutting. Use a steel cross-cut blade which is recommended for finish cuts on plywood, veneers, laminates, etc. The blade should have at least 6 teeth per inch. All the teeth should be of the same shape, height, and point to point distance. Set the blade height just a little above the thickness of the steel to prevent chipping. Hold Plexiglas down firmly when cutting. Do not force feed. Follow edge finishing instructions in Section 4.

3. Drilling by Hand with Standard Twist Drills. Standard twist drills commonly used for metals can be used to drill Plexiglas if reasonable care is exercised. Back Plexiglas with wood, clamp or hold firmly, use a sharp drill, very slow speed, and minimum pressure.

Caution: If too much speed is used, Plexiglas will tend to climb the drill. If too much pressure is used, chipping will occur on the back side of the hole. (See instructions for through fastening).

4. Edge Finishing. Sawed edges and other tool marks should be removed by scraping the edge with a sharp knife, filing with a fine tooth file, and/or sanding with medium grit (60–80) paper. This will insure maximum breakage resistance of the Plexiglas part. To further improve the appearance of the surface or edge, follow the initial finishing with "wet or dry" (150–220) grit sand paper. For a transparent edge, follow this step with grits to 400 and buff with a clean muslin wheel dressed with a good grade of fine grit buffing compound. Finish up with a clean soft cotton flannel wheel.

5. Cementing. Capillary cementing with a solvent such as methylene chloride ("MDC") or ethylene dichloride ("EDC") or 1,1,2-trichloroethane is an easy method of joining two pieces of Plexiglas. Sand the surfaces to be cemented; do not polish. Remove the protective masking paper. Hold pieces together with strips of masking tape.

Apply the solvent to the joint with syringe, oil can with a very fine spout, eye dropper, or small paint brush. Let joint dry thoroughly. Caution: solvents may be toxic if inhaled for extended periods of time or if swallowed; many are also flammable. Use in a well ventilated area, keep away from children.

6. Through-Fastening of Plexiglas. Drill oversize holes (allow 1/16" per foot of length of the Plexiglas to provide for expansion and contraction), following drilling instructions in Section 3. Holes should be located to provide at least 1/4" solid material from edge of hole to edge of sheet. Smoothing hole surface with a round file should provide maximum resistance to breakage. Bring the screws up just snugly and back off 1/4 turn to provide freedom of movement for expansion or contraction of the Plexiglas.

7. Cleaning of Plexiglas. Wash the Plexiglas with a mild soap and lukewarm water solution. Use a clean soft cloth or sponge and as much of the solution as possible. Rinse well. Dry by blotting with a damp cloth or chamois. A periodic waxing with a good grade of hard automobile paste wax (not a cleaner-wax combination) will fill in minor surface scratches and help maintain the lustre. Apply sparingly and buff lightly with clean cotton flannel or jersey. Sanding and buffing as described in #4 (Edge Finishing) will remove deeper scratches.

Aluminum Alloy and Temper Designations

Aluminum Alloy Designations

The aluminum industry uses a four-digit index system for the designation of its wrought aluminum alloys.

As outlined below, the first digit indicates the alloy group according to the major alloying elements.

1xxx Series. In this group, minimum aluminum content is 99 percent, and there is no major alloying element.

The second digit indicates modifications in impurity limits. If the second digit is zero, there is no special control on individual impurities. Digits 1 through 9, which are assigned consecutively as needed, indicate special control of one or more individual impurities.

The last two digits indicate specific minimum aluminum content. Although the absolute minimum aluminum content in this group is 99 percent the minimum for certain grades is higher than 99 percent, and the last two digits represent the hundredths of a per cent over 99.

Thus, 1030 would indicate 99.030 percent minimum aluminum, without special control on individual impurities. The designations 1130, 1230, 1330, etc. indicate the same purity with special control on one or more impurities. Likewise, 1100 indicates minimum aluminum content of 99.00 percent with individual impurity control.

2xxx through 9xxx Series. The major alloying elements are indicated with the first digit, as follows:

2xxx	Copper
3xxx	Manganese
4xxx	Silicon
5xxx	Magnesium
6xxx	Magnesium and silicon
7xxx	Zinc
8xxx	Other element
9xxx	Unused series

The second digit indicates alloy modification. If the second digits is zero, it indicates the original alloy: digits 1 through 9, which are assigned consecutively, indicate alloy modifications. The last two digits have no special significance, serving only to identify the different alloys in the group.

Experimental Alloys. Experimental alloys are designated according to the four digit system, but they are prefixed by the letter X. The prefix is dropped when the alloy becomes standard. During development, and before they are designated as experimental, new alloys are identified by serial numbers assigned by their originators. Use of the serial numbers is discontinued when the X number is assigned.

Aluminum Temper Designations

Temper designations of wrought aluminum alloys consist of suffixes to the numeric alloy designations. For example, in 3003-H14, 3003 denotes the alloy and "H14" denotes the temper, or degree of hardness. The temper designation also reveals the method by which the hardness was obtained. Temper designations differ between nonheat-treatable alloys and heat-treatable alloys, and their meanings are given below:

Nonheat-treatable Alloys. The letter "H" is always followed by 2 or 3 digits. The first digit indicates the particular method used to obtain the temper, as follows:

- H1 means strain hardened only
- H2 means strain hardened, then partially annealed
- H3 means strain hardened, then stabilized.

The temper is indicated by the second digit as follows:

- 2 1/4 hard
- 4 1/2 hard
- 6 3/4 hard
- 8 full hard
- 9 extra hard

Added digits indicate modification of standard practice.

Heat-treatable Alloys.

- F As fabricated
- O Annealed
- T Heat treated

The letter "T" is always followed by one or more digits. These digits indicate the method used to produce the stable tempers, as follows:

- T3 Solution heat treated, then cold worked.
- T351 Solution heat treated, stress-relieved stretched, then cold worked.
- T36 Solution heat treated, then cold worked (controlled).
- T4 Solution heat treated, then naturally aged.
- T451 Solution heat-treated, then stress relieved stretched.
- T5 Artificially aged only.
- T6 Solution heat treated, then articicially aged.
- T61 Solution heat treated (boiling water quench), then artificially aged.
- T651 Solution heat treated, stress-relieved by compression, then artificially aged.
- T7 Solution heat treated, then stabilized.
- T8 Solution heat treated, cold worked, then artifically aged.
- T81 Solution heat treated, cold worked (controlled), then artificially aged.
- T851 Solution heat treated, cold worked, stress-relieved stretched, then artificially aged.
- T9 Solution heat treated, artifically aged, then cold worked
- T10 Artificially aged, then cold worked.

Added digits indicate modification of standard practice.

Comparison of Modern & Old Systems of Aluminum Alloy Designation

Although the old system of aluminum identification has been obsolete for many years, stock with the old markings is still occasionally found. The following comparison is presented as an aid in identifying such materials in terms of the modern system.

In the old system, alloy composition was indicated by a one- or two-digit number followed by the letter "S" to indicate that it was a wrought alloy, i.e., an alloy that could be shaped by rolling, drawing, or forging. Any variation in the basic composition was indicated by a letter preceding the numerical alloy designation. For example, A17S was a modification of the basic alloy 17S. In modern terminology these two alloys are designated by a second letter: "O" for soft (annealed), "H" for strain hardness of nonheat-treatable alloys, and "T" for hardness of heat-treatable alloys. Degree of hardness of nonheat-treatable alloys was indicated by a fraction preceding the letter "H". For example, 3S1/4H would be quarter-hard 3S alloy.

Modern System	Old System
1100	2S
3003	3S
3003-0	3SO
2014	14S
2017	17S
2117	A17S
2018	18S
2218	B18S
2024T	24ST
5052	52S
7075T6	75ST6

Aluminum—The Most Common Grades

1100. This grade is commercially pure aluminum. It is soft and ductile and has excellent workability. It is ideal for applications involving intricate forming because it work hardens more slowly than other alloys. It is the most weldable of aluminum alloys, by any method. It is nonheat-treatable. It has excellent resistance to corrosion and is widely used in the chemical and food processing industries. It responds well to decorative finishes which make it suitable for giftware.

2011. This the most free-machining of the common aluminum alloys. It also has excellent mechanical properties. Thus, it is widely used for automatic screw machine products in parts requiring extensive machining.

2014, 2017. The 2017 alloy combines excellent machinability and strength with the result that it is one of the most widely used alloys for automatic screw machine work. It is a tough, ductile alloy suitable for heavy-duty structural parts. It strength is slightly less than that of 2014.

2024. This is one of the best known of the high strength aluminum alloys. With its high strength and excellent fatigue resistance, it is used to advantage on structures and parts where good strength-to-weight ratio is desired. It is readily machined to a high finish. It is readily formed in the annealed condition and may be subsequently heat treated. Arc or gas welding is generally not recommended, although this alloy may be spot, seam or flash welded. Since corrosion resistance is relatively low, 2024 is commonly used with an anodized finish or in clad form ("Alclad") with a thin surface layer of high purity aluminum. Applications: aircraft structural components, aircraft fittings, hardware, truck wheels and parts for the transportation industry.

3003. This is the most widely used of all aluminum alloys. It is essentially commercially pure aluminum with the addition of manganese which increases the strength some 20% over the 1100 grade. Thus, it has all the excellent characteristics of 1100 with higher strength. It has excellent corrosion resistance. It has excellent workability and it may be deep drawn or spun, welded or brazed. It is nonheat-treatable. Applications: cooking utensils, decorative trim, awnings, siding, storage tanks, chemical equipment.

5005. This alloy is generally considered to be an improved version of 3003. It has the same general mechanical properties as 3003 but appears to stand up better in actual service. It is readily workable. It can be deep drawn or spun, welded or brazed. It has excellent corrosion resistance. It is nonheat-treatable. It is well suited for anodizing and has less tendency to streak or discolor. Applications same as 3003.

5052. This is the highest strength alloy of the more common nonheat-treatable grades. Fatigue strength is higher than most aluminum alloys. In addition this grade has particularly good resistance to marine atmosphere and salt water corrosion. It has excellent workability. It may be drawn or formed into intricate shapes and its slightly greater strength in the annealed condition minimizes tearing that occurs in 1100 and 3003. Applications: Used in a wide variety of applications from aircraft components to home appliances, marine and transportation industry parts, heavy duty cooking utensils and equipment for bulk processing of food.

5083, 5086. For many years there has been a need for aluminum sheet and plate alloys that would offer, for high strength welded applications, several distinct benefits over such alloys as 5052 and 6061. Some of the benefits fabricators have been seeking are greater design efficiency, better welding characteristics, good forming properties, excellent resistance to corrosion and the same economy as in other nonheat-treatable alloys. Metallurgical research has developed 5083 and 5086 as superior weldable alloys which fill these needs. Both alloys have virtually the same characteristics with 5083 having slightly higher mechanical properties due to the increased manganese content over 5086. Applications: unfired pressure vessels, missile containers, heavy-duty truck and trailer assemblies, boat hulls and superstructures.

6061. This is the least expensive and most versatile of the heat-treatable aluminum alloys. It has most of the good qualities of aluminum. It offers a range of good mechanical properties and good corrosion resistance. It can be fabricated by most of the commonly used techniques. In the annealed condition it has good workability. In the T4 condition fairly severe forming operations may be accomplished. The full T6 properties may be obtained by artificial aging. It is welded by all methods and can be furnace brazed. It is available in the clad form ("Alclad") with a thin surface layer of high purity aluminum to improve both appearance and corrosion resistance. Applications: this grade is used for a wide variety of products and applications from truck bodies and frames to screw machine parts and structural components. 6061 is used where appearance and better corrosion resistance with good strength are required.

6063. This grade is commonly referred to as the architectural alloy. It was developed as an extrusion alloy with relatively high tensile properties, excellent finishing characteristics and a high degree of resistance to corrosion. This alloy is most often found in various interior and exterior architectural applications, such as windows, doors, store fronts and assorted trim items. It is the alloy best suited for anodizing applications–either plain or in a variety of colors.

7075. This is one of the highest strength aluminum alloys available. Its strength-to-weight ratio is excellent and it is ideally used for highly stressed parts. It may be formed in the annealed condition and subsequently heat treated. Spot or flash welding can be used, although arc and gas welding are not recommended. It is available in the clad ("Alclad") form to improve the corrosion resistance with the over-all strength being only moderately affected. Applications: used where highest strength is needed.

How Aluminum is Made

Aluminum is the world's most abundant metal. Though some is present in practically any handful of dirt, it never appears in nature as free metal. The only practical source is bauxite ore with 45 to 60% aluminum oxide. Ground and kiln-dried bauxite is mixed with caustic soda solution and heated to dissolve the aluminum oxide. This forms sodium aluminate that is drawn off as a clear "green liquor" after most iron oxides and other impurities have been removed by controlled settling and filtration. Complex precipitation, washing and high temperature dehydration produce the fine white powdered aluminum oxide known as alumina. By weight it is half oxygen, half aluminum. Alumina, cryolite (sodium aluminum fluoride) and aluminum fluoride are mixed in a carbon-lined cell or "reduction pot" where direct current is applied through a carbon electrode. This is essentially an electric smelting furnace where the aluminum is freed as the oxygen from the alumina joins carbon in the anode as carbon dioxide. Between the cell's carbon lining and the floating crust of alumina, molten aluminum collects and is tapped or siphoned to crucibles, tested, and conveyed to alloying furnaces. Ingots or billets are then cast in suitable forms for further processing.

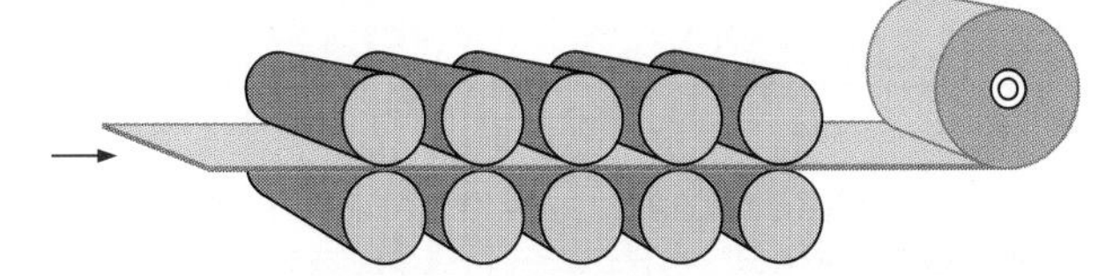

Hugh rolling ingot is reduced through hot and cold rolling mills to desired thickness. Annealing and other thermal treatments before, between, and/or after rolling must be carefully controlled to produce specified quality.

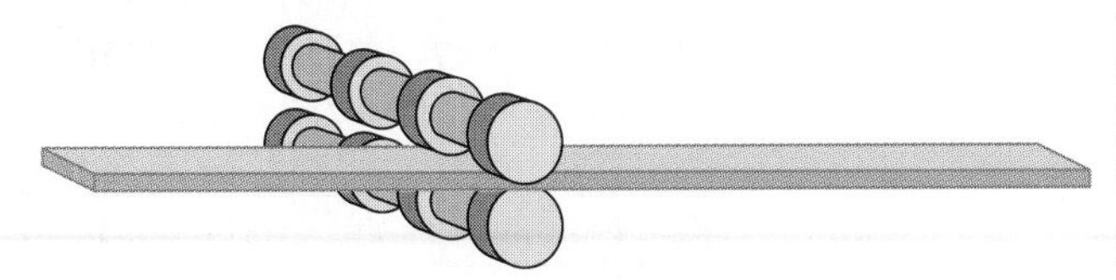

Wrought forms of rod and bar are rolled to specified dimensions.

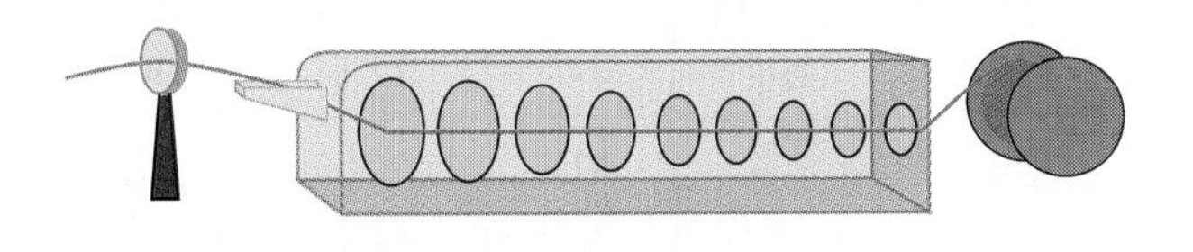

Rod is drawn through a series of dies into wire of various forms.

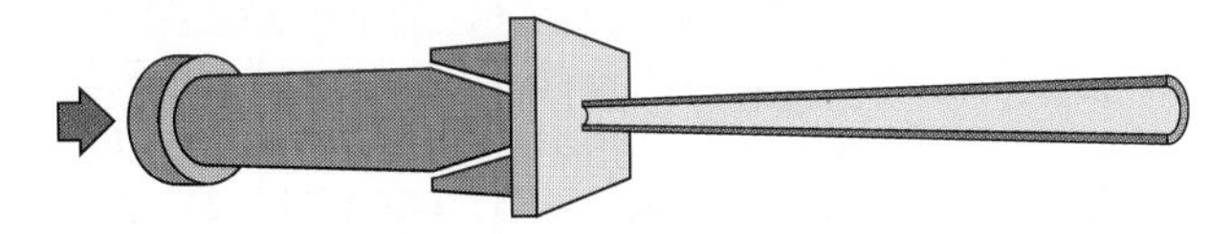

Extrusion ingot (billet) is heated and forced under tremendous pressure through dies to form extruded rod, bar, pipe, tub, and various shapes.

Sheet Gauges

Gauge No.	Steel Sheets		Galvanized Sheets		Stainless Steel Sheets			Aluminum Sheets	
					Wt. lbs. per Sq. Ft.				
	Weight Lbs. per Square Foot	Thick-ness in Inches	Weight Lbs. per Square Foot	Thick-ness in Inches	Straight Chromium (400 Series)	Chromium Nickel (300 Series)	Approx. Thick-ness in Inches	Weight Lbs. per Sq. Ft. (1100)	Thick-ness in Inches
38	.25000	.0060						.0558	.00396
37	.26562	.0064						.0627	.00445
36	.28125	.0067						.0705	.00500
35	.31250	.0075						.0791	.00561
34	.34375	.0082						.0888	.00630
33	.39500	.0090						.0998	.00708
32	.40625	.0097	.56250	.0134	.3708	.3780	.010	.1121	.00795
31	.43750	.0105	.59375	.0142	.4506	.4594	.011	.1259	.00893
30	.50000	.0120	.65625	.0157	.5150	.5250	.013	.1410	.0100
29	.56250	.0135	.71875	.0172	.5794	.5906	.014	.1593	.0113
28	.62500	.0149	.78125	.0187	.6438	.6562	.016	.1777	.0126
27	.68750	.0164	.84375	.0202	.7081	.7218	.017	.2002	.0142
26	.75000	.0179	.90625	.0217	.7725	.7875	.019	.2242	.0159
25	.87500	.0209	1.03125	.0247	.9013	.9187	.022	.2524	.0179
24	1.0000	.0239	1.15625	.0276	1.0300	1.0500	.025	.2834	.0201
23	1.1250	.0269	1.28125	.0306	1.1587	1.1813	.028	.3187	.0226
22	1.2500	.0299	1.40625	.0336	1.2875	1.3125	.031	.3567	.0253
21	1.3750	.0329	1.53125	.0366	1.4160	1.4437	.034	.4019	.0285
20	1.5000	.0359	1.65625	.0396	1.5450	1.5750	.038	.4512	.0320
19	1.7500	.0418	1.90625	.0456	1.8025	1.8375	.044	.5062	.0359
18	2.0000	.0478	2.15625	.0516	2.0600	2.1000	.050	.5682	.0403
17	2.2500	.0538	2.40625	.0575	2.3175	2.3625	.056	.6387	.0453
16	2.5000	.0598	2.65625	.0635	2.5750	2.6250	.063	.7163	.0508
15	2.8125	.0673	2.96875	.0710	2.8968	2.9531	.070	.8051	.0571
14	3.1250	.0747	3.28125	.0785	3.2187	3.2812	.078	.9038	.0641
13	3.7500	.0897	3.90625	.0934	3.8025	3.9375	.094	1.015	.0720
12	4.3750	.1046	4.53125	.1084	4.5063	4.5937	.109	1.139	.0808
11	5.0000	.1196	5.15625	.1233	5.1500	5.2500	.125	1.279	.0907
10	5.6250	.1345	5.78125	.1382	5.7937	5.9062	.141	1.437	.1019
9	6.2500	.1495	6.40625	.1532	6.4375	6.5625	.156	1.613	.1144
8	6.8750	.1644	7.03125	.1681	7.0813	7.2187	.172	1.812	.1285
7	7.5000	.1793						2.035	.1443
6	8.1250	.1943						2.284	.1620
5	8.7500	.2092						2.565	.1819
4	9.3750	.2242						2.881	.2043
3	10.000	.2391						3.235	.2294

Metal Hardness and Finishing Data

Heat Treatment of Steel

By thermal treatment, steel may be made harder or softer, stresses induced or relieved, mechanical properties increased or decreased, crystalline structure changed, machinability enhanced, etc. The terms used to describe such heat treatments and their effects are listed below.

Normalize. Normalizing consists of uniform heating to a temperature slightly above the point at which grain structure is affected (known as the critical temperature), followed by cooling in still air to room temperture. This produces a uniform structure and hardness throughout.

Anneal. When not preceded by a descriptive adjective, annealing consists of heating to and holding at a suitable temperature, then allowing to cool slowly. Annealing removes stresses, reduces hardness, increases ductility and produces a structure favorable for formability.

1. Full Anneal—This term is synonymous with annealing and is used to differentiate anneal from bright anneal, stress relief anneal, etc.
2. Spherodize Anneal—This treatment is similar to full annealing except the steel is held at an elevated temperature for a prolonged period of time, followed by slow cooling in order to produce a microstructure where carbides exist in a globular or spheroidal form.
3. Soft Anneal—When maximum softness and ductility are required without change in grain structure, steel should be ordered soft annealed. This process consists of heating to a temperature slightly below the critical temperature and cooling in still air.
4. Stress Relief Anneal—Stress relieving is intended to reduce the residual stresses imparted to the steel in the drawing operation. It generally consists of heating the steel to a suitable point below the critical temperature followed by slow cooling.

Relationship of Hardness to Tensile Strength of Carbon and Alloy Steel

Brinell Indentation Diameter mm	Brinell Hardness Number		Rockwell Hardness Number		Tensile Strength (Approximate) 1000 psi
	Standard Ball	Tungsten-Carbide Ball	B Scale	C Scale	
2.45	—	627	—	58.7	347
2.50	—	601	—	57.3	328
2.55	—	578	—	56.0	313
2.60	—	555	—	54.7	298
2.65	—	534	—	53.5	288
2.70	—	514	—	51.1	274
2.75	—	495	—	51.0	264
2.80	—	477	—	49.6	252
2.85	—	461	—	48.5	242
2.90	—	444	—	47.1	230
2.95	429	429	—	45.7	219
3.00	415	415	—	44.5	212
3.05	401	401	—	43.1	202
3.10	388	388	—	41.8	193
3.15	375	375	—	40.4	184

Brinell Indentation Diameter mm	Brinell Hardness Number		Rockwell Hardness Number		Tensile Strength (Approximate) 1000 psi
	Standard Ball	Tungsten-Carbide Ball	B Scale	C Scale	
3.20	363	363	—	39.1	177
3.25	352	352	—	37.9	171
3.30	341	341	—	36.6	164
3.35	331	331	—	35.5	159
3.40	321	321	—	34.3	153
3.45	311	311	—	33.1	149
3.50	302	302	—	32.1	146
3.55	293	293	—	30.9	141
3.60	285	285	—	29.9	138
3.65	277	277	—	28.8	134
3.70	269	269	—	27.6	130
3.75	262	262	—	26.6	127
3.80	255	255	—	25.4	124
3.85	248	248	—	24.2	120
3.90	241	241	100.0	22.8	116
3.95	235	235	99.0	21.7	114
4.00	229	229	98.2	20.5	111
4.05	223	223	97.3	—	104
4.10	217	217	96.4	—	103
4.15	212	212	95.5	—	100
4.20	207	207	94.6	—	99
4.25	201	201	93.8	—	97
4.30	197	197	92.8	—	94
4.35	192	192	91.9	—	92
4.40	187	187	90.7	—	90
4.45	183	183	90.0	—	89
4.50	179	179	89.0	—	88
4.55	174	174	87.8	—	86
4.60	170	170	86.8	—	84
4.65	167	167	86.0	—	83
4.70	163	163	85.0	—	82
4.80	156	156	82.9	—	80
4.90	149	149	80.8	—	73
5.00	143	143	78.7	—	71
5.10	137	137	76.4	—	67
5.20	131	131	74.0	—	65
5.30	126	126	72.0	—	63
5.40	121	121	69.0	—	60
5.50	116	116	67.6	—	58
5.60	111	111	65.7	—	56

*This table, which is based on ASTM A 370-68, Table III, lists the approximate relationship of hardness values to corresponding approximate tensile strength values of steels. Some compositions and processing histories may deviate from these relationships. The data in this table do not represent hardness-to-tensile strength conversions for austenitic, ferritic, and martensitic stainless steel. If more precise conversions are required, they should be developed for each specific composition and heat treatment. Related Rockwell superficial hardness numbers, if of interest, may be found in ASTM A 370-68.

5. Bright Anneal—This process consists of annealing in a closely controlled furnace atmosphere which will permit the surface to remain relatively bright.

Quench. Quenching consists of heating steel above the critical range, then hardening by immersion in an agitated bath of oil, water, brine or caustic. Quenching increases tensile strength, yield point and hardness. It reduces ductility and impact resistance. By subsequent tempering some ductility and impact resistance may be restored, but at some sacrifice of tensile strength, yield point and hardness.

Temper. Tempering is the reheating of steel, after quenching, to the specified temperature below the critical range, then air cooling. It is done in furnaces, oil or salt baths, at temperatures varying from 300 to 1200° F. Low tempering temperatures give maximum hardness and wear resistance. Maximum toughness is achieved at the higher temperatures.

Metal Finishing Data

Cadmium Plating. A nonporous electrolytically deposited layer of cadmium that offers better corrosion resistance for steel than zinc coating. Plating is per specification MIL-P-416A (or equivalent commercial specification QQ-P-416A). Three types of cadmium plating are considered in this specification:
1. Type I—Pure silver-colored cadmium plate, without supplementary treatment. This type of cadmium coating was used on all steel aircraft hardware in the past.
2. Type II—This consists of Type I plating followed by a chromate treatment. Type II plating is a light to dark gold color. It has improved corrosion resistance. Procurement specifications for aircraft hardware now specify Type II plating.
3. Type III—This is Type I coating followed by a phosphate treatment. It is used mainly as a paint base.

In addition to the type of plating, MIL-P-416A also defines the plating thickness in terms of the following classes:

- Class 1—.0005" minimum
- Class 2—.0003" minimum
- Class 3—.0002" minimum

Steel parts with a Rockwell hardness greater than R 40 (approx. 180,000 PSI tensile strength) must be stress relieved before cleaning and plating, and if they are subject to flexure (springs, etc.) they must be baked at 375° F within 30 minutes after plating to prevent hydrogen embrittlement.

Anodizing. This finish, applied to aluminum by an acid plating process, hardens the surface, reduces porosity, increases abrasion resistance and has high dielectric strength. Anodized aluminum can be dyed almost any color. Specification MIL-A-8625B covers three types of anodizing:
1. Type I—Chromic anodize coating will vary from a light to a dark gray color, depending on the alloy. Coating is given a chromate treatment to seal surface.
2. Type II—Sulfuric anodize coating is the best coating for dying (Class 2). Nondyed (Class 1) coatings will have a dull yellow-green (gold) appearance when sealed with a chromate treatment.
3. Type III—Hard anodize coating can be used as an electrical insulation coating or as an abrasion resisting coating on devices such as hydraulic cylinders, wear surfaces and actuating cams.

Steel Alloy Designation

Standard AISI and SAE Steels

Studies have been made in the steel industry for the purpose of establishing certain "standard" steels and eliminating the manufacture of other steels which vary only slightly in composition from the standard steels, as much as possible. These standard steels are selected on the basis of serving the significant metallurgical and engineering needs of fabricators and users of steel products.

Standard Carbon Steels

Definition. By common custom, steel is considered to be carbon steel when no minimum content is specified or required for aluminum, boron, chromium, cobalt, columbium, molybdenum, nickel, titanium, tungsten, vanadium or zirconium, or for any other element added to obtain a desired alloying effect; when the specified minimum for copper does not exceed 0.40 %; or when the maximum content specified for any of the following elements does not exceed the percentages noted: manganese 1.65, silicon 0.60, copper 0.60.

Numbering System. In the AISI system of identification, the prefix "B" is used to designate acid bessemer steel. The letter "L" within the grade number is used to identify leaded steels.

A four-numeral series is used to designate graduations of chemical composition of carbon steel, the last two numbers of which are intended to indicate the approximate middle of the carbon range. For example, in the grade designation 1035, 35 represents a carbon range of 0.32 to 0.38 %.

It is necessary, however, to deviate from this rule and to interpolate numbers in the case of some carbon ranges and for variations in manganese, phosphorus or sulphur with the same carbon range.

The first two digits of the four-numeral series of the various grades of carbon steel and their meanings are as follows:

- 10xx Nonresulphurized carbon steel grades
- 11xx Resulphurized carbon steel grades
- 12xx Rephosphorized and resulphurized carbon steel guides
- 15xx Nonresulphurized high manganese carbon steels.

Standard Alloy Steels

Definition. Steel is considered to be alloy steel when the maximum of the range given for the content of alloying elements exceeds one or more of the following limits: manganese, 1.65%; silicon, 0.60%; copper, 0.60%; or in which a definite range or a definite minimum quantity of any of the following elements is specified or required within the limits of the recognized field of constructional alloy steels: aluminum, boron, chromium up to 3.99%, cobalt, columbium, molybdenum, nickel, titanium, tungsten, vanadium, zirconium or any other alloying element added to obtain a desired alloying effect.

Numbering System. In the AISI numbering system, the prefix letter E is used to designate steels normally made only by the basic electric furnace process. Steels without a prefix letter are normally manufactured by the basic open hearth or basic oxygen processes, but may be manufactured by the basic electric furnace process with adjustments in phosphorus and sulphur limits.

The last two digits of the four-numeral series are intended to indicate the approximate middle of the carbon range. For example, in the grade designation 4142, 42 represents a carbon range of 0.40 to 0.45%. (Where a five-numeral series occurs, the last three digits indicate the carbon content.) It is necessary, however, to deviate from this rule and to interpolate numbers in the case of some carbon ranges, and for variations in manganese, sulphur, chromium, or other elements.

The first two digits indicate the type of alloy according to alloying elements as follows:

- 13xx Manganese 1.75%
- 40xx Molybdenum 0.20 or 0.25%
- 41xx Chromium 0.50, 0.80 or 0.95%—Molybdenum 0.12, 0.20 or 0.30%
- 43xx Nickel 1.83%—Chromium 0.50 or 0.80%—Molybdenum 0.25%
- 44xx Molybdenum 0.53%
- 46xx Nickel 0.85 or 1.83%—Molybdenum 0.20 or 0.25%
- 47xx Nickel 1.05%, Chromium 0.45%
- 48xx Nickel 3.50%, Molybdenum 0.25%
- 50xx Chromium 0.40%
- 51xx Chromium 0.80, 0.88, 0.93, 0.95 or 1.00%
- 5xxxx Carbon 1.04%—chromium 1.03 or 1.45%
- 61xx Chromium 0.60 or 0.95%—Vanadium 0.13% or 0.15% min.
- 86xx Nickel 0.55%—Chromium 0.50%—Molybdenum 0.25%
- 87xx Nickel 0.55%—Chromium 0.50%—Molybdenum 0.35
- 88xx Nickel 0.55%—Chromium 0.50%—Molybdenum 0.35
- 92xx Silicon 2.00%

Effects of Common Alloying Elements in Steel

By definition, steel is a combination of iron and carbon. Steel is alloyed with various elements to improve physical properties and to produce special properties, such a resistance to corrosion or heat. Specific effects of the addition of such elements as outlined below:

Carbon (C), although not usually considered as an alloying element, is the most important constituent of steel. It raises tensile strength, hardness and resistance to wear and abrasion. It lowers ductility, toughness and machinability.

Manganese (Mn) is a deoxidizer and degasifier and reacts with sulphur to improve forgeability. It increases tensile strength, hardness, hardenability and resistance to wear. It decreases tendency toward scaling and distortion. It increases the rate of carbon-penetration in carburizing.

Phosphorus (P) increases strength and hardness and improves machinabilty. However, it adds marked brittleness or cold-shortness to steel.

Sulphur (S) improves machinability in free-cutting steels, but without sufficient manganese it produces brittleness at red heat. It decreases weldability, impact toughness and ductility.

Silicon (Si) is a deoxidizer and degasifier. It increases tensile and yield strength, hardness, forgeability and magnetic permeability.

Chromium (Cr) increases tensile strength, hardness, hardenability, toughness, resistance to wear and abrasion, resistance to corrosion and scaling at elevated temperatures.

Nickel (Ni) increases strength and hardness without sacrificing ductility and toughness. It also increases resistance to corrosion and scaling at elevated temperatures when introduced in suitable quantities in high chromium (stainless) steels.

Molybdenum (Mo) increases strength, hardness, hardenability and toughness, as well as creep resistance and strength at elevated temperatures. It improves machinability and resistance to corrosion and it intensifies the effects of other alloying elements. In hot-work steels, it increases red-hardness properties.

Tungsten (W) increases strength, hardness and toughness. Tungsten steels have superior hot-working and greater cutting efficiency at elevated temperatures.

Vanadium (V) increases strength, hardness and resistance to shock impact. It retards grain growth, permitting higher quenching temperatures. It also enhances the red hardness properties of high speed metal cutting tools and intensifies the individual effects of other major elements.

Cobalt (Co) Increases strength and hardness and permits higher quenching temperatures. It also intensifies the individual effects of other major elements in more complex steels.

Aluminum (Al) is a deoxidizer and degasifier. It retards grain growth and is used to control austenitic grain size. In nitriding steels it aids in producing a uniformly hard and strong nitrided case when used in amounts 1.00%–1.25%.

Lead (Pb), while not strictly an alloying element, is added to improve machining characteristics. It is almost completely insoluble in steel, and minute lead particles, well dispersed, reduce friction where the cutting edge contacts the work. Addition of lead also improves chip-breaking formations.

Steel—Some Common Types

1010. This is one of the most widely used low carbon steels for low strength applications. It is best suited for parts whose fabrication involves moderate to severe forming and some machining. Its weldability is excellent and it can be case hardened for wear resistance by cyaniding.

1018. This is a popular carburizing grade of steel. It can be strengthened by cold working or surface hardened by carburizing or cyaniding. It is relatively soft and has good weldability and formability.

1020. This is a general-purpose low-carbon "mild" steel. It is easy to fabricate by the usual methods such as mild cold or hot forming and welding. It is weldable by all processes and the resulting welds are of extremely high quality.

4130. This chromium-molybdenum alloy is one of the most widely used aircraft steels because of its combination of weldability, ease of fabrication and mild hardenability. In relatively thin sections, it may be heat treated to high strength levels. In the normalized condition it has adequate strength for many applications. It may be nitrided for resistance to wear and abrasion.

4140. This chromium-molybdenum alloy is a deep hardening steel used where strength and impact toughness are required. It has high fatique strength making it suitable for critical stressed applications. It may be nitrided for increased resistance to wear and abrasion.

4340. This chromium-nickel-molybdenum alloy is a widely used deep-hardening steel. It possesses remarkable ductility and toughness. With its high alloy content uniform hardness is developed by heat treatment in relatively heavy sections. Its high fatique strength makes it ideal for highly stressed parts.

4620. This nickel-molybdenum alloy is a carburizing steel capable of developing high case hardness and core toughness. It can be forged similarly to the other carburizing grades. Because of its relatively high nickel content, it is not as readily cold-formed.

5160. This carbon-chromium grade of spring steel has a high yield/tensile strength ratio, excellent toughness and high ductility. It is very difficult to machine in the as-rolled condition and should be annealed prior to machining. It is not readily welded, but it can be welded by either gas or arc welding processes if the section involved is preheated and stress relieved after welding.

52100. This high carbon-high chromium alloy is produced by the electric furnace process and then vacuum degassed to meet the rigid standards of the aircraft industry for bearing applications. It develops high hardness and has exceptional resistance to wear and abrasion.

6150. This chromium-vanadium alloy steel is similar to 4340. It has good hardenability, good fatique properties and excellent resistance to impact and abrasion.

8620. This is a "triple alloy" chromium-nickel-molybdenum steel. It is readily carburized. It may be heat treated to produce a strong, tough core and high case hardness. It has excellent machinability and responds well to polishing operations. It is easily welded by any of the common welding processes, although the section should be heated and stress relieved after welding.

9310. This chromium-nickel-molybdenum alloy is a carburizing steel capable of attaining high case hardness with high core strength. It has excellent toughness and ductility.

Bending of 4130 Steel

Specification MIL-SF18729C states that 4130 steel .749 inch and less in thickness can withstand bending without cracking at room temperature, with the axis of bending transverse to the direction of rolling, through and angle as indicated in the table. Condition N materials will be bent around a diameter three times the thickness of the material. Test samples are bent cold either by pressure or blows. In the event of dispute, bending will be by pressure. Paragraph 4.5.3 of the specification states that the formation of cracks not over 1/16" in aggregate lengths at the corners on the outside of the bend will not be cause for rejection.

Thickness of Material Inches	Minimum Angle of Bend Cond. N Degrees
Under .090	180
.090 to .187, Incl.	135
Over .187 to .249, Incl.	90
Over .249 to .749, Incl.	90

Modern Steelmaking

Modern Steelmaking

Steel is essentially a combination of iron and carbon, the carbon content of common grades ranging from a few hundredths to about one per cent. All steels also contain varying amounts of other elements, principally manganese, phosphorus, sulfur, and silicon, which are always present if only in trace amounts. The presence and amounts of these and some 20 other alloying elements, which are added in various combinations as desired, determine to a great extent the ultimate properties and characteristics of the particular steel.

Raw Materials

The principal raw materials of the steel industry are iron ore, iron and steel scrap, coal, and limestone. Iron ore is a natural combination of iron oxides and other materials, such as silicon and phosphorus. Until recently, the industry's main sources of iron were the high-grade ores, containing from 55 to 65% iron, which were mined and sent directly to the steel plants. Today, the most available domestic iron ore is taconite, which contains a lesser amount of iron, making its use uneconomical without some kind of beneficiation, a process in which the material is upgraded and formed into high-iron-bearing pellets. Nearly one-half of the iron ore produced on this continent is now used in this pellet form.

A second source of iron is scrap. Most of this comes from the steel plant itself: only about two-thirds of the steel produced by steel plants is shipped as product, the remainder being discarded during processing and returned to the furnaces as scrap. Other scrap, if needed, comes from outside the plant from such sources as old automobiles, worn out railway cars and rails, obsolete machinery, and cuttings from metalworking shops.

Coal is converted in coke, gas, and chemicals in the coke ovens. The coke is used in the blast furnace as a fuel and reducing agent, the gas is burned in heating units, and the chemicals are processed into various organic materials.

Limestone is employed as a flux in both the blast furnace and steelmaking furnace where it serves to remove impurities from the melt. It is used either as crushed stone direct from the quarry or, after calcining, as burnt lime.

Blast Furnace

The principal charging material used in making steel is molten pig iron, the product of the blast furnace. To produce it, iron ore, coke, and limestone are charged into the top of the furnace. A continuous blast of preheated air, introduced near the bottom of the furnace, reacts with the coke to form carbon monoxide gas which then combines with the oxygen in the iron oxides, thereby reducing them to metallic iron. The molten iron is tapped into a ladle for transportation to the steel producing unit.

Pig iron contains considerable amounts of carbon, manganese, phosphorus, sulfur, and silicon. In the solid form, it is hard and brittle and therefore unsuitable for applications where ductility is important.

Steelmaking Methods

Steelmaking may be described as the process of refining pig iron or ferrous scrap by removing the undesirable elements from the melt and then adding the desired elements in predetermined amounts. These additions are often in the same elements which were originally removed, the difference being that the elements present in the final steel product are in the proper proportion to produce the desired properties.

The open-hearth, the basic oxygen, and the electric-arc processes account for nearly all the steel tonnage produced in this country today. The open-hearth furnace was the nation's major source of steel until 1969, when this role was assumed by the relatively new basic oxygen process. Together, these two methods account for over 80% of the steel made in America. The remainder is made up of electric furnace steels.

Open-Hearth Furnace. The open-hearth furnace has the ability to produce steels in a wide range of compositions. The process can be closely controlled, yielding steels of high quality from charges which need be only nominally restrictive in their analyses. Most modern open-hearth furnaces are lined with a chemically basic material, such as magnesite, and use a basic refining slag. Furnace capacities range from 100 to 500 tons per melt, or heat, each heat requiring from 4 to 10 hours furnace time.

To begin the process, the basic open-hearth furnace is charged with scrap, limestone, and iron ore. This initial charge lies on an "open" hearth, where it is melted by exposure to flames sweeping over its surface. The pig iron, which may constitute as much as 75% of the charge, is added in the molten state after the scrap is partially melted. During the subsequent refining of the heat—a process which is frequently accelerated by the introduction of oxygen through roof lances—nearly all of the manganese, phosphorus, and silicon are oxidized and retained by the slag, which floats on the heavier molten metal. Appreciable percentages of sulfur can also be taken into the slag.

The heat is allowed to react until its carbon content has been reduced by oxidation to approximately that desired in the finished steel. The furnace is then tapped, allowing the molten metal to flow into a ladle. To obtain the desired analysis, appropriate quantities of needed elements, usually in the form of ferroalloys, are added to the heat as it pours into the ladle, or, in the case of some elements, added to the furnace just prior to tapping. A deoxidizer, such as aluminum or ferrosilicon, is also normally added to control the amount of gas evolved during solidification. The heat is then usually poured into ingot molds where it solidifies into steel ingots.

Basic Oxygen Furnace. The "BOF" involves the same chemical reactions as the open-hearth, but uses gaseous oxygen as the oxidizing agent to increase the speed of these reactions and thereby reduce the time of the refining process. Although the advantages of the use of oxygen were obvious to steelmakers a hundred years ago, only in recent years has the pure gas become commercially available in the vast quantities required to make the BOF feasible. Heats of steel as large as 300 tons can be made in less than an hour, several times faster than the average open-hearth can operate. The steel is of excellent quality, equivalent to open-hearth steel in every respect.

The basic oxygen furnace, a closed-bottom, refractory-lined vessel, is charged with molten pig iron and scrap. During the oxygen blow, burnt lime and fluorspar, which form the slag, are charged into the furnace. A high-velocity stream of oxygen is directed down onto the charge through a water-cooled lance, causing the rapid oxidation of carbon, manganese, and silicon in the melt. These reactions provide the heat required for scrap melting, slag formation, and refining. Additions of deoxidizers and any required alloying elements are made as the steel is tapped from the vessel into the ladle. It is then usually poured into ingot molds, as with other steelmaking processes.

Electric-arc Furnace. Special steels, such as the high-alloy, stainless, and tool steels, are normally made in electric-arc furnaces. The primary advantage of this type furnace is that it permits the extremely close control of temperature, heat analysis, and refining conditions required in the production of these complex steels. As another advantage, these furnaces can be operated efficiently on a cold metal charge, thereby eliminating the need for blast furnaces and associated facilities. For this reason, electric furnaces are today being used with increasing frequency for the production of standard carbon and alloy steels.

The furnace proper is round or elliptical, with carbon or graphite electrodes extending through the roof. In operation, the electrodes are lowered to a point near the charge, which is melted by the heat of the electricity arcing between the electrodes and the charge. When the charge of carefully selected steel scrap is about 70% molten, iron ore and burnt lime are added. Alloying elements are added during a later stage of the refining process. Some 3 to 7 hours are required for each heat, depending mostly on the type of steel being produced. Furnace capacity can vary from a few hundred pounds to 200 tons or more.

Composite Materials

Foams

Five different types of rigid, closed cell foams are currently being used in proven designs.

Styrofoam FB. Low density (2 lb./ft.^{3}), large cell, fire retardant. Varies in color from white to blue. The large-cell type provides better protection from delamination than the more commonly used small-cell blue styrofoam. Cuts smoothly with a hot wire for airfoil shapes. Do not confuse styrofoam with expanded polystyrene which is the type seen in the average picnic cooler. The compression strength of polystyrene is too low and it dissolves in most solvents and fuel.

Styrofoam. Blue-Low Density (2 lb./ft.^{3}), used for years in flotation for boat docks and other marine uses. Commonly called a Bouyancy Billet, this foam has applications and costs similar to Styrofoam FB. The prime advantage is the availability in the large billet sizes. This foam is now used in the Q2 design, and should be used with Safe-T-Poxy only. RAE type epoxy tends to bleed into the foam which can result in a dry layup.

Caution: Use only epoxy with styrofoam. Polyester will dissolve the foam.

Urethane. Low density (2 lb./ft.^{3}), small cell, colored green or tan. Used extensively in the fuselage and fuel tanks as it is completely fuel proof. Easy to carve and contour with a large knife. Do not hot-wire urethane foam as a hazardous gas is discharged.

Urethane Polyester. Medium and high density, white (4 to 18 lb./ft.^{3}), this is a small cell rigid urethane-polyester foam which can be readily cut and carved and has limited heat forming characteristics. Its fuel compatability makes it suitable for sandwich tank construction. The uniform surface, excellent compressive strength and low cost of this white foam makes it an ideal replacement for some types of PVC foam.

PVC. Medium and high density, tan or blue (3 to 15 lb./ft.3), used in fuselage bulkheads and other areas where high compressive strength is required. Some PVCs have recently been replaced by white Urethane/Polyester foams which combine comparable qualities at much lower cost.

Caution: Either epoxy or polyester systems may be used with urethane and PVC. **Do not store foams in sunlight.**

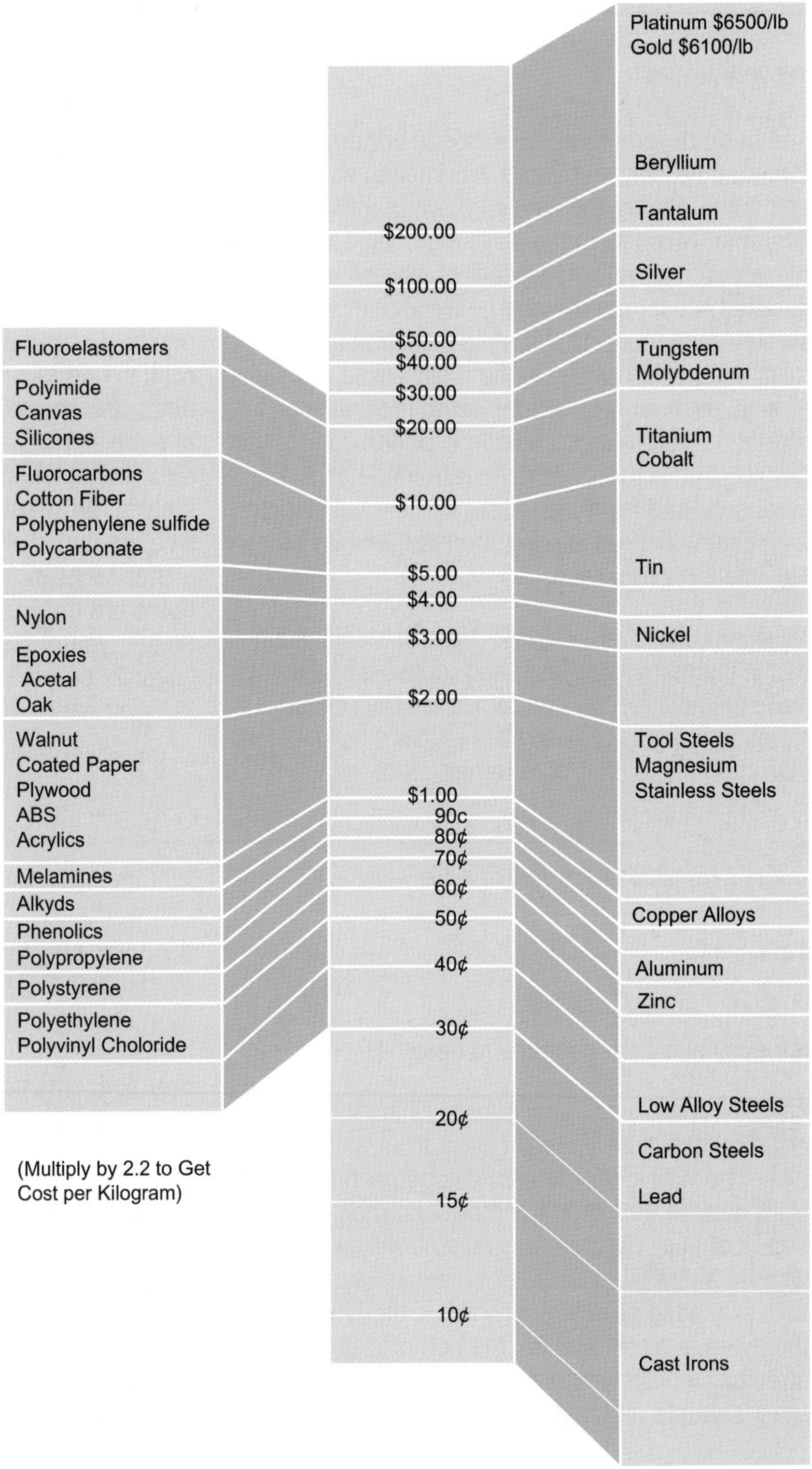

Platinum $6500/lb
Gold $6100/lb
Beryllium
Tantalum
$200.00
Silver
$100.00
Fluoroelastomers
$50.00
$40.00
Tungsten
Molybdenum
Polyimide
Canvas
Silicones
$30.00
$20.00
Titanium
Cobalt
Fluorocarbons
Cotton Fiber
Polyphenylene sulfide
Polycarbonate
$10.00
Tin
$5.00
$4.00
Nylon
$3.00
Nickel
Epoxies
Acetal
Oak
$2.00
Walnut
Coated Paper
Plywood
ABS
Acrylics
Tool Steels
Magnesium
Stainless Steels
$1.00
90c
80¢
70¢
Melamines
60¢
Alkyds
50¢
Copper Alloys
Phenolics
Polypropylene
40¢
Aluminum
Polystyrene
Zinc
Polyethylene
Polyvinyl Choloride
30¢
Low Alloy Steels
20¢
Carbon Steels
(Multiply by 2.2 to Get Cost per Kilogram)
Lead
15¢
10¢
Cast Irons

Relative cost of material per pound

Appendix C material courtesy of Aircraft Spruce and Specialty Company

Index

Note: Page numbers in **BOLD** type reference non-text material.